Sports Injuries and Prevention

Kazuyuki Kanosue
Editor in Chief

Tetsuya Ogawa • Mako Fukano
Toru Fukubayashi
Editors

Sports Injuries and Prevention

Springer

Editor in Chief
Kazuyuki Kanosue
Faculty of Sport Sciences
Waseda University
Saitama, Japan

Editors
Tetsuya Ogawa
Faculty of Sport Sciences
Waseda University
Saitama, Japan

Toru Fukubayashi
Faculty of Sport Sciences
Waseda University
Saitama, Japan

Mako Fukano
Faculty of Sport Sciences
Waseda University
Saitama, Japan

ISBN 978-4-431-55317-5 ISBN 978-4-431-55318-2 (eBook)
DOI 10.1007/978-4-431-55318-2

Library of Congress Control Number: 2015944319

Springer Tokyo Heidelberg New York Dordrecht London

Printed on acid-free paper

Springer Japan KK is part of Springer Science+Business Media (www.springer.com)

Introduction to the Series

Waseda University of Japan has a tradition of producing great athletes among its graduates, such as Mikio Oda, the first Japanese ever to win an Olympic gold medal. Waseda University strongly supports coaching techniques that embody a practical application of the knowledge gained from the fundamental research findings of sports science. Waseda University also takes pride in providing athletes with medical care that utilizes leading-edge sports medicine and formulates management strategies that combine all these elements. This approach has led to a strong tradition of sports-based research—what we like to call "Waseda Sports"—which has enjoyed an unprecedented level of success. This tradition was enhanced by the Faculty of Sport Sciences in Waseda University in 2009 when they initiated the Global COE (Center of Excellence) Program, entitled "Sport Sciences for the Promotion of Active Life." The Global COE Program is under the aegis of the Japanese Ministry of Education, Culture, Sports, Science, and Technology; this Ministry supports the development of international centers of education and research excellence.

While life expectancy in Japan is the highest in the world, large-scale societal changes here and elsewhere have led to an increase in health problems due to a decrease in activity and physical fitness. In the aging population, there has been a deterioration of overall health, much of which can be attributed to inactivity and excess body weight. It is especially troubling that similar problems are increasing among children and are associated with severe physical and mental disabilities. The international scope of the above problems provided the impetus for Waseda University to form the Global COE Program. This effort involved the construction of an international hub of education and research specifically designed to develop and encourage talented researchers to create sports programs that would contribute to an active and vital lifestyle. The program emphasizes the development of specialist knowledge in conjunction with a broad understanding and awareness of the diverse world of sports. One of our goals was to focus not just on improving the individual health of mind and body but also to develop an understanding of the conditions present in regions and societies that facilitate such improvements in the lifestyle of individuals.

The sports sciences have created and are extending an important body of knowledge. It is critical that this information be utilized to produce an active, two-way interaction between the investigators and the active participants of sporting events. In order to provide a focus for developing this reciprocal intercommunication, the Global COE Program identified three strategic project themes: (1) Active Children Project, (2) Active Elderly Project, and (3) Elite Athlete Project. The COE Program was proactive in seeking out mature graduate students who were returning to higher education after a period of work, thereby facilitating a meaningful contribution to the formation of academic careers for specialists who were active in the practical domain of sports. Many graduate students from abroad, especially from Asian countries, joined the program and have contributed to our goals via both the creation of academic knowledge and direct participation in the sports relevant to their area of investigation.

The formal funding for the Global COE Program came to an end in March 2014, but the projects initiated by the program and the activities of the graduates continue unabated. The accomplishments made during the 5 years of the program have been documented in a series of four books with the overall theme of "Sports Science and an Active Life." We are proud to present this substantial body of research in the following series of books: Vol. 1: *Sports Management and Sports Humanities* (Kohei Kogiso, Daichi Oshimi, Munehiko Harada, Eds.), Vol. 2: *Physical Activity, Exercise, Sedentary Behavior, and Promoting Health* (Satomi Oshima, Zhen-Bo Cao, Koichiro Oka, Eds.), Vol. 3: *Sports Performance* (Tomoyuki Nagami, Jun Tsuchiya, Eds.), and Vol. 4: *Sports Injuries and Prevention* (Tetsuya Ogawa, Mako Fukano, Toru Fukubayashi, Eds.). The series was written by the dedicated faculty members and young graduate students and postdoctoral researchers under the guidance of investigators who took part in the Global COE Program. The series was also contributed to by leading researchers around the world, most of who belong to Waseda University's research institute or university partners. I appreciate their contributions as well as their participation in the Global COE Program. During the 5 years of the program, an international network of individuals and universities doing active research in the area of sports sciences has been established. I expect this network to grow wider and stronger in the future and to contribute to the solution of many of the health problems that plague modern societies. We will all continue to work hard to involve sports activities in the solutions to these problems and in the process aid in advancing the sports activities themselves.

Finally, I express my appreciation to the editors of each volume, who not only did a fine job of organizing the volumes but also wrote chapters that were important scientific contributions to the overall effort. We would also like to thank the Global COE staff for their efficient work and the kind support they extended to the graduate students. Drs. Larry Crawshaw and Candace S. O'Connor are thanked for their enthusiastic editorial assistance.

Program Leader Kazuyuki Kanosue
Global COE "Sport Sciences for the
Promotion of Active Life"
Waseda University

Preface

As contemporary humans, we benefit from sports in many ways. These include promotion and maintenance of both physical and mental well-being, enhancement of communication with others, and the fostering of a sense of community in the overall society. Moreover, international competition in sports utilizing commonly agreed upon rules can result in a deepening of mutual understanding between people all over the world. This awareness and tolerance can extend beyond cultures and languages. Furthermore, through sports, we can address our ultimate possibilities as human beings.

In opposition to these possible advantages, on the other hand is an important matter that must be taken into account. Participating in sports can lead to injuries, and it is necessary to develop an awareness of sports-induced injury incidence. Moreover, the fact remains that such injuries can be prevented or minimized through appropriate interventions. As a representative example of efforts toward sports injury prevention, over the last decade the International Olympic Committee (IOC) has been placing a particular emphasis on the maintenance of health and injury prevention among Olympic athletes. In 2011, the IOC for the first time hosted The IOC World Conference on Prevention of Injury and Illness in Sport and declared a focus of their support on prevention of injury and illness among the athletes.

In another example on the international level, in 1993 the International Federation of Football Associations (FIFA) established the FIFA Medical Assessment and Research Centre (F-Mark) and have been working on the assessment and analysis of injury occurrences in World Cup and other international games. It is considered as crucially important to address injury occurrence based on a common standard worldwide. For prevention, understanding the source and frequency of injuries on the basis of a valid, complete statistical analysis can play a significant role.

Since its inception, preventive medicine has brought about a transformation of modern medicine and has been making significant contributions toward prevention of infectious diseases, primarily through vaccination. For the prevention of sports-related injuries and illnesses, on the other hand, the effectiveness of some approaches such as applications of particular training programs and the use of medical checkups before competitions have been recognized as important interventions. Application of appropriate training programs has been found to significantly decrease the incidence of sports-related injuries. In addressing the construction of strategies for sports-related injury prevention in the future, it is essential to promote a better understanding of the underlying mechanisms and risk factors.

Throughout this book, the authors provide such information. Current data are presented on the incidence of sports-related injuries, the types of injuries specific to particular sports, and the importance of factors such as age and gender. Possible injury mechanisms and risk factors are presented based on an analysis involving recent scientific findings.

The book is composed of seven parts. In the first part, current situations and the general characteristics of sports-related injuries are outlined on the basis of an investigation utilizing statistical data involving a large number of populations. In the following parts, detailed information on the injuries in terms of the types of sports activities, body sites, symptoms, and the relationships among these factors are discussed. Part II, for example, deals with topics on concussion and severe head–neck injuries which occur frequently in rugby and judo.

In Parts III and IV, as one of the major sports-related injuries, anterior cruciate ligament (ACL) injuries are discussed. Beginning with the underlying mechanisms as assessed by using the latest measuring techniques, characteristic features of their occurrence are described. This particular type of sports injury is particularly dependent on gender and occurs much more frequently in females than in males. The background and possible interventions for prevention will be discussed, particularly among soccer and basketball players. Further, Part IV deals with topics on postoperative (ACL reconstruction) aspects of ACL injuries, especially those related to muscle functions and tendon regeneration in the hamstring muscles.

Part V deals with muscle strain and focuses particularly on those occurring in the hamstring muscles as this muscle group is known as one of the most frequent sites of muscle strain. The occurrence mechanisms in relation to the characteristic features of the muscle group as well as possible intervention for prevention are explained.

In Part VI, disorders related to the ankle and foot are introduced. Precise movements of the foot bones and their relationship to the incidence of injuries as well as possible interventions for prevention are discussed. Finally, Part VII provides information on lower back disorders. Included are detailed mechanisms of their incidence, epidemiology, and implications for their prevention.

A variety of sports are included to allow the reader to better generalize the results as well as to apply appropriate procedures to specific sports. The authors have emphasized basic scientific findings to help the reader gain a broad knowledge of sports injuries. The potential audience includes medical doctors, physical therapists, athletic trainers, coaches, and interested parents. This book is expected to play a prominent role in the construction of training programs for both healthy and injured players. The focus on junior athletes will aid in their education, injury prevention, and increased performance. It will also benefit instructors at the junior and senior high school levels.

Saitama, Japan Tetsuya Ogawa
Saitama, Japan Mako Fukano
Saitama, Japan Toru Fukubayashi

Contents

Part I
Sports Injury Surveillance

Chapter 1
Sports Injury Surveillance in Japan (from Sports Safety Association)

Toru Fukubayashi, Tetsuya Ogawa, and Mako Fukano

Abstract The chapters of this book focus on the development of a better understanding of the mechanisms and risk factors involved in sports-related injuries. Implicit in this understanding is the attempt, also covered in this book, to create intervention programs which minimize or prevent sports related injuries In this initial chapter we provide a foundation for the ensuing chapters by assessing the current incidence and general features of sports-related injuries in Japan. We herein supply the latest statistics on the incidence of sports-related injuries in Japan on the basis of surveillance conducted by the Sports Safety Association in the fiscal years from 2000 to 2011. We hope our organization of the statistical material, and our structured arrangement by anatomical site, type and severity of disorder, type of sport, and gender will allow readers to easily access needed information as well as better understand the contents in the later chapters of this book.

It is important to note that the incidence of sports-related injuries in junior high school and high school students during their official school activities (e.g. physical education class and club activities) are not included in this analysis since they are covered by accident insurance provided by the Japan Sports Council. This information will be discussed separately in a subsequent chapter.

Keywords Sports Injury Surveillance • Sports Safety Association • Sports Accident Insurance

1.1 Introduction

The statistics presented below on the incidence of sports-related injuries are based on the record of payment for insurance claims to the sport safety association accident insurance fund provided by the Sports Safety Association of Japan. This information was previously introduced in our bulletin reports (Fukubayashi 2010,

T. Fukubayashi (✉) • T. Ogawa • M. Fukano
Faculty of Sport Sciences, Waseda University, Saitama, Japan
e-mail: fukubayashi@waseda.jp; t.ogawa@aoni.waseda.jp; mako.fukano@aoni.waseda.jp

K. Kanosue et al. (eds.), *Sports Injuries and Prevention*,
DOI 10.1007/978-4-431-55318-2_1

2011, 2012a, b). Sport safety association accident insurance has approximately 10 million subscribers, who are made up of people engaging in amateur sports and a variety of other activities which include cultural events, volunteering, community events, and coaching activities. This insurance and has been widely used by group members and coaches in sporting activities and social educational activities. The data provided focus mainly on those reported in 2011, in which year the total incidence was 183,399 cases.

1.2 Statistics

1.2.1 Overview of Incidence

The number of subscribers over the past decade has stayed relatively constant. The number of people insured in 2011 was 9,786,467. In this population, there was a total incidence of 183,399 cases (117,517 males and 64,794 females). For reference, Fig. 1.1 describes the overall incidence numbers on an annual basis for the three most recent years available (2009–2011). The total yearly incidences, illustrated in Fig. 1.1a, were 163,803 (2009), 193,802 (2010), and 183,399 (2011). The corresponding incidence rates are given in Fig. 1.1b and are 1,656, 1,959, and 1,874 cases per 100,000 people, respectively. Figure 1.1c depicts the incidence by gender.

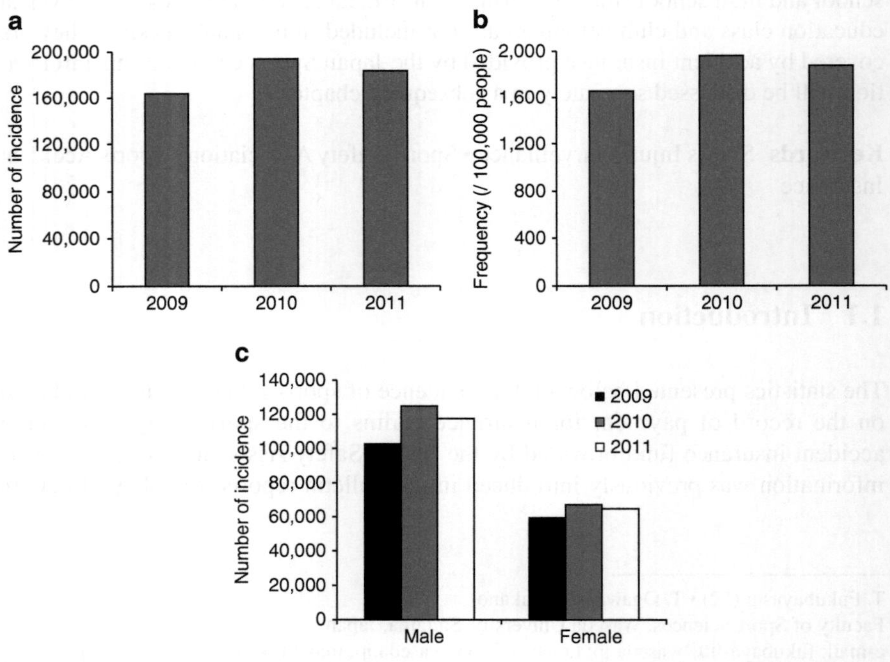

Fig. 1.1 Incidence of sports related injury (annually) from 2009 to 2011. (**a**) Numbers of incidence, (**b**) incidence rate (per 100,000 people), and (**c**) numbers of incidence by sex

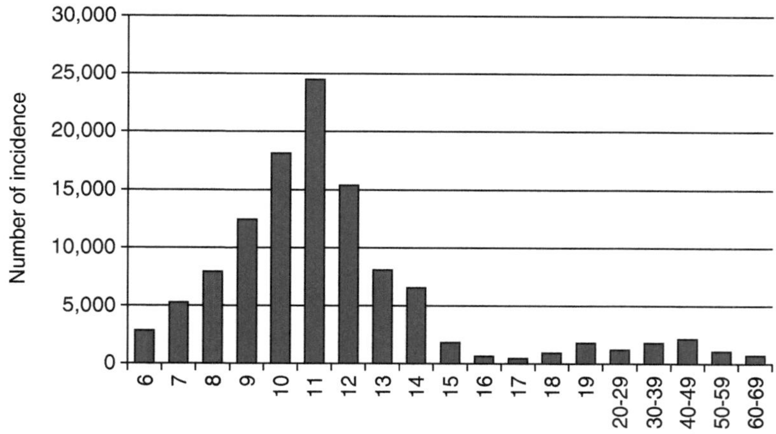

Fig. 1.2 Incidence by age. Numbers for the ages of 20–29, 30–39, 40–49, 50–59, and 60–69 represent the average in the respective age groups

For males, the figures were 102,546 (2009), 124,720 (2010), and 117,517 (2011). The female incidence was 59,545 (2009), 67,162 (2010), and 64,794 (2011). There were slightly more female than male subscribers.

1.2.2 Incidence by Age Bracket

The injury incidence by age bracket is described in Fig. 1.2. The incidence increases abruptly with increase in age up to 11. As mentioned above, since the statistics provided here do not include the incidence among junior high school and high school students during their official school activities (physical education class and club activities), the values reported here do not cover the actual incidence for those age groups (12–18). After high school (19 and higher), the numbers show a slight increase. This does not necessarily mean there was an increase in incidence rate. Rather, the increase likely occurred because Japan Sports Council insurance cannot be applied for after high school. In addition, there is an apparent decrease in incidence over the age of 70 (data not shown in the figure), but this is due to a considerable decrease of participation in sports activities in this age group and therefore also in subscribers.

1.2.3 Incidence by Month

Figure 1.3 portrays the injury incidence numbers on a monthly basis. From April when school year begins in Japan, there is a considerable increase that continues through November. During the winter (December through March), although the

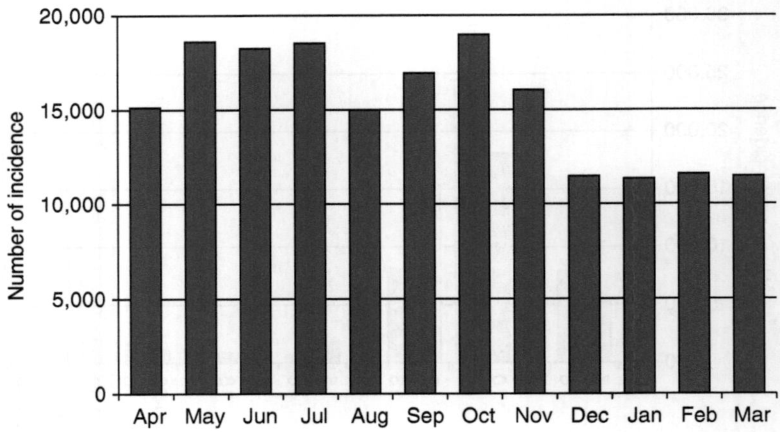

Fig. 1.3 Incidence by month

incidence still exceeds 10,000, the numbers decrease abruptly in comparison to the rest of the year since sports activities, except for winter sports ,are less frequent during this time.

1.2.4 Most Frequent Anatomical Sites

Injury incidence by anatomical site is shown in Fig. 1.4. The hand maniphalanx region had the highest incidence, accounting for 20.8 % of the total. For other sites, the corresponding percentage incidences (in decreasing order) were: ankle, 15.9 %; head and neck, 11.9 %; knee, 10.2 %; and foot toe regions, 9.0 %. With a regional categorization, the incidences were: upper extremity 36.3 %, lower extremity 41.8 %, head and neck 11.9 %, and trunk 6.5 %. It is clear that the incidence of sports-related injuries is predominantly focused on the limbs.

1.2.5 Major Injury (and by Gender)

Incidence of injury by sprain, bone fracture, contusion bruise, and "other" is described in Fig. 1.5 for the overall population. The breakdown by gender is also shown, and indicatest that the injury distribution differs considerably for the two sexes. For example, in contrast to the incidence of sprain in males, where it accounts for approximately 30 % of the injuries, in females sprains account for nearly half of the total. On the other hand, while the, incidence of bone fracture is as high as that of sprain in males (31.8 %), bone fractures in females account for only 11.1 % of the total. Given the prominent difference between males and females, we will next

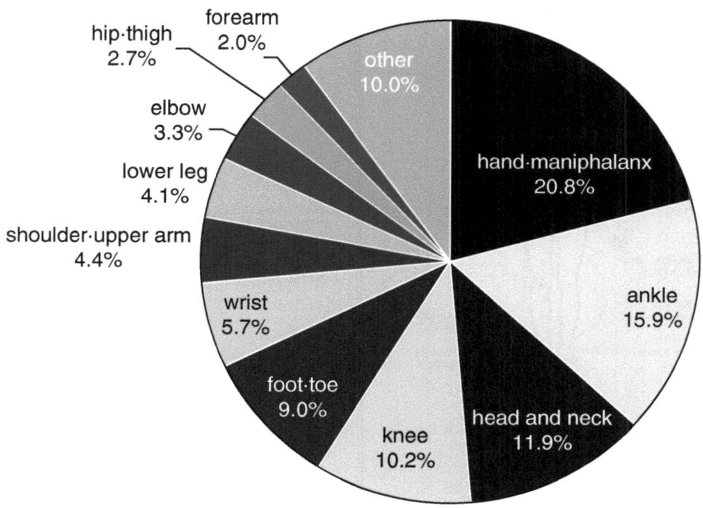

Fig. 1.4 Injury incidence by anatomical site

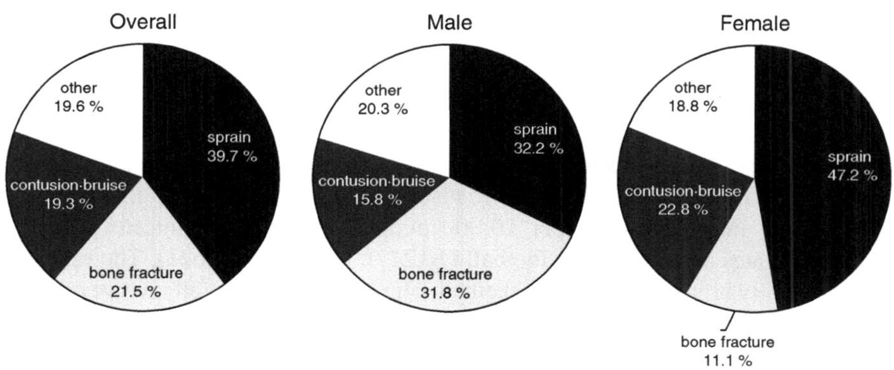

Fig. 1.5 Types of incidence overall and by sex

look into the specific injury, particularly from the perspective of gender. Figure 1.6 shows the ten most frequent injuries for males and females. Generally, the incidence rate is high in the regions of the hand maniphalanx and also for joints in the lower extremity such as the ankle and knee. In males, but not in females, contusion bruise and wound in the head and neck region are frequent, which is considered to be due to a difference in the type of sports males engage in. Although for both males and females the most frequently injured anatomical the region is hand maniphalanx, the types of injury in males and females differs. Bone fracture dominates the incidence in males (bone fracture in hand maniphalanx (11.9 %) and wrist (3.9 %),while in females the incidence of sprain is relatively high. In other regions

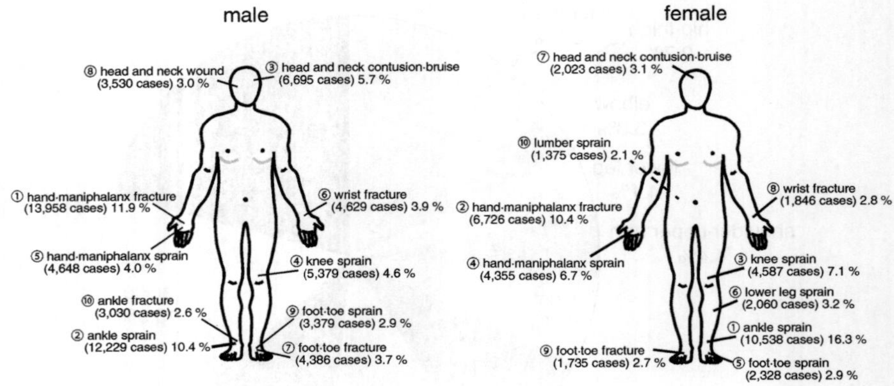

Fig. 1.6 Types and body site of incidence by sex

such as the knee and ankle, there are considerably higher rates of incidence of sprain in females than in males.

1.2.6 Types of Sports (Overview)

Figure 1.7 represents the injury incidence for the 12 sports with the highest number of injuries. Some sports such as soccer, volleyball, baseball, and basketball have high numbers of injuries simply because many subscribers play these sports. The yearly incidence rate (per 10,000 people) tells a different story. Injury incidence rates are highest for football (10,247), and rugby (5,621). The rates are also relatively high for judo and basketball. The foregoing sports all involve contact, and with the exception of judo, also involve high running speeds as well as changes in direction and occasional jumping. Non-contact sports such as track and field and swimming, despite a large number of subscribers, have a low overall incidence and an extremely low incidence rate.

1.2.7 Injury by Types of Sports Activities

Lastly, in Fig. 1.8 we will look at major sports injuries for the 12 sports described above. In soccer (Fig. 1.8a), with the highest number of cases (39,298), ankle sprain is the most frequent injury with 5,930 cases, and accounts for 15.1 % of the total. Surprisingly, injury of the upper extremity, such as fracture and sprain in the hand maniphalanx regions and the wrist fracture are relatively frequent, despite the fact that the lower extremities are predominantly used in soccer. This may be attributed

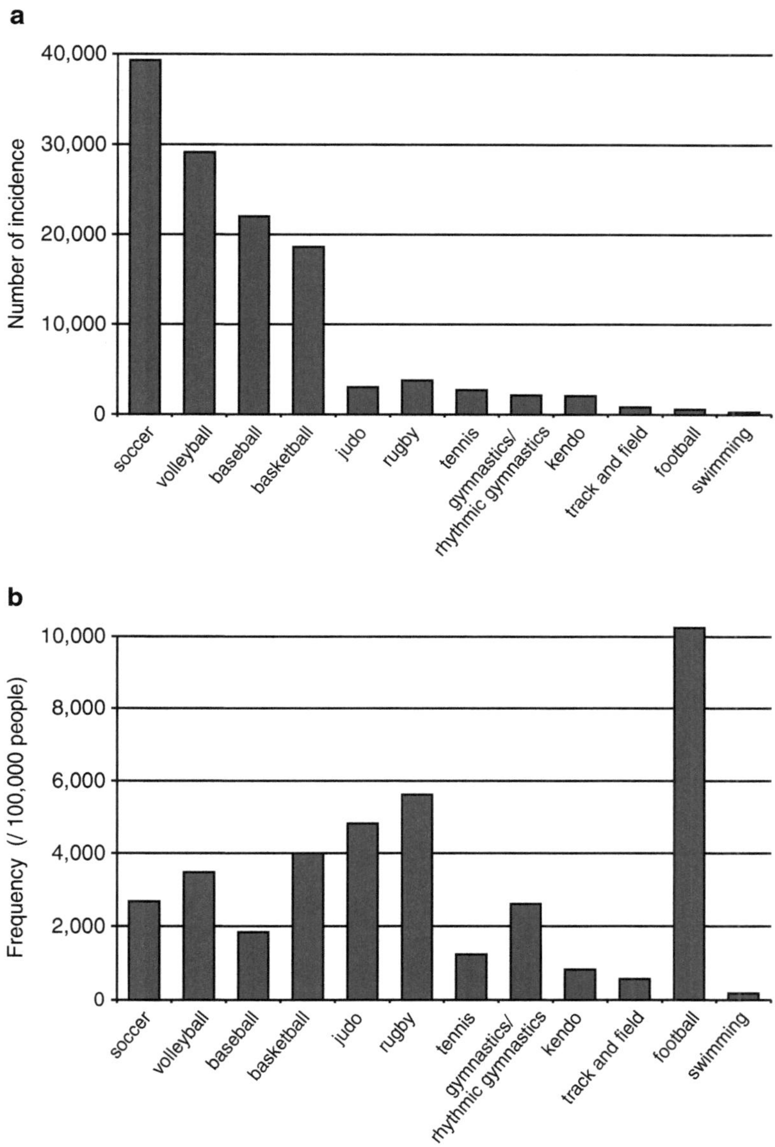

Fig. 1.7 (**a**) Incidence by sports, (**b**) frequency by sports

to the fact that soccer is a readily accessible sports from a young age, and children in elementary and middle school are included. For each sport we also detail the incidence of cases involving severe trauma. Severe cases are defined as those that involve insurance payments of more than 100,000 Japanese yen per case. For soccer, there were 1,917 such cases, which represented 5.0 % of the total.

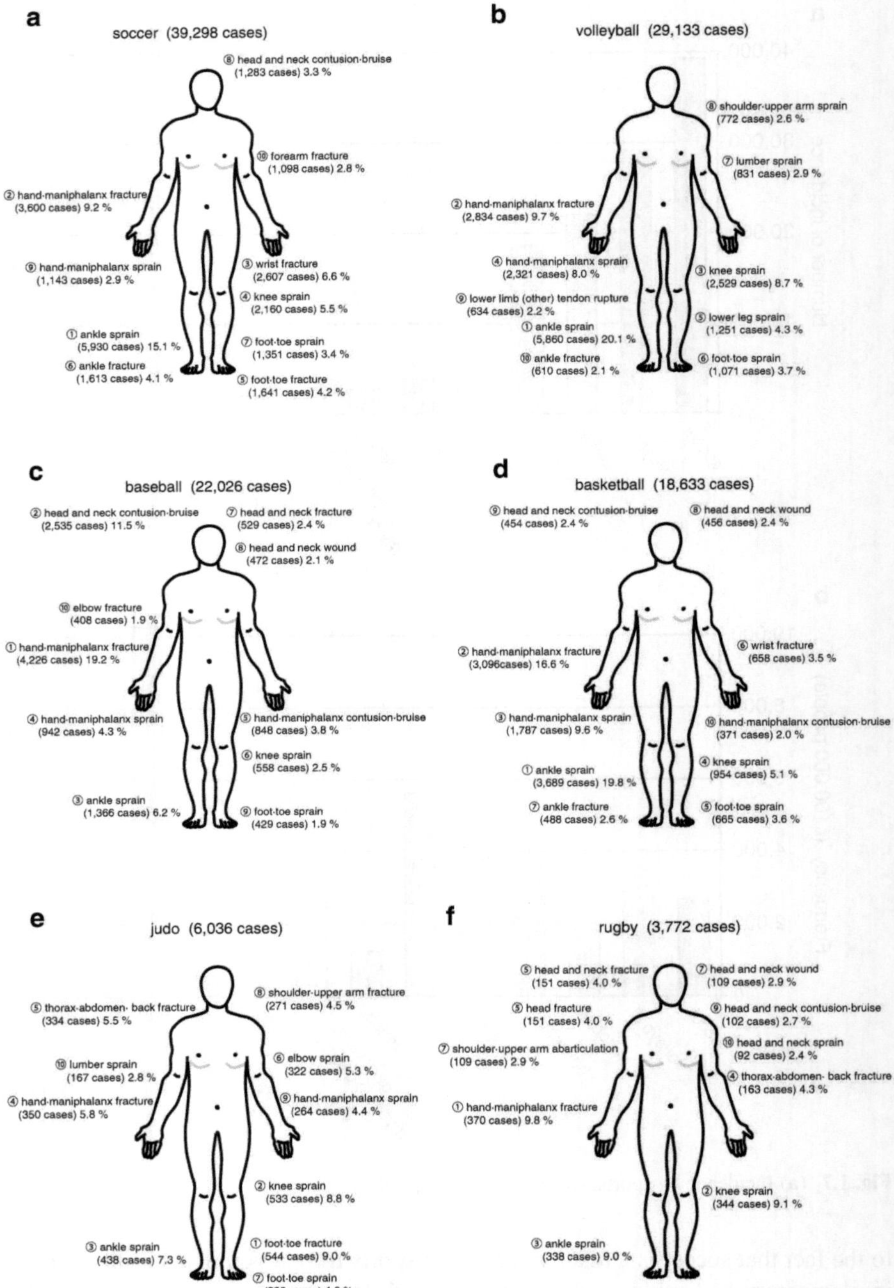

Fig. 1.8 Major type and body site of injury by sports. (**a**) Soccer, (**b**) volleyball, (**c**) baseball, (**d**) basketball, (**e**) judo, (**f**) rugby, (**g**) gymnastics/rhythmic gymnastics, (**h**) tennis, (**i**) kendo, (**j**) track and field, (**k**) american football, (**l**) swimming

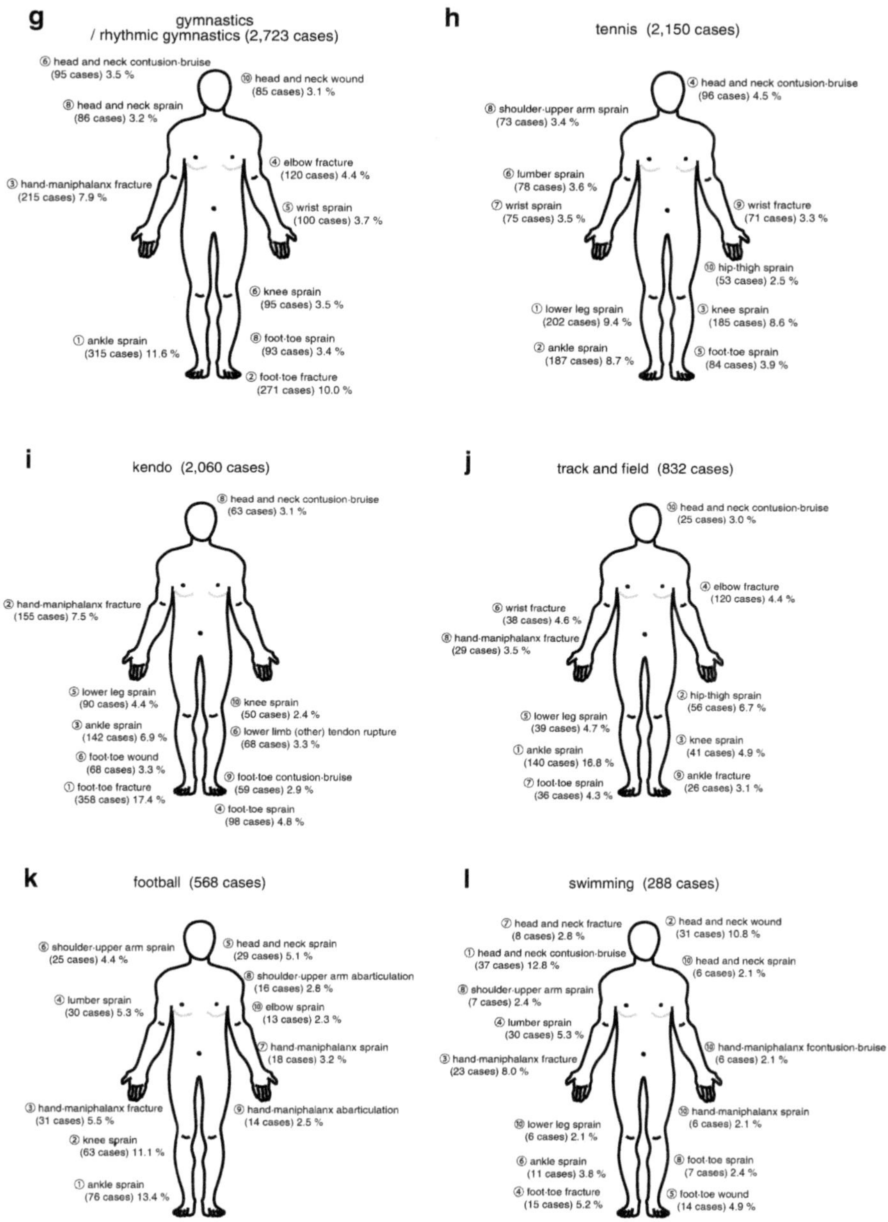

Fig. 1.8 (continued)

Among these cases, knee sprain (ligament injury) had the highest in the incidence (263 cases) and accounted for 21.7 % of the total.

There were 29,133 injury cases in volleyball (Fig. 1.8b). Ankle sprain (ligament injury) was the most common injury (20.1 %), followed by hand maniphalanx fracture (9.7 %), knee sprain (8.7 %), and hand maniphalanx sprain (8.0 %). Injuries in the upper extremity, typically included finger sprains and fractures resulting from exposure to hit balls, while those in the lower extremity, such as knee and ankle sprains, were more likely to be due to stresses resulting from landing after a jump. There were 2,061 severe trauma cases, which accounted for 7.1 % of the cases. Overall, there were 476 cases (23.1 %) of knee sprain (ligament injury), 411 cases (19.9 %) of Achilles tendon rupture, 150 cases of ankle sprain, and 135 cases of hand maniphalanx fracture.

Data under the baseball heading (Fig. 1.8c) include both hardball and softball. There were for a total of 22,026 baseball related incidences. The most numerous were hand maniphalanx fractures (19.2 %), followed by head and neck contusion bruise (11.5 %), ankle sprain (6.2 %), and hand maniphalanx sprain (4.3 %). There were occasional bruises in the ophthalmic region and fractures of facial bones attributable to hit balls. The incidence of severe trauma cases was relatively low, with 734 cases (3.3 %).

There were 18,633 injuries for basketball (Fig. 1.8d). The most frequent major injuries were ankle sprain (19.8 %), hand maniphalanx fracture (16.6 %), hand maniphalanx sprain (9.6 %), and knee sprain (ligament injury) (5.1 %). Although the general features of basketball injuries are comparatively similar to those found in volleyball, basketball injuries are more likely to result from collisions with other players upon landing, rapid changes in direction, or loss of balance. High school female athletes have an especially high incidence of anterior cruciate ligament (ACL) injury. There were 596 severe traumatic cases which account for 3.2 % of the cases. Of these, 187 cases were knee sprain (ligament injury) and 68 cases involved Achilles tendon ruptures.

In judo (Fig. 1.8e), 6,036 injuries were documented, the most frequent being foot toe fracture (544 cases, 9.0 %), followed by knee sprain (ligament injury) (533 cases, 8.8 %), ankle sprain (438 cases, 7.3 %), hand maniphalanx fracture (350 cases, 5.8 %), thorax abdomen back fracture (334 cases, 5.5 %), elbow sprain (322 cases, 5.3 %), and foot toe sprain (288 cases, 4.8 %). In judo, injuries of the shoulder and elbow were more common than in other sports. There were 416 severe trauma cases, which accounted for 6.9 % of the total. This is relatively high. Among these severe cases, there were 71 cases of knee sprain, 16 cases of elbow joint problems, 28 cases of, shoulder fracture, 28 cases of clavicle fracture, and 14 cases of abarticulation in the shoulder joint.

Rugby had 3,772 injury cases (Fig. 1.8f). Given the small number of rugby players, the number of cases is quite large (incidence rate: 5,621 cases per 100,000 people).The most frequent injuries were hand maniphalanx fracture (370 cases, 9.8 %), followed by knee sprain (344 cases, 9.1 %), and ankle sprain (338 cases, 9.0 %). As is clear from the figure, the major injuries are concentrated in the upper extremity in comparison with the other sports. It is noteworthy that there were 151 cases (4.0 %) of head and neck fracture resulting from collision with other players during scrum, ruck, and tackling, which are common activities in rugby. Careful attention to players with these injuries is required, because some of the head and neck injuries are accompanied by the risk of concussion, and in severe cases, brain trauma and intracranial bleeding. There were 377 severe cases in rugby players which accounted for 10.0 % of the injuries. Again, this figure is higher than that found in other sports. Of these cases, there were 94 knee sprains, 46 head and neck fractures, and 20 Achilles tendon ruptures.

In tennis (Fig. 1.8h), the injury incidence was low in comparison to other sports. There were 2,150 injury cases. Most frequent were lower leg sprains (202 cases, 9.4 %), ankle sprains (187 cases, 8.7 %), and knee sprains (185 cases, 8.6 %). There were 156 severe injuries (7.3 %). Thirty of these were Achilles tendon ruptures, which represented the largest number of severe injuries.

Figure 1.8 also illustrates the injury incidence for other sports activities such as gymnastics / rhythmic gymnastics, (8G) kendo (8I), track and field (8 J), football (8 K), and swimming (8 L).

As the statistical summary provided above makes clear, the incidence of sports-related injuries differ for each type of sport. Since each sport is characterized by a unique set of physical, anatomical, physiological, and environmental challenges, the development of intervention programs to prevent injuries should tailored to address the specific needs of each activity.

References

Fukubayashi T (2010) A survey on incidence of sports-related trauma in the sports safety insurance. Jpn Sports Assoc Sports Med Sci Res Rep 2:12–26 (in Japanese)

Fukubayashi T (2011) A survey on incidence of sports-related trauma in the sports safety insurance. Jpn Sports Assoc Sports Med Sci Res Rep 2:17–32 (in Japanese)

Fukubayashi T (2012a) A survey on incidence of sports-related trauma in the sports safety insurance; statistical report of year 2011. Jpn Sports Assoc Sports Med Sci Res Rep 2:34–47 (in Japanese)

Fukubayashi T (2012b) A survey on incidence of sports-related trauma in the sports safety insurance; summary of three year from 2009 to 2011. Jpn Sports Assoc Sports Med Sci Res Rep 2:48–53 (in Japanese)

Chapter 2
Sports Injury Surveillance in Japanese Junior and Senior High School Students

Toru Okuwaki

Abstract The number and rate of sports-related injuries in junior and senior high school sports clubs were investigated for 11 sports, including soccer, using the records of the Injury and Accident Mutual Aid Benefit System from 2009 through 2011. The number and rate of sports-related injuries has increased in each sport. More than 500 cases of sports-related head and neck injuries were reported each year. Of the sports surveyed, , anterior cruciate ligament injuries were one of the most common sports-related injuries, occurred more frequently in women than in men, and were most frequent in basketball players. Establishment of a Sports Injury Surveillance system based on the Injury and Accident Mutual Aid Benefit System will be invaluable for the prevention of sports-related injuries in athletes, including junior and senior high-school students.

Keywords Sports-related injury • Junior and senior high school sports clubs • Head and neck injury • Anterior cruciate ligament injury • Sports Injury Surveillance system

2.1 Introduction

The number and rate of sports-related injuries in students in Japan have been recorded based on the payment of insurance claims by the Injury and Accident Mutual Aid Benefit provided by the Japan Sports Council. The Injury and Accident Mutual Aid Benefit was established by the association for Japan School Safety in

---

The structure of this study
This study was based on "Sports medicine and science report-- the establishment of the Sports Injury Surveillance system in Japan –first through third reports" published from 2009 to 2011 by the Japan Amateur Sports Association, with permission from the Japan Sport Council (JSC).

T. Okuwaki (✉)
Japan Institute of Sports Sciences, Tokyo, Japan
e-mail: toru.okuwaki@jpnsport.go.jp

K. Kanosue et al. (eds.), *Sports Injuries and Prevention*,
DOI 10.1007/978-4-431-55318-2_2

1960. The association has investigated the annual incidence of injury during school-supervised activities from 1965 through the present.

The Injury and Accident Mutual Aid Benefit covers approximately 17 million students, which is about 97 % of all students in Japan. Over 99 % of elementary and junior high school students are affiliated with this program. This high affiliation rate allows for a simple and accurate determination of the number and rate of incidents by consulting the records of payment of insurance claims by the Injury and Accident Mutual Aid Benefit. However, this system does not cover medical expenses of less than 5,000 yen under insurance benefits, so minor injuries have been excluded. One can also assume that not all cases are claimed; therefore, the numbers and rates of injuries in this study are estimates.

Approximately 1.1 million incidents (including sports-related injuries) occurred under school supervision in 2008 (the rate of incidents = the number of claims/the number of affiliated students × 100 = 6.3 %). About one-fourth of the 1.1 million injuries, or 0.3 million, were sports-related (overall rate = 2.1 % (0.1 % for elementary students)). The timing of the incidents varied with the age of the students: more than half of all injuries to elementary students occurred during recess, whereas injuries to junior and senior high school students frequently occurred during school sports club activities. Being able to defining the injuries that occurred during junior and senior high school sports clubs as sports-related injuries would improve our understanding of the circumstances of all sports-related injuries in this age group; therefore, we investigated the number and rate of sports-related injuries in 11 typical junior and senior high school sports for which the number of students participating was known.

2.2 Subjects and Methods

The survey material included the Injury and Accident Mutual Aid Benefit System records from April to March for 3 years (2009, 2010, and 2011). The number of sports-related injuries was defined as the number of incidents reported to have occurred during school club activities. The school club sports included in this study were soccer, baseball, volleyball, basketball, rugby, tennis, Kendo, Judo, gymnastics, swimming, and track and field.

The numbers of students belonging to the school sports clubs were extracted from the records of the Nippon Junior High School Physical and Culture Association, All Japan High School Athletic Federation, and Japan High School Baseball Federation. The rate of sports-related injuries was shown as the number of injuries per 100,000 students per year (/100,000 per year). Severe head and neck injuries were defined as those incurring medical expenses of greater than 100,000 yen during the first month after the first hospital visit. The data were collected with reference to the amount of reimbursement for medical costs, regardless of the symptoms or severity of the injury; therefore, these data cannot be directly compared with other sports injury surveillance results.

2.3 Results

2.3.1 *The Characteristics of the 11 Sports*

The Number and Rate of Sports-Related Injuries

The number and rate of injuries increased slightly over the course of the study period (Table 2.1, Fig. 2.1). This trend indicates that there is an urgent need to promote and execute a program for preventing such injuries.

Table 2.1 The number and rate of sports-related injuries from all 11 sports

Year	Injuries, no.	Injury rate
2009	256,369	8,864
2010	264,369	9,294
2011	268,608	9,452

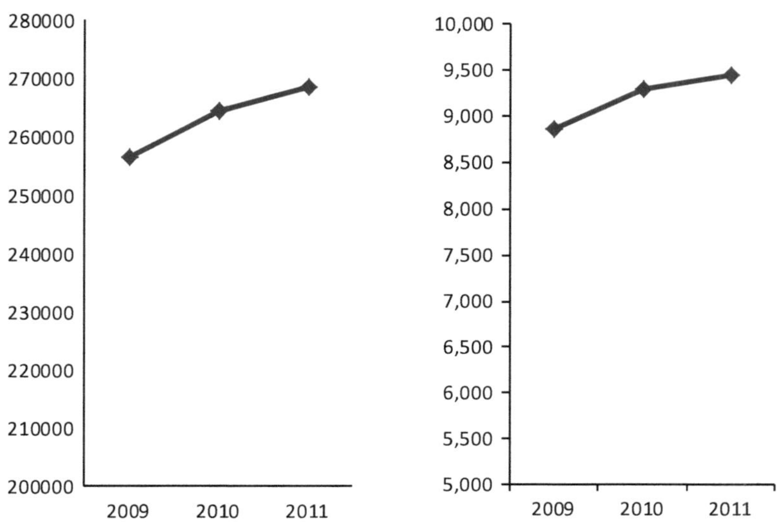

Fig. 2.1 The number (left) and rate (right) of sports-related injuries from all 11 sports

The Number and Rate of Sports-Related Injuries in Male and Female Students

Although the number of injuries for male students was more than twice that of female students, the rate of injury did differ minimally with respect to sex. The number of injuries increased slightly each year for both sexes (Table 2.2, Fig. 2.2).

Table 2.2 The number and rate of sports-related injuries in male and female students

Sex	Year	Injuries, no.	Injury rate
Female	2009	80,440	8,188
	2010	82,574	8,639
	2011	84,651	8,914
Male	2009	175,929	9,211
	2010	181,795	9,625
	2011	183,957	9,723

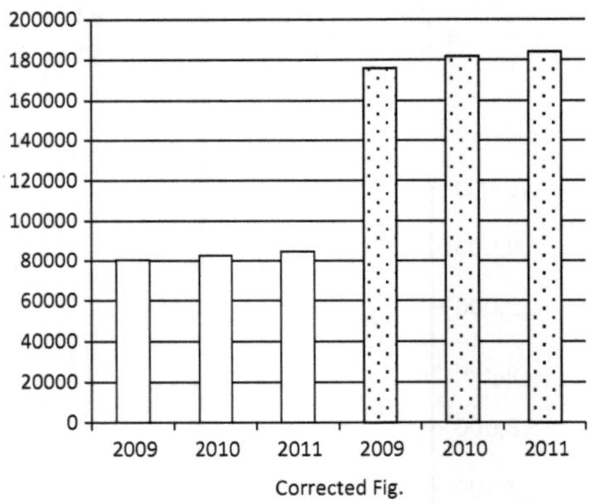

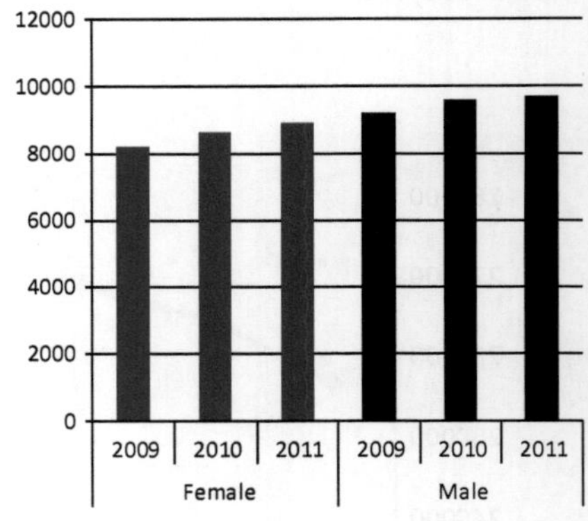

Fig. 2.2 The number (left: corrected Fig.) and rate (right) of sports-related injuries in male and female students

The Number and Rate of Sports-Related Injuries in Junior and Senior High School Students

The number of injuries was 1.5-fold higher in junior high school students than in senior high school students and increased slightly each year for both junior and senior high school students. The rate of injury was slightly higher in senior high school than in junior high school students (Table 2.3, Fig. 2.3).

Table 2.3 The number and rate of sports-related injuries in junior and senior high school students

	Junior high school			Senior high school		
Injury	2009	2010	2011	2009	2010	2011
Injuries, no.	157,347	161,817	161,545	99,022	102,552	107,063
Injury rate	8,351	8,790	8,723	9,669	10,973	10,639

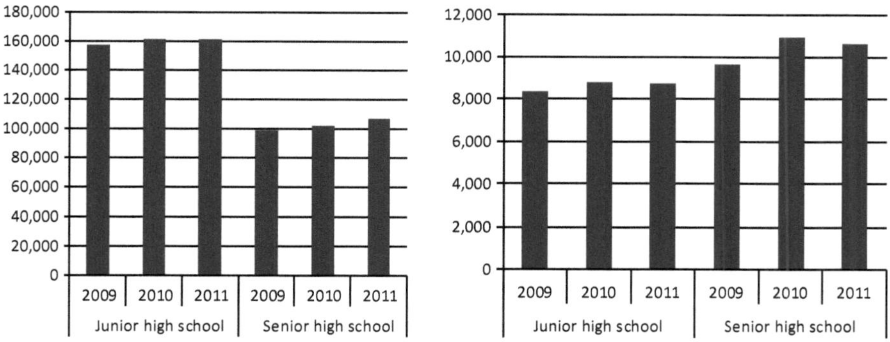

Fig. 2.3 The number (left) and rate (right) of sports-related injuries in junior and senior high school students

The Number of Sports-Related Injuries by Month

The number of injuries demonstrated a similar trend during the year for each of the three years of the study period: the greatest number of sports-related injuries occurred in the beginning of the school term, especially in May. This shows the importance of executing the prevention program beginning at the start of the school term and continuing it through the summer (Table 2.4, Fig. 2.4).

Table 2.4 The number of sports-related injuries by month

Month\year	2009	2010	2011
April	28,313	26,965	27,553
May	29,389	30,569	29,828
June	27,404	27,649	27,367
July	27,401	27,259	28,370
August	24,821	25,209	26,623
September	22,762	23,859	23,852
October	16,293	20,399	21,681
November	14,107	16,629	17,645
December	15,821	16,300	17,718
January	17,232	17,807	18,414
February	13,096	12,515	12,571
March	19,730	19,209	16,986

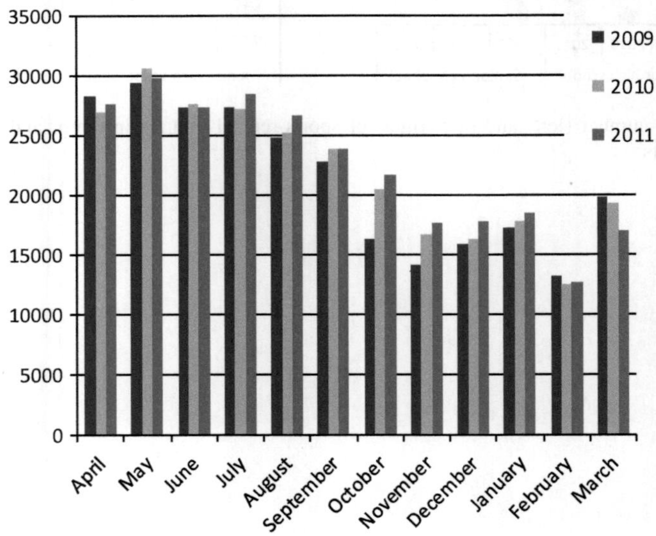

Fig. 2.4 The number and rate of sports-related injuries by month

The Number of Sports-Related Injuries by Body Part

The body part most frequently injured was the ankle joint, followed by the hand and finger, head, and knee joint (Table 2.5, Fig. 2.5).

Table 2.5 The number of sports-related injuries by body part

Body part\year	2009	2010	2011
Head	41,326	42,160	41,366
Neck	2,566	2,701	2,424
Shoulder complex	8,957	9,055	9,275
Upper arm	2,120	2,123	2,523
Elbow	6,711	6,805	6,710
Forearm	8,466	8,491	8,681
Wrist	9,735	9,958	10,511
Hand finger	48,001	48,529	48,315
Thoracoabdominal	4,379	4,495	4,094
Lumbar	10,684	10,934	11,634
Buttock	1,030	3,056	1,027
Hip and thigh	9,165	7,442	10,515
Knee	22,633	23,409	23,711
Shank	11,479	11,676	12,006
Ankle	51,922	55,172	57,731
Foot toe	15,384	15,467	15,107
Other	1,811	2,896	2,978

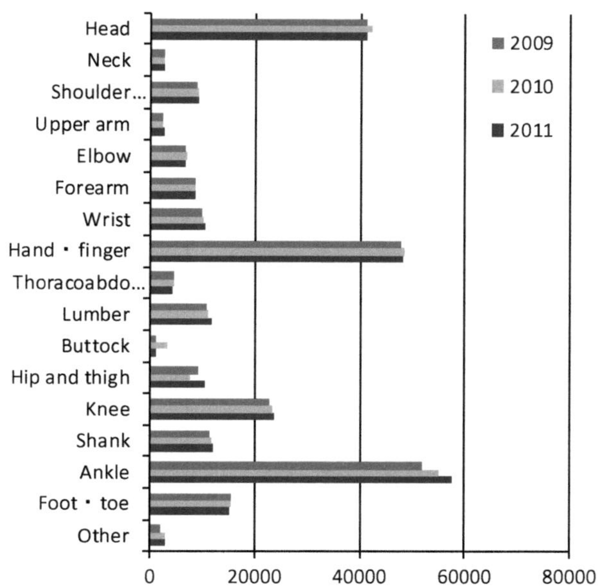

Fig. 2.5 The number of sports-related injuries by body part

The Number of Sports-Related Injuries by Type

The most frequent type of injury was bone fracture, followed by joint sprain and contusion/bruise (Table 2.6, Fig. 2.6)

Table 2.6 The number of sports-related injuries by type

Type\year	2009	2010	2011
Bone fracture	71,058	73,363	75,066
Joint sprain	64,131	63,450	65,996
Contusion bruise	60,260	60,745	66,046
Muscle tendon strain	24,575	29,211	24,522
Joint trauma	7,664	6,137	6,138
Bone trauma	5,716	6,141	6,047
Laceration abrasion	9,939	10,144	10,005
Dislocation	5,893	5,872	7,492
Tooth fracture	1,581	1,733	1,845
Eye trauma	1,299	1,537	866
Heat stroke	1,137	2,331	2,080
Skin trauma	712	624	736
Internal trauma	962	542	406
Neurologic trauma	525	956	602
Ear trauma	427	578	260
Burn	115	105	71
Food poisoning other poisoning	75	78	59
Other	300	822	371

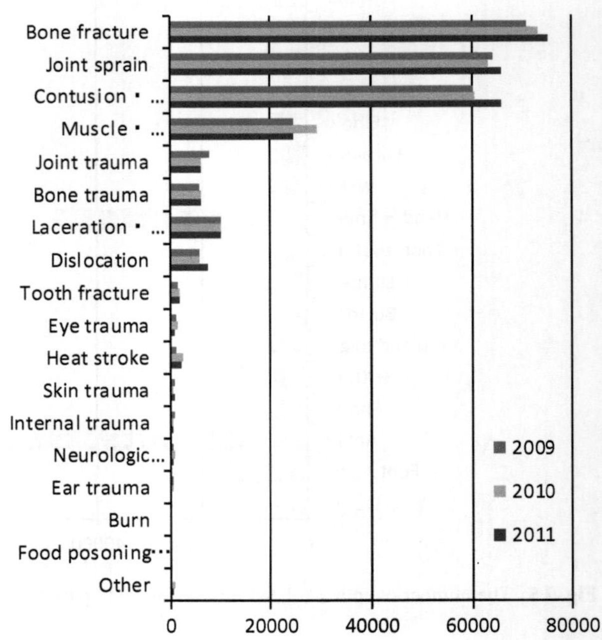

Fig. 2.6 The number of sports-related injuries by type

The Number and Rate of Sports-Related Injuries by Sport

The sport with the largest total number of injuries was basketball, followed by soccer and baseball (Table 2.7, Fig. 2.7). The sport with the highest frequency of injury was rugby, followed by Judo and basketball (Table 2.8, Fig. 2.8). Notably, these results showed a trend towards a gradual increase in the number and rate of sports-related injuries for many sports.

Table 2.7 The number of sports-related injuries by sport

Sport\year	2009	2010	2011
Basketball	64,448	67,495	69,335
Soccer	48,304	50,594	53,465
Baseball	48,603	48,400	45,580
Volleyball	33,722	33,789	32,824
Tennis	17,757	17,733	17,283
Track and field	11,804	13,547	17,226
Judo	11,803	11,864	11,749
Rugby	10,057	10,320	10,881
Kendo	6,563	6,943	6,684
Gymnastics	2,040	2,315	2,601
Swimming	1,268	1,369	980

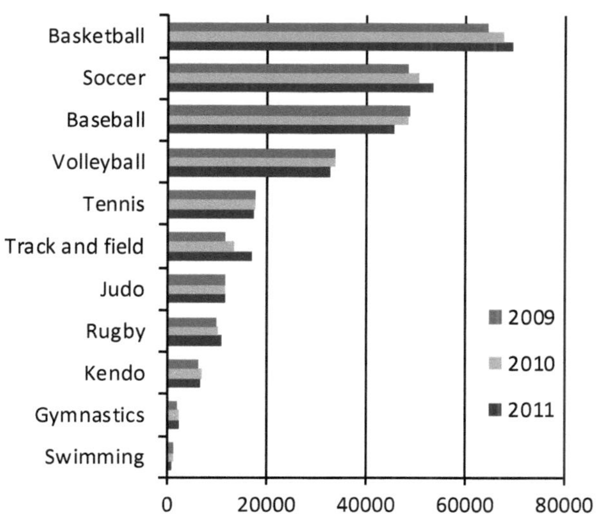

Fig. 2.7 The number of sports-related injuries by sport

Table 2.8 The rate of sports-related injuries by sport

Sport\year	2009	2010	2011
Rugby	29,204	32,051	33,923
Judo	16,592	17,536	17,954
Baseball	13,411	14,029	14,542
Soccer	10,172	13,240	13,330
Gymnastics	10,388	11,857	13,299
Volleyball	10,293	10,851	10,697
Baseball	12,663	10,499	10,140
Kendo	4,506	4,755	4,579
Tennis	3,125	3,235	5,393
Track and field	3,799	4,268	3,164
Swimming	1,688	1,756	1,232

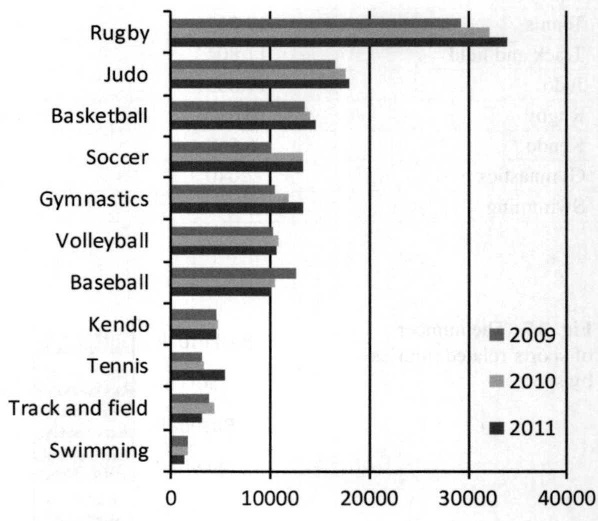

Fig. 2.8 The rate of sports-related injuries by sport

2.3.2 Analysis of Individual Sports-Related Injuries

Severe Head and Neck Injury

More than 500 cases of severe head and neck injury were reported every year, and the number increased each year (Table 2.9, Fig. 2.9). When assessed separately, the numbers of both head injuries and neck injuries also increased every year.

Table 2.9 The number (left) of severe head and neck injuries and the rate (right) of severe head injuries by year

	Injury			Rate		
Region\year	2009	2010	2011	2009	2010	2011
Head	434	461	490	18	19	20
Neck	73	86	90			

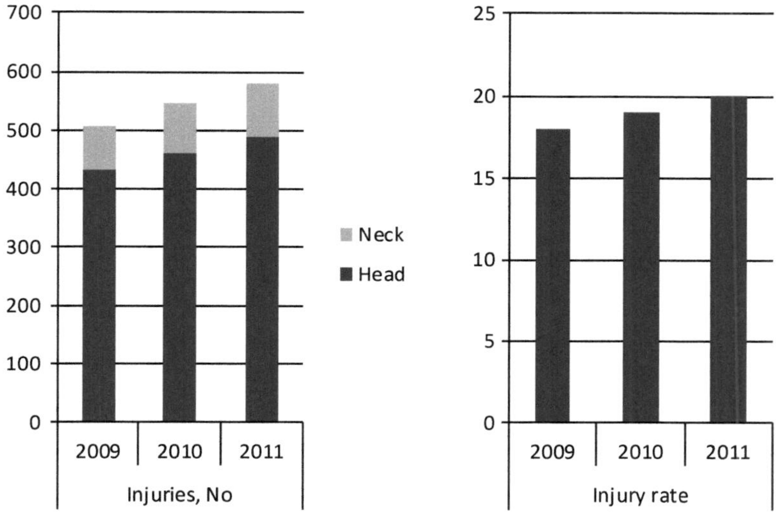

Fig. 2.9 The number (left) of severe head and neck injuries and the rate (right) of severe head injuries by year

Details of the Head Injuries

Although concussion and head contusion were the most common head injuries for all years, acute subdural hematoma or acute extradural hematoma represent approximately 50 cases each year (Table 2.10, Fig. 2.10).

Table 2.10 Details of the numbers of severe head injuries

Type\year	2009	2010	2011
Concussion	166	183	180
Head contusion	124	102	130
Acute subdural extradural hematoma	42	67	51
Bruise	36	40	47
Skull fracture (includes facial fracture)	36	21	43
Subarachnoid hemorrhage other	30	46	39

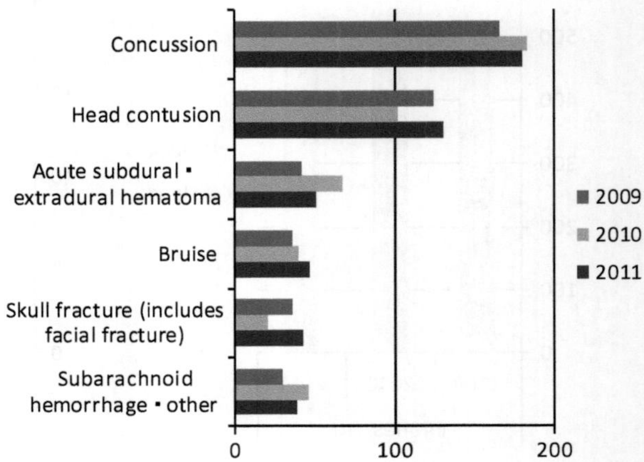

Fig. 2.10 Details of the numbers of severe head injuries

Trends in Severe Head and Neck Injuries by Sport

The sport with the largest number of severe head and neck injuries was baseball, followed by soccer and rugby. The number of severe head and neck injuries tended to increase over the study period, especially in soccer and rugby (Table 2.11, Fig. 2.11). The sport with the highest rate of severe head and neck injuries was rugby, followed by Judo and gymnastics (Table 2.12, Fig. 2.12).

Table 2.11 The number of severe head and neck injuries by sport

Sport\year	2009	2010	2011
Baseball	151	141	143
Soccer	122	132	140
Rugby	74	86	97
Judo	42	59	50
Basketball	44	46	61
Track and field	11	25	26
Volleyball	13	18	13
Kendo	14	13	18
Tennis	18	13	15
Gymnastics	9	7	11
Swimming	9	7	6

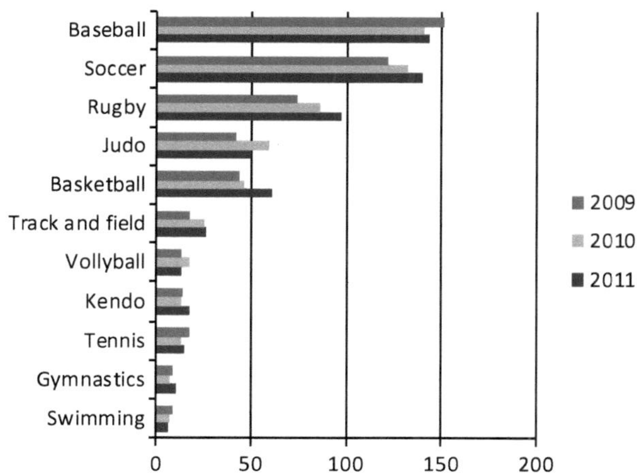

Fig. 2.11 The number of severe head and neck injuries by sport

Table 2.12 The rate of severe head and neck injury by sport

Sport\year	2009	2010	2011
Rugby	215	267	302
Judo	59	87	76
Gymnastics	46	36	56
Soccer	32	35	35
Baseball	32	31	32
Basketball	9	10	13
Kendo	10	9	3
Swimming	3	9	8
Track and field	4	8	8
Volleyball	4	6	4
Tennis	3	2	12

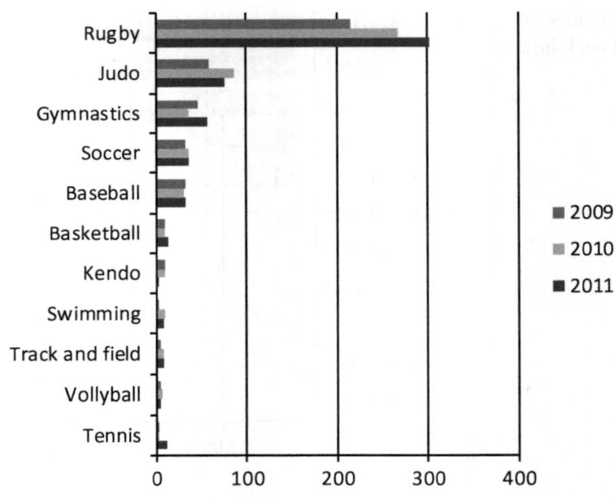

Fig. 2.12 The rate of severe head and neck injury by sport

Four Types of Specific Sports-Related Injury

Four common specific sports-related injuries (ankle sprain, anterior cruciate ligament tear, shoulder dislocation, and fifth metatarsal fracture) were investigated over the 3 years. Ankle sprain was the most frequently reported of these injuries. The number and rate of ankle sprain and shoulder dislocation were both lower in 2010 than in 2009 or 2011, but the reason for this is unknown (Table 2.13, Fig. 2.13).

First, we shows the number and rate of anterior cruciate ligament tears that is the severe injury for the athletes. Secondly, the rate of shoulder dislocation, thirdly that of ankle sprain and fourthly fifth metatarsal fractures.

Table 2.13 The number and rate of four common specific sports-related injuries

	Injuries, no.			Injury rate		
Type\year	2009	2010	2011	2009	2010	2011
Ankle sprain	44,267	39,812	47,839	1,530	1,443	1,683
Anterior cruciate ligament tear	2,439	2,577	2,642	84	91	93
Shoulder dislocation	1,289	829	1,253	45	30	44
Fifth metatarsal fracture	1,128	1,280	1,504	39	46	53

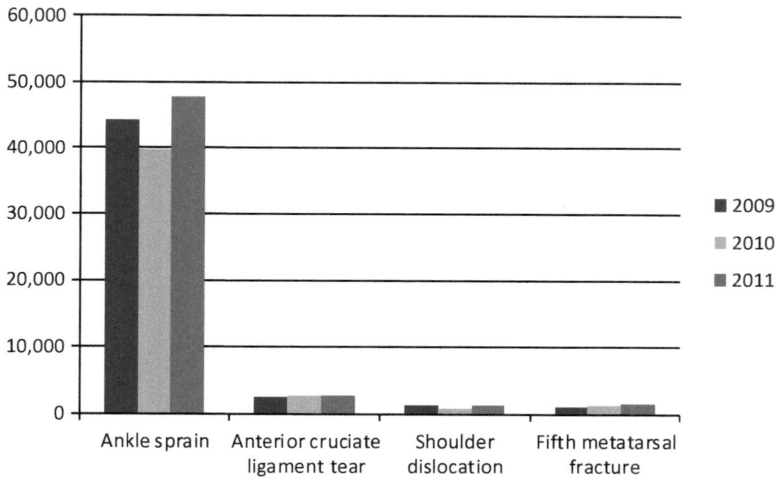

Fig. 2.13 The number and rate of four common specific sports-related injuries

Anterior Cruciate Ligament Tear

The number and rate of anterior cruciate ligament tears increased over the 3 years (Table 2.14, Fig. 2.14 and Table 2.15, Fig. 2.15). Both the number and the rate of an anterior cruciate ligament tear were higher in female than in male students for all years. The sport with the greatest number of anterior cruciate ligament tears was basketball, which accounted for half of the total caseload. The sport with the highest rate of this injury was rugby, followed by Judo and basketball.

Table 2.14 The number of anterior cruciate ligament tears by sport

Sport\year	2009	2010	2011
Basketball	1,205	1,224	1,307
Soccer	370	453	386
Volleyball	334	300	318
Judo	178	225	220
Rugby	154	153	166
Baseball	91	130	97
Track and field	28	54	51
Tennis	39	31	41
Gymnastics	29	32	41
Kendo	11	9	15
Swimming	0	2	0

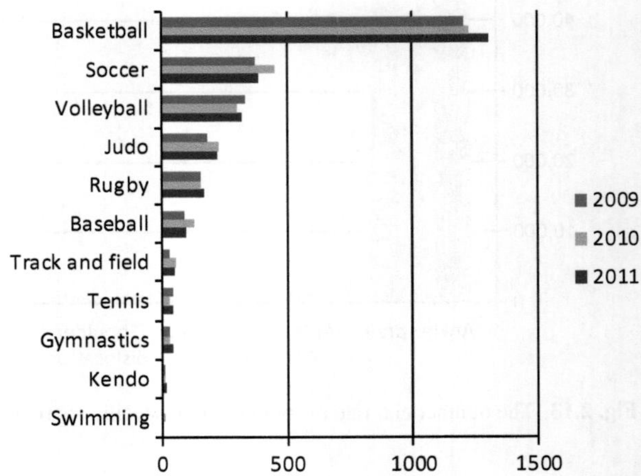

Fig. 2.14 The number of anterior cruciate ligament tears by sport

Table 2.15 The rate of anterior cruciate ligament tear by sport

Sport\year	2009	2010	2011
Rugby	447	474	518
Judo	250	333	336
Basketball	251	254	274
Gymnastics	148	164	210
Volleyball	102	96	104
Soccer	97	119	96
Baseball	19	28	22
Track and field	9	17	16
Tennis	7	7	8
Kendo	8	6	10
Swimming	0	3	0

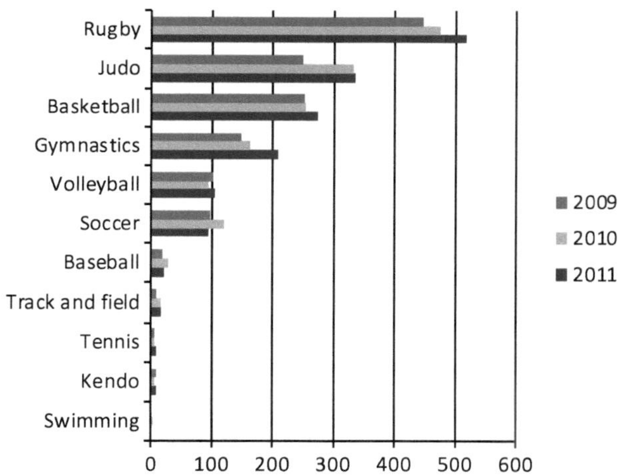

Fig. 2.15 The rate of anterior cruciate ligament tear by sport

Shoulder Dislocation

The number and rate of shoulder dislocations were higher in male than in female students. The sport with the largest number of shoulder dislocations was rugby, followed by baseball and basketball. The sport with the highest rate was also rugby, for which the rate was more than four-fold higher than for the sport with the second-highest frequency, Judo (Table 2.16, Fig. 2.16). The number and rate of shoulder dislocation were lower in 2010 than in 2009 or 2011 for many sports, but the reason for this is unknown.

Table 2.16 The rate of shoulder dislocation by sport

Sport\year	2009	2010	2011
Rugby	810	651	995
Judo	218	148	196
Gymnastics	56	31	56
Baseball	42	34	52
Basketball	35	25	42
Soccer	33	27	40
Volleyball	32	24	36
Swimming	15	7	10
Tennis	9	9	9
Track and field	9	4	7
Kendo	6	5	8

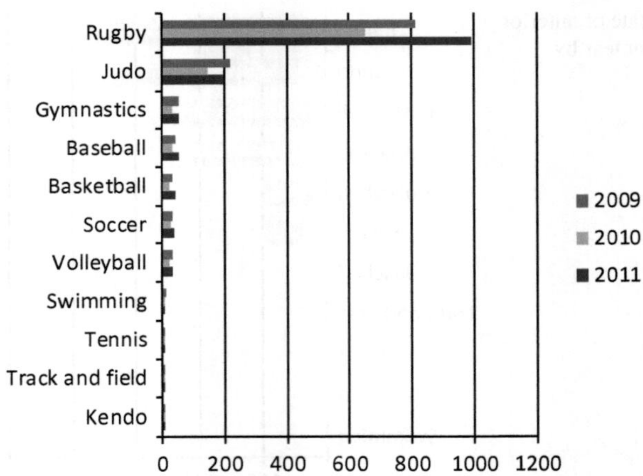

Fig. 2.16 The rate of shoulder dislocation by sport

Ankle Sprain

While there was no sex difference in the total number of ankle sprains, the rate of ankle sprain was almost twice as high in female as in male students. The number and rate of ankle sprains tended to be higher in ball sports, such as basketball and volleyball. Both basketball and volleyball produced more than 10,000 cases each year, and both sports showed an injury rate of greater than 3,000 per 100,000 participants per year (Table 2.17, Fig. 2.17).

Table 2.17 The rate of ankle sprain by sport

Sport\year	2009	2010	2011
Basketball	3,475	3,149	3,772
Volleyball	3,153	2,980	3,452
Rugby	2,625	2,105	2,996
Soccer	1,845	1,706	2,014
Gymnastics	1,492	1,547	2,081
Judo	1,113	986	1,241
Baseball	654	558	890
Track and field	589	768	491
Tennis	458	519	620
Kendo	382	349	416
Swimming	129	132	99

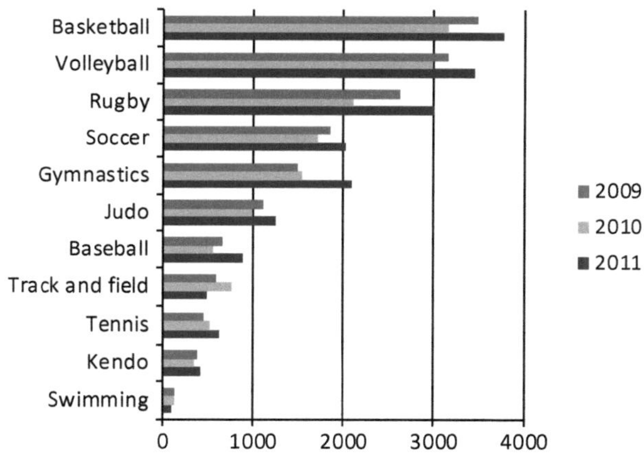

Fig. 2.17 The rate of ankle sprain by sport

Fifth Metatarsal Fracture (Including Stress Fracture)

The number and rate of fifth metatarsal fractures were higher in male than in female students. Although the number and rate of this injury showed no trend across the 3 years, they tended to be higher for gymnastics, soccer, Kendo, Rugby, and volleyball (Table 2.18, Fig. 2.18).

Table 2.18 The rate of fifth metatarsal fractures by sport

Sport\year	2009	2010	2011
Gymnastics	117	92	123
Kendo	105	93	35
Rugby	93	68	100
Soccer	87	95	104
Judo	62	71	84
Basketball	56	65	67
Volleyball	42	42	103
Baseball	31	25	46
Track and field	21	28	15
Swimming	15	14	4
Tennis	14	16	37

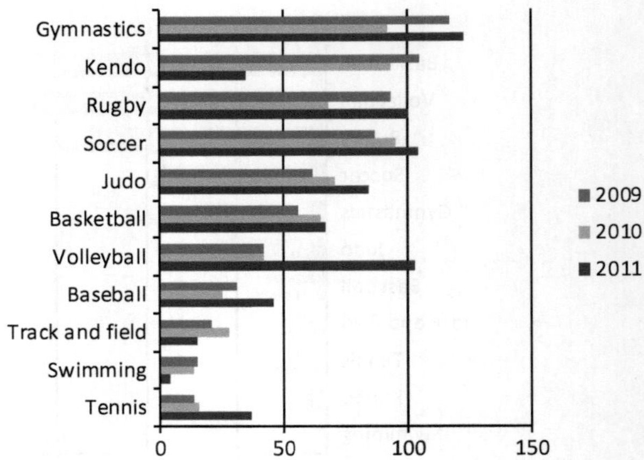

Fig. 2.18 The rate of fifth metatarsal fractures by sport

2.3.3 *Conclusion*

The Sports Injury Surveillance system has not yet been established in Japan, and most of the reported statistics on sports-related injuries reflect estimated values from a limited number of previous studies. The present study reveals the characteristics of sports-related injuries that occurred during school sports club activities at junior and senior high schools.

Continued investigation of the characteristics of sports-related injuries using the Sports Injury Surveillance system will be needed in the future on order to reveal the mechanisms and causes of such injuries and to promote a program for their prevention. The Sports Injury Surveillance system will also be necessary for valid statistical analysis of the effects of such a prevention program (i.e., an accurate recording of the number and rate of sports-related injuries). We anticipate that the establishment of the Sports Injury Surveillance system in Japan will help all athletes, including junior and senior high school students, due to an improved ability to prevent sports-related injuries.

The Numbers of Students Affiliated with the 11 Clubs

The numbers of students were extracted from the affiliation numbers reported by the Nippon Junior High School Physical and Culture Association, All Japan High School Athletic Federation, and Japan High School Baseball Federation. However, the numbers of female baseball and rugby athletes were excluded because they were unknown.

2009

Sport	Junior high school			Senior high school			Total		Grand total
(school club)	Male	Female	Total	Male	Female	Total	Male	Female	
Baseball	307,053	1,333	308,386	169,449	*	169,449	476,502	1,333	477,835
Soccer	223,951	3,429	227,380	145,522	8,551	154,073	369,473	11,980	381,453
Tennis	173,514	203,385	376,899	114,769	76,513	191,282	288,283	279,898	568,181
Basketball	172,342	154,917	327,259	89,786	63,506	153,292	262,128	218,423	480,551
Track and field	122,512	90,294	212,806	61,867	36,060	97,927	184,379	126,354	310,733
Kendo	62,095	34,501	96,596	33,256	15,798	49,054	95,351	50,299	145,650
Volleyball	51,958	171,263	223,221	39,594	64,807	104,401	91,552	236,070	327,622
Judo	33,604	9,461	43,065	22,832	5,238	28,070	56,436	14,699	71,135
Swimming	28,050	17,070	45,120	18,254	11,738	29,992	46,304	28,808	75,112
Rugby	7,748	119	7,867	26,570	*	26,570	34,318	119	34,437
Gymnastics	2,133	9,212	11,345	3,083	5,210	8,293	5,216	14,422	19,638
Total	1,184,960	694,984	1,879,944	724,982	287,421	1,012,403	1,909,942	982,405	2,892,347

2010

Sport (school club)	Junior high school			Senior high school			Total		Grand total
	Male	Female	Total	Male	Female	Total	Male	Female	
Baseball	291,015	1,505	292,520	168,488	*	168,488	459,503	1,505	461,008
Soccer	221,407	3,538	224,945	148,764	8,421	157,185	370,171	11,959	382,130
Tennis	167,674	193,279	360,953	113,545	73,653	187,198	281,219	266,932	548,151
Basketball	174,443	153,046	327,489	91,034	62,598	153,632	265,477	215,644	481,121
Track and field	124,611	90,833	215,444	65,234	36,716	101,950	189,845	127,549	317,394
Kendo	60,881	36,312	97,193	32,672	16,138	48,810	93,553	52,450	146,003
Volleyball	50,621	160,867	211,488	38,335	61,575	99,910	88,956	222,442	311,398
Judo	31,434	9,207	40,641	21,795	5,220	27,015	53,229	14,427	67,656
Swimming	29,229	16,876	46,105	19,997	11,876	31,873	49,226	28,752	77,978
Rugby	6,775	45	6,820	25,379	*	25,379	32,154	45	32,199
Gymnastics	2,447	9,071	11,518	2,946	5,061	8,007	5,393	14,132	19,525
Total	1,160,537	674,579	1,835,116	728,189	281,258	1,009,447	1,888,726	955,837	2,844,563

2011

Sport (school club)	Junior high school			Senior high school			Total		Grand total
	Male	Female	Total	Male	Female	Total	Male	Female	
Baseball	280,917	1,658	282,575	166,925	*	166,925	447,842	1,658	449,500
Soccer	237,783	3,946	241,729	150,655	8,713	159,368	388,438	12,659	401,097
Tennis	166,815	196,129	362,944	110,375	71,088	181,463	277,190	267,217	544,407
Basketball	178,468	146,601	325,069	89,510	62,225	151,735	267,978	208,826	476,804
Track and field	127,248	91,168	218,416	65,646	36,403	102,049	192,894	127,571	320,465
Kendo	61,113	36,974	98,087	31,484	16,412	47,896	92,597	53,386	145,983
Volleyball	50,299	161,691	211,990	35,721	59,151	94,872	86,020	220,842	306,862
Judo	30,936	9,001	39,937	20,638	4,864	25,502	51,574	13,865	65,439
Swimming	30,276	17,438	47,714	20,040	11,763	31,803	50,316	29,201	79,517
Rugby	7,054	40	7,094	24,982	*	24,982	32,036	40	32,076
Gymnastics	2,232	9,468	11,700	2,910	4,948	7,858	5,142	14,416	19,558
Total	1,173,141	674,114	1,847,255	718,886	275,567	994,453	1,892,027	949,681	2,841,708

Chapter 3
Injury and Illness Surveillance Among Olympic Athletes: Summary of the 2010 Winter, and the 2008 and 2012 Summer Olympic Games

Kathrin Steffen and Lars Engebretsen

Abstract Protection of the athletes' health is a clearly articulated objective of the International Olympic Committee (IOC). Longitudinal surveillance of injuries and illnesses can provide valuable data that may identify high-risk sports and disciplines. Such surveillance would build a foundation for introducing tailored preventive measures. During the XXIX (Beijing) and XXX (London) Summer Olympic Games and the XXI (Vancouver) Winter Olympic Games, comprehensive injury and illness recording by the medical staff of the participating National Olympic Committees and the sports medicine clinics at the different Olympic venues revealed that between 7 and 11 % of all athletes incurred an injury or suffered from at least one occurrence of illness during the Games. The incidence of injuries and illnesses varied substantially between sports. In the future, risk factor and injury mechanism analyses in high-risk Olympic sports are essential to better direct injury prevention strategies. Concomitantly, Periodic Health Evaluations of athletes would be instrumental in optimizing health protection.

Keywords Olympic athlete • Health protection • Surveillance • Prevention

3.1 Introduction

With almost 11,000 athletes from more than 200 countries, the XXX London 2012 Olympic Games were one of the largest sports events ever. More than 2,500 athletes participated in the XXI Winter Olympic Games held in Vancouver in 2010. The protection of the athlete's health is an important task for the International Olympic Committee (IOC) (Ljungqvist et al. 2009). Systematic injury and illness

K. Steffen (✉) • L. Engebretsen
Department of Sports Medicine, Oslo Sports Trauma Research Center, Norwegian School of Sport Sciences, Oslo, Norway
e-mail: kathrin.steffen@nih.no

K. Kanosue et al. (eds.), *Sports Injuries and Prevention*,
DOI 10.1007/978-4-431-55318-2_3

surveillance monitoring over long periods of time, and the identification of high-risk sports, including the most common and severe injuries and illnesses, is a starting place from which to address injury prevention (Junge et al. 2008, 2009; Engebretsen et al. 2010, 2013). Following the six stage TRIPP model of Finch (2006), analysing the extent of a problem such as a high injury and/or illness risk in a specific population is the first step in the development of an effective prevention strategy.

The IOC injury and illness surveillance system, developed in cooperation with the International Sports Federations (IFs) and National Olympic Committees (NOCs), was successfully implemented in the 2008 Beijing Olympics (injury surveillance only) (Junge et al. 2009) and expanded to also include the registration of illness cases in the 2010 Vancouver Olympics (Engebretsen et al. 2010). The system was further developed and optimized for the 2012 London Olympic Games (Engebretsen et al. 2013) as well as the 2014 Sochi Olympic Games. As early as 1998, the Fédération Internationale de Football Association (FIFA) initiated a survey that included all injuries incurred during their competitions (Junge and Dvorak 2007, 2010, 2013; Junge et al. 2004; Dvorak et al. 2007, 2011). Other major sports federations followed FIFA's Medical Assessment and Research Centre (F-MARC) lead (Bahr and Reeser 2003; Alonso et al. 2009, 2010, 2012; Mountjoy et al. 2010). Based on the experience of these federations, a group of experts, convened by the IOC, developed an injury surveillance program for multi-sport events (Junge et al. 2008). During the 2008 Beijing Olympic Games, the IOC performed surveillance on all sports for the first time (Junge et al. 2009).

Continuous injury and illness surveillance during major sporting events will build a foundation for providing evidence useful for the development of injury prevention programs (Ljungqvist et al. 2009). The aim of this book chapter is to summarize the injuries and illnesses that occurred in the previous three Olympic Games. This will enable the National Olympic Committee (NOC) physicians and medical personnel to better prepare for up-coming Olympic Games. In order to better protect the athletes' health, practical implications and suggestions for further research are provided.

3.2 Methods

In Beijing, Vancouver, and London, all National Olympic Committees' head physicians participated in the Olympic surveillance studies and reported daily on the occurrence (or non-occurrence) of newly sustained injuries and illnesses (only injuries in Beijing) on a standardized reporting form. The injury definition and data collection procedures are described comprehensively elsewhere (Junge et al. 2008, 2009; Engebretsen et al. 2010, 2013). In addition, information on all athletes treated for injuries and illnesses by the Local Organizing Committee (LOC) medical services were retrieved from available medical centers located at select venues. Athletes seen for an injury or illness in the venue medical stations or central clinics were assessed through the central clinic database.

3.2.1 Injury and Illness Report Form

The report form on injuries and illnesses required broad documentation of the respective injury and illness. A booklet was provided with detailed instructions, as well as examples about how to fill out the form correctly. Daily injury information was also received from polyclinics in the Olympic Villages. Injury and illness report forms were distributed to the NOCs in all required languages (Junge et al. 2008, 2009; Engebretsen et al. 2010, 2013).

3.2.2 Definition of Injury and Illness

An athlete was defined as injured or ill if he/she received medical attention, regardless of the consequences with respect to absence from competition or training. Following the IOC injury surveillance system, an injury was expected to be reported if it fulfilled the following criteria: (1) musculo-skeletal complaint or concussion, (2) newly incurred (pre-existing, not fully rehabilitated should not be reported) or re-injuries (if the athlete has returned to full participation after the previous injury), (3) occurred in competition or training, and (4) occurred during the XXIX Summer Olympic Games 2008 (August 9–24, 2008), the XXI Winter Olympic Games 2010 (February 12–28, 2010), and the XXX Summer Olympic Games 2012 (July 27-August 12, 2012). An illness was defined as any physical complaint (not related to injury) newly occurred during the Games that received medical attention regardless of the consequences with respect to absence from competition or training. All information was handled confidentially and injury reports were anonymized after the Olympic Games. Ethical approval was obtained by the Regional Committee for Medical Research Ethics, Region Øst-Norge, Norway.

3.3 Results

3.3.1 Injury Risk in Different Sports

All NOCs with more than 30 (Summer Games) or 10 registered athletes (Winter Games) were included in the analysis of response-rate, and these countries represented more than 94 % of all participating athletes (Junge et al. 2009; Engebretsen et al. 2010, 2013). In the three Olympics, the overall injury rate was similar at

Table 3.1 Comparison between injuries sustained by athletes participating in Summer (Beijing 2008, London 2012) and Winter Olympic Games (Vancouver 2010)

	Beijing 2008	London 2012	Vancouver 2010
Participating athletes	10,977	10,568	2,567
Injuries (per 1,000 athletes)	1,055 (96.1)	1,361 (128.8)	287 (111.8)
Most common diagnosis	Ankle sprains (7 %), thigh strains (7 %)	Most severe injuries: shoulder, elbow, and knee dislocations, muscle strains and ruptures, fractures/stress fractures, ligament sprains and ruptures, incl ACL, tendon ruptures	Concussions (7 %)
Most affected locations	Trunk (13 %), thigh (13 %), head/neck (12 %), knee (12 %)		Head/neck (16 %), knee (14 %), thigh (7 %)
Most common mechanisms	Non-contact (20 %)	Contact with another athlete (14 %)	Contact with another athlete (15 %)
	Overuse (22 %)	Non-contact (20 %)	Contact with a stationary object (22 %)
	Contact with another athletes (33 %)	Overuse (25 %)	Non-contact (23 %)
Expected time-loss injuries	50 %	35 %	23 %[a]
Competition – training injuries	73–27 %	55–45 %	46–54 %
High risk sports (injuries per 100 athletes)	Football, taekwondo, field Hockey, handball, weightlifting	Football, taekwondo, BMX, handball, MTB, athletics	Snowboard cross, freestyle aerials and cross, bobsleigh, ice hockey
Low risk sports (injuries per 100 athletes)	Canoeing/kayaking, diving, rowing, sailing, synchronized swimming, fencing	Archery, canoe slalom and sprint, track cycling, rowing, shooting, equestrian	Nordic skiing disciplines, curling, speed skating

[a]This figure may underestimating the number of time-loss injuries as the response rate to this information was low and many of the injuries were of severe outcome, without estimated time-loss registered (more details in the Vancouver paper) (Engebretsen et al. 2010)

around 11 % of registered athletes sustaining at least one injury (Table 3.1). In Beijing, a total of 1,055 injuries were reported among 10,977 athletes, equivalent to an incidence of 96.1 injuries per 1,000 registered athletes (Junge et al. 2009) Among the 2,567 registered athletes (1,045 females, 1,522 males) in Vancouver, a total of 287 injuries were reported resulting in an injury rate of 111.8 injuries per 1,000 registered athletes (Engebretsen et al. 2010). In London, 10,568 athletes (4,676 females, 5,892 males) sustained 1,361 injuries, equalling 128.8 injuries per 1,000 participating athletes (Engebretsen et al. 2013).

The incidence of injuries varied substantially among the different sports in all three Games (Table 3.2). The risk of sustaining an injury was highest for football (soccer), taekwondo, BMX and MTB cycling, field hockey, handball, weightlifting, and boxing (all ≥15 % of the athletes) (Junge et al. 2009; Engebretsen et al. 2013). In Vancouver, injury risk was highest for bobsleigh, ice hockey, short track, alpine, and for freestyle and snowboard cross (15–35 % of registered athletes were affected in each sport) (Engebretsen et al. 2010).

Table 3.2 Injury and illness distribution (percentage of participating athletes) for selected sports on the program for the 2008 Beijing, 2010 Vancouver, and 2012 London Olympic Games

	Injuries			Illnesses	
Olympic sports	Beijing 2008	Vancouver 2010	London 2012	Vancouver 2010	London 2012
Alpine skiing		14.9		4.2	
Aquatics					
Diving	2.1		8.1		5.1
Swimming	3.4		5.4		7.3
Synchronized swimming	1.9		13.5		12.5
Water polo	9.7		13.1		
Archery	7.0		1.6		7.8
Athletics	11.3		17.7		10.5
Badminton	4.7		15.9		3.0
Baseball	11.1		–		
Basketball	13.2		11.1		3.1
Beach volleyball	8.3		12.5		18.8
Biathlon		1.5		11.4	
Bobsleigh		20.0		4.4	
Boxing	14.9		9.2		6.4
Canoeing/kayaking	1.2		5.2		10.4
Cross country skiing		3.1		6.8	
Curling		4.0		10.0	
Cycling	5.8				
BMX			31.3		4.2
MTB			21.1		6.6
Road			9.0		3.3
Track			3.0		19.6
Equestrian	5.2		4.5		5.5
Fencing	2.4		9.3		5.3
Field hockey	20.4		17.0		7.5
Figure skating		14.3	12.3		
Football	31.5		35.2		12.2

(continued)

Table 3.2 (continued)

Olympic sports	Injuries Beijing 2008	Injuries Vancouver 2010	Injuries London 2012	Illnesses Vancouver 2010	Illnesses London 2012
Freestyle					
Aerials		19.1		2.1	
Cross		19.0		2.9	
Moguls		1.8		0	
Gymnastics	7.5				
Artistic			7.7		2.6
Rhythmic			7.3		1.0
Trampoline			6.3		3.1
Handball	17.4		21.8		4.9
Ice hockey		18.5		5.6	
Judo	11.2		12.3		4.2
Luge		1.9		6.5	
Modern pentathlon	5.6		8.3		1.4
Nordic combined		1.9		7.8	
Rowing	1.8		3.3		7.3
Sailing	0.8		14.7		10.0
Shooting	7.8		3.8		4.4
Short track		9.0		9.2	
Skeleton		6.4		10.6	
Ski jumping		4.5		1.5	
Snowboard					
Cross		35.0		10.5	
Half pipe		13.0		5.8	
Slalom		6.8		6.8	
Softball	13.4		–		–
Speed skating		2.8		12.5	
Table tennis	5.2		6.3		6.9
Taekwondo	27.0		39.1		10.9
Tennis	5.9		11.4		2.2
Triathlon	9.2		14.5		6.4
Volleyball	8.0		6.9		2.8
Weightlifting	16.9		17.5		4.0
Wrestling	9.4		12.0		4.7
Total					

3.3.2 *Injury Location and Type*

In the Summer Olympic Games of Beijing and London, the distribution of injuries was as follows: About half of the diagnoses affected the lower extremity, with knee and shoulder dislocations, thigh strains, fractures/stress fractures and tendon ruptures the most severe injuries (Beijing, London) (Junge et al. 2009; Engebretsen et al. 2013). In Vancouver, the face, head, and cervical spine (21 %) and knee (14 %) were the most prominent injury locations. In alpine, freestyle and snowboarding, 22 out of 102 injuries (22 %) affected the head/cervical spine, and one quarter of all injuries affected the knee (24 %). Twenty concussions were reported, affecting 7 % of the registered athletes. A death occurred in luge (Engebretsen et al. 2010).

3.3.3 Injury Mechanism, Circumstance, and Severity

The four most commonly reported injury mechanisms for summer sports were overuse, non-contact trauma, contact with another athlete, and contact with a stationary object (Junge et al. 2009; Engebretsen et al. 2013). In Vancouver, the three most commonly reported injury mechanisms were non-contact trauma (23 %), contact with a stagnant object (22 %), and contact with another athlete (15 %) (Table 3.1) (Engebretsen et al. 2010). In Beijing, about half of the injuries were expected to prevent the athletes from further training or competition (50 %). In London, these numbers were 35 %. Physicians estimated that 13–33 % of the injuries would result in an absence from sport of up to 1 week (Junge et al. 2009; Engebretsen et al. 2013). In Vancouver, of the 287 injuries, 65 (23 %) were expected to result in a time-loss situation for the athlete (Engebretsen et al. 2010).

3.3.4 Incidence and Distribution of Illnesses

Illnesses were reported from the participants of a variety of sports (Table 3.2), and mostly affected the respiratory system. In Vancouver (Engebretsen et al. 2010) as in London (Engebretsen et al. 2013), the most frequent diagnosis was an upper respiratory tract infection (pharyngitis, sinusitis, tonsillitis).

3.4 Discussion

This book chapter summarizes the first three IOC surveillance projects and provides a direct comparison of the injury and illness occurrence of athletes participating in the Summer and Winter Olympic Games. The findings reveal that of all participating athletes in the three most recent Olympic Games (2008–2012), about 11 % suffered an injury and 7 % an illness. Certain team sports (such as soccer, ice hockey, field hockey, handball, and basketball), martial arts or weight class sports (such as Taekwondo, boxing, and weightlifting), and speed sports (such as BMX, MTB, bobsled, and the skiing and snowboard disciplines) emerge as sports with high injury risk (Junge et al. 2009; Engebretsen et al. 2010, 2013). Upper respiratory tract infection was the major cause of illness during the Vancouver and London Games.

In both summer and winter sports, many of the injuries occur as a result from athlete-to-athlete contact, typical for team sports characterized by tackling or checking. Many injuries that occur during the Winter Olympics involve high speeds. In freestyle and snowboard cross, for example, athletes race while passing challenges such as turns, jumps, and waves. Combined with the speed component, competing in heats may promote an additional risk-taking attitude for the athletes (Bakken et al. 2011; Randjelovic et al. 2014) Parallels can be drawn with velodrome and road cycling when the athletes position themselves for the final sprint. Overuse injuries constitute a high proportion of injuries in sports as swimming (Mountjoy et al. 2010), athletics events (Alonso et al. 2012; Jacobsson et al. 2013), beach volleyball (Bahr and Reeser 2003), and cycling (Clarsen et al. 2010). Identifying overuse injuries, including their injury mechanisms, through the current injury surveillance methodology is still a challenge and a new standardized methodology to quantify overuse injuries in surveillance studies needs to be developed (Bahr 2009; Clarsen et al. 2013).

The lowest injury risk during the Summer Olympics was observed for water sports such as canoeing/kayaking, rowing, equestrian events, and swimming (Junge et al. 2009; Engebretsen et al. 2013). The low injury risk for athletes competing in the Nordic Skiing disciplines, as compared to the alpine events of freestyle and snowboard athletes, is not surprising as Nordic competitors are not exposed to high speeds on icy surfaces with minimal protection (Bakken et al. 2011; Bere et al. 2011a, b; Randjelovic et al. 2014).

In Vancouver, a major concern was that every 5th injury affected the head, neck, and cervical spine. These injuries were mainly diagnosed as abrasions, skin lesions, contusions, fractures or concussions. The figures for concussions were twice as high as those reported from the Summer Olympic Games (Engebretsen et al. 2010). Consequently, for the relevant sports events, identifying the underlying mechanisms for concussion incidents would be of major importance. In many cases, head and knee injuries result in a long absence from training and competition, and thus the prevention of concussions and severe knee ligament sprains, including anterior cruciate ligament ruptures, is important.

A change in injury incidence can be the result of changes in competition rules, in equipment or in other environmental factors. Changes in injury rates can also follow an increased or reduced awareness among both the athletes and their medical staff in recognizing and reporting even minor incidents (broad injury and illness definitions are utilized in the IOC surveillance studies). In certain sports, changes may also be attributable to more comprehensive and accurate data reporting by team physicians who over time have been trained as injury and illness recorders through the implementation of surveillance systems by their own Federation. Also, rate differences (lower or higher) may simply be the result of the natural fluctuation (variability) of an athlete's exposure to risk, an observation that emphasizes the value of on-going surveillance to monitor trends over time, e.g. the effect of rule or equipment changes in the period between major sports events.

3.4.1 Incidence and Distribution, Type and Cause of Illnesses

The incidence figures of illnesses from Vancouver and London is comparable for data from athletics (7 %) (Alonso et al. 2012; Jacobsson et al. 2013), aquatics (7 %) (Mountjoy et al. 2010), and football (12 %) (Dvorak et al. 2011). Almost two thirds of the illnesses affected the respiratory system were caused by infections, which is a higher rate than reported in swimming (respiratory system 50 %, infection 49 %) (Fitch 2012). Airway inflammation has been shown to often affect elite swimmers, ice-hockey players, and cross-country skiers (Kippelen et al. 2012; Fitch 2012). It has been documented that good sanitation, early recognition and the timely isolation of ill players can successfully reduce infections and illness in a team setting (Kippelen et al. 2012).

3.4.2 Practical Implications and Further Research

Introducing and implementing successful preventive measures relies, in part, on the proper characterization of risk factors and mechanisms (Finch 2006). Two recent reports describing situations leading to serious injuries in World Cup alpine skiing and snowboarding revealed that individual technical errors and inappropriate tactical choices and technical errors at take-off for jumping were primary causes of the injuries (Bakken et al. 2011; Bere et al. 2011b).

Analysing the inciting event of an injury, the moment when the injury actually occurs, will also be crucial for a better understanding of high-risk injury situations in summer sports. Two main injury mechanisms have been identified for anterior cruciate ligament injury; one is during plant and cut and 1-legged jump shot landings in team handball and basketball; the other occurs when a large valgus force is combined with internal tibial rotation with the knee close to full extension (Olsen et al. 2004; Koga et al. 2010, 2011). Hamstring strain injuries, typically occur during maximal acceleration and deceleration movements. A recent paper showed that hamstring muscles are most susceptible to strain injuries during the late swing phase of sprinting (Chumanov et al. 2012). This suggests that preventive measures should focus on eccentric hamstring strengthening exercises (Petersen et al. 2011).

As the causes and mechanisms of an injury vary substantially between sports, successful preventive strategies need to be tailored to the respective sport and athlete at risk. The effect of potential measures to reduce injury risk will be monitored in up-coming Games.

As a new initiative to address the next generation of future Olympic athletes in a timely manner, the IOC has created a new sporting event for young athletes. The first Summer Youth Olympic Games (YOG) were held in Singapore in August 2010, and the first Winter Youth Olympic Games in Innsbruck, Austria in January 2012. These events each brought together around 5,000 athletes, aged 14–18, from all over the world to participate in high-level Olympic competitions. Little is known about the injury risk of the young athlete when competing at high level sports (Steffen and Engebretsen 2010). Consequently, a comprehensive injury and illness surveillance, based on the IOC model for previous Olympic Games, was initiated during the 9 days of the 2012 first Winter Youth Olympic Games (Ruedl et al. 2012).

The IOC and other major international Sports Federations, such as the International Football Association (FIFA) (Dvorak et al. 2011), the International Aquatic Federation (FINA) (Mountjoy et al. 2010), and the International Association of Athletics Federations (IAAF) (Alonso et al. 2012) have taken a second step and extended their injury surveillance to also include illness monitoring. In addition, the IOC is currently looking into developing an Electronic Health Record (EHR) system, to be offered to the NOCs to facilitate Periodic Health Evaluations (PHE). This will increase the pre- and in-Games knowledge base on both injuries and illnesses, and will help the IOC and NOCs to maximize the health protection of their elite athletes (Ljungqvist et al. 2009).

In Sochi 2014, the IOC with the NOCs and IFs will continue running the injury and illness surveillance system. Following the data from Beijing (Junge et al. 2009), Vancouver (Engebretsen et al. 2010), and London (Engebretsen et al. 2013), the monitoring of the athletes' health will enable researchers and clinicians to follow the injury and disease trends in the various sports and further the work of the IOC Medical Commission on protecting the health of the athletes. The message from this and other long-term projects initiated by the IOC and the IFs is that we need to monitor the development of injury and illness rates over several years in order to properly identify potential risk factors and mechanisms for injury and illnesses in disciplines and sports. By acquiring new knowledge on injury and illness trends, we can optimise and target future research on risk factors, mechanisms and finally, prevention. The key to a meaningful study of epidemiology lies in a well-organised procedure for data collection with coordinated efforts from sports medicine professionals, coaches and athletes, combined with systematic subsequent analyses.

3.5 Conclusion

The present data collection procedures were well-accepted by the medical staff of the National Olympic Committees as demonstrated by the high response rates of completion for the injury and illness forms. At least one in ten athletes incurred an injury during the XXIX and XXX Summer or XXI Winter Games, and 7 % of the athletes suffered from at least one illness. The incidence of injuries and illnesses varied substantially between sports. Periodic Health Evaluations will be instrumental in preventing injuries and illnesses, and hence, in protecting the health of the athletes.

References

Alonso JM, Junge A, Renstrom P, Engebretsen L, Mountjoy M, Dvorak J (2009) Sports injuries surveillance during the 2007 IAAF World Athletics Championships. Clin J Sport Med 19:26–32

Alonso JM, Tscholl PM, Engebretsen L, Mountjoy M, Dvorak J, Junge A (2010) Occurrence of injuries and illnesses during the 2009 IAAF World Athletics Championships. Br J Sports Med 44:1100–1105

Alonso JM, Edouard P, Fischetto G, Adams B, Depiesse F, Mountjoy M (2012) Determination of future prevention strategies in elite track and field: analysis of Daegu 2011 IAAF Championships injuries and illnesses surveillance. Br J Sports Med 46:505–514

Bahr R (2009) No injuries, but plenty of pain? On the methodology for recording overuse symptoms in sports. Br J Sports Med 43:966–972

Bahr R, Reeser JC (2003) Injuries among world-class professional beach volleyball players. The Federation Internationale de Volleyball beach volleyball injury study. Am J Sports Med 31:119–125

Bakken A, Bere T, Bahr R, Kristianslund E, Nordsletten L (2011) Mechanisms of injuries in World Cup Snowboard Cross: a systematic video analysis of 19 cases. Br J Sports Med 45:1315–1322

Bere T, Florenes TW, Krosshaug T, Koga H, Nordsletten L, Irving C, Muller E, Reid RC, Senner V, Bahr R (2011a) Mechanisms of anterior cruciate ligament injury in World Cup alpine skiing: a systematic video analysis of 20 cases. Am J Sports Med 39:1421–1429

Bere T, Florenes TW, Krosshaug T, Nordsletten L, Bahr R (2011b) Events leading to anterior cruciate ligament injury in World Cup Alpine Skiing: a systematic video analysis of 20 cases. Br J Sports Med 45:1294–1302

Chumanov ES, Schache AG, Heiderscheit BC, Thelen DG (2012) Hamstrings are most susceptible to injury during the late swing phase of sprinting. Br J Sports Med 46:90

Clarsen B, Krosshaug T, Bahr R (2010) Overuse injuries in professional road cyclists. Am J Sports Med 38:2494–2501

Clarsen B, Myklebust G, Bahr R (2013) Development and validation of a new method for the registration of overuse injuries in sports injury epidemiology: the Oslo Sports Trauma Research Centre (OSTRC) overuse injury questionnaire. Br J Sports Med 47:495–502

Dvorak J, Junge A, Grimm K, Kirkendall DT (2007) Medical Report from the 2006 FIFA World Cup GermanyTM. Br J Sports Med 41:578–581

Dvorak J, Junge A, Derman W, Schwellnus M (2011) Injuries and illnesses of football players during the 2010 FIFA World Cup. Br J Sports Med 45:626–630

Engebretsen L, Steffen K, Alonso JM, Aubry M, Dvorak J, Junge A, Meeuwisse W, Mountjoy M, Renstrom P, Wilkinson M (2010) Sports injuries and illnesses during the Winter Olympic Games 2010. Br J Sports Med 44:772–780

Engebretsen L, Soligard T, Steffen K, Alonso JM, Aubry M, Budgett R, Dvorak J, Jegathesan M, Meeuwisse WH, Mountjoy M, Palmer-Green D, Vanhegan I, Renstrom PA (2013) Sports injuries and illnesses during the London Summer Olympic Games 2012. Br J Sports Med 47:407–414

Finch C (2006) A new framework for research leading to sports injury prevention. J Sci Med Sport 9:3–9

Fitch KD (2012) An overview of asthma and airway hyper-responsiveness in Olympic athletes. Br J Sports Med 46:413–416

Jacobsson J, Timpka T, Kowalski J, Nilsson S, Ekberg J, Dahlstrom O, Renstrom PA (2013) Injury patterns in Swedish elite athletics: annual incidence, injury types and risk factors. Br J Sports Med 47:941–952

Junge A, Dvorak J (2007) Injuries in female football players in top-level international tournaments. Br J Sports Med 41(Suppl 1):i3–i7
Junge A, Dvorak J (2010) Injury risk of playing football in Futsal World Cups. Br J Sports Med 44:1089–1092
Junge A, Dvorak J (2013) Injury surveillance in the World Football Tournaments 1998–2012. Br J Sports Med 47:782–788
Junge A, Dvorak J, Graf-Baumann T (2004) Football injuries during the World Cup 2002. Am J Sports Med 32:23S–27S
Junge A, Engebretsen L, Alonso JM, Renstrom P, Mountjoy M, Aubry M, Dvorak J (2008) Injury surveillance in multi-sport events: the International Olympic Committee approach. Br J Sports Med 42:413–421
Junge A, Engebretsen L, Mountjoy ML, Alonso JM, Renstrom PA, Aubry MJ, Dvorak J (2009) Sports injuries during the Summer Olympic Games 2008. Am J Sports Med 37:2165–2172
Kippelen P, Fitch KD, Anderson SD, Bougault V, Boulet LP, Rundell KW, Sue-Chu M, McKenzie DC (2012) Respiratory health of elite athletes – preventing airway injury: a critical review. Br J Sports Med 46:471–476
Koga H, Nakamae A, Shima Y, Iwasa J, Myklebust G, Engebretsen L, Bahr R, Krosshaug T (2010) Mechanisms for noncontact anterior cruciate ligament injuries: knee joint kinematics in 10 injury situations from female team handball and basketball. Am J Sports Med 38:2218–2225
Koga H, Bahr R, Myklebust G, Engebretsen L, Grund T, Krosshaug T (2011) Estimating anterior tibial translation from model-based image-matching of a noncontact anterior cruciate ligament injury in professional football: a case report. Clin J Sport Med 21:271–274
Ljungqvist A, Jenoure P, Engebretsen L, Alonso JM, Bahr R, Clough A, De BG, Dvorak J, Maloley R, Matheson G, Meeuwisse W, Meijboom E, Mountjoy M, Pelliccia A, Schwellnus M, Sprumont D, Schamasch P, Gauthier JB, Dubi C, Stupp H, Thill C (2009) The International Olympic Committee (IOC) Consensus Statement on periodic health evaluation of elite athletes March 2009. Br J Sports Med 43:631–643
Mountjoy M, Junge A, Alonso JM, Engebretsen L, Dragan I, Gerrard D, Kouidri M, Luebs E, Shahpar FM, Dvorak J (2010) Sports injuries and illnesses in the 2009 FINA World Championships (Aquatics). Br J Sports Med 44:522–527
Olsen OE, Myklebust G, Engebretsen L, Bahr R (2004) Injury mechanisms for anterior cruciate ligament injuries in team handball: a systematic video analysis. Am J Sports Med 32:1002–1012
Petersen J, Thorborg K, Nielsen MB, Budtz-Jørgensen E, Hölmich P (2011) Preventive effect of eccentric training on acute hamstring injuries in men's soccer: a cluster-randomized controlled trial. Am J Sports Med 39(11):2296–2303. doi:10.1177/0363546511419277, Epub 2011 Aug 8
Randjelovic S, Heir S, Nordsletten L, Bere T, Bahr R (2014) Injury situations in Freestyle Ski Cross (SX): a video analysis of 33 cases. Br J Sports Med 48:29–35
Ruedl G, Schobersberger W, Pocecco E, Blank C, Engebretsen L, Soligard T, Steffen K, Kopp M, Burtscher M (2012) Sport injuries and illnesses during the first Winter Youth Olympic Games 2012 in Innsbruck, Austria. Br J Sports Med 46:1030–1037
Steffen K, Engebretsen L (2010) More data needed on injury risk among young elite athletes. Br J Sports Med 44:485–489

Part II
Concussion and Severe Head–Neck Injury

Chapter 4
Concussions in Junior Rugby Football Athletes and Their Prevention

Mana Otomo and Toru Fukubayashi

Abstract Sports are second only to motor vehicle crashes as a cause of traumatic brain injury among 15–24 year olds. A wide range of head and neck injury risks are incurred in sports, and some involve serious injury. Among other head injury in sports, Concussions have recently attracted much attention, and are a hot topic in sports medicine. Concussions are caused by a direct impact to the head and face or other part of the body that includes the transfer of an impact force to the brain. Full-contact sports are known for their high incidence of such injuries. Although rugby athletes can sustain injuries to all body parts, concussions are the most common. Tackling is the phase of play that produces the highest proportion of injuries in rugby union football. Helmets and headgear have been shown to reduce the risk of severe head and facial injury, but have been of littie or no help in reducing the rate of concussions. Concussion injury prevention must be multifaceted, addressing relevant factors of the environment as well as those of training and supervision.

Keywords Concussions • Rugy football • Injury prevention

4.1 Introduction

In recent years, concussions occurring during in sports activities have become recognized as a central issue in sports medicine in Europe and the United States and the development of concussion awareness and management programs is underway in many sports medicine institutions. The first conference of the International Consensus Conference on Concussion (ICC) was held in Vienna, Austria in 2001, the second in 2004, the third in 2008, and the fourth in 2012. An important function of the ICC has been to increase the worldwide awareness of concussions in sports

M. Otomo (✉)
Graduate School of Sport Sciences, Waseda University, Saitama, Japan
e-mail: mana1325@gmail.com

T. Fukubayashi
Faculty of Sport Sciences, Waseda University, Saitama, Japan

K. Kanosue et al. (eds.), *Sports Injuries and Prevention*,
DOI 10.1007/978-4-431-55318-2_4

McCrory et al. (2013). A particularly dangerous aspect of concussions incurred in sports is the high likelihood of repeat injury. A repeat injury during the acute stage of a concussion can result in second impact syndrome, which may result in fatal brain damage. In addition, acute subdural hematoma can be a complication that occurs immediately after a concussion, and repeated concussions may result in chronic brain injury which negatively impacts cognitive function. It was stated in the ICC that junior athletes who incur a concussion should be managed particularly carefully. Sports concussions in junior athletes are particularly dangerous, and involve, a long recovery time (Gary S. Solomon et al. 2011) as well as a large number of symptoms and signs after the concussion (Alexis Chiang Clovin et al. 2009). Unfortunately, a standard method for preventing sports-related concussions has yet to be developed. Therefore, the ICC has stated that all people working with junior athletes must learn about the signs of concussions. Athletes, referees, parents, coaches, and healthcare providers must be educated about concussions, their clinical features and assessment techniques, and the principles to be followed in order to have a safe return to play.

4.1.1 Definition of Concussion

The ICC (2013) officially identified concussion as a brain injury and defined it as a complex pathophysiological process induced by biomechanical forces that affects the brain. Concussion is caused by a direct impact to the head and face or other part of the body that includes the transfer of an impact force to the brain. Concussion is included among a handful of short-term disorders of neurological function that resolve naturally. Their symptoms and signs evolve over the course of a few minutes to a few hours. Acute clinical symptoms mainly reflect functional disability and abnormal findings such as anatomical disorders that are not always seen on magnetic resonance imaging or computed tomography. Concussion severity results in correspondingly serious clinical symptoms. Loss of consciousness is one clear mark of a more severe concussion.

4.2 Injury Incidence

Sports are second only to motor vehicle crashes as a cause of traumatic brain injury among people aged 15–24 years. Stephanie J. Hollis et al. (2009) studied 26 non-professional rugby clubs and eight schools in Australia. A total of 3,207 male players with a mean age of 22.7 (5.5) years (range, 15–49) were recruited for the study. Of the players, 36 % (n = 1,034) reported wearing protective headgear and 80 % (n = 1,816) reported wearing a mouth guard during the game. Almost 15 % (n = 326) of the players had sustained a concussion in the 12 months prior to recruitment, and 25 % (n=81) of these players had sustained a concussion in the previous 3 months. Interestingly, 64 % of the recent concussions occurred in players with >8 years of playing experience. This study concluded that nonprofessional rugby has a high incidence of mild traumatic brain injury (mTBI) and that the absence of headgear and a recent history of mTBI are associated with an increased risk of subsequent mTBI. These findings highlight the fact that the use of headgear and the proper management of prior mTBI would likely be beneficial in reducing the likelihood of mTBI among nonprofessional rugby players, who comprise >99 % of the rugby union players in Australia.

Gessel et al. (2007) compared the rates of concussion among high school and collegiate athletes in a variety of sports and found that concussions represented 8.9 % (n=396) of all high school injuries and 5.8 % (n=482) of all collegiate injuries. Bleakley et al. (2011) calculated that the incidence of concussion in adolescent rugby ranged from 0.1920 to 1.4523/1,000 playing hours and from 3.826 to 5.728/1,000 athlete exposures.

4.2.1 mTBI

The reported incidence of mTBI in rugby varies, partially because studies do not always evaluate the players' exposure but also because of variations in injury definition and study design. Rugby studies that have evaluated exposure and, more specifically, performed systematic reviews of the incidence of mTBI in a number of contact sport studies report that rugby has a high incidence of mTBI (0.62–9.05 per 1,000 player game hours). The level of competition affects mTBI. For professional rugby players, it is 2.9–9.1 per 1,000 player game hours, whereas mTBI is lower in nonprofessional players, at 0.6–5.0 per 1,000 player game hours. Hollis et al. (2009) found that for school-aged players, the incidence of mTBI was approximately 1.03 per 1,000 player game hours, but found no conclusive evidence which indicated that prior concussion was a precursor to future mTBI. While many studies have indicated that prior concussions are not a precursor to mTBI, other studies have found that high school and collegiate athletes experiencing ≥3 concussions are more vulnerable to a subsequent concussion note. The literature highlighting recent mTBI as a precursor to subsequent mTBI indicates that incomplete recovery from a prior injury may be responsible for this phenomenon. For example, it has been postulated that an increase in lactate production after brain injury leads to secondary ischemic injury, which may predispose the brain to a repeat injury (Hollis et al. 2009) ***document?***. Unfortunately, in this study, we were unable to determine whether an inadequate recovery contributed to the repeat incidence of concussion. A recent study of mTBI found no relationship between initial and subsequent concussions on mTBI (Hollis et al. 2009) ***document?***.

4.2.2 Rugby Football

Rugby football is one of the most popular sports in the world. In the America, the rugby union includes approximately 25,000 participants at the high school, collegiate, and club levels, representing a 25 % increase in recent years (Kaplan et al. 2008). In Australia, the number of nonprofessional registered players has increased by 22 % in recent years, and almost half of the participants are school-aged players. Rugby football is a full-contact (collision) sport. Full-contact sports are known for their high incidence of injury. Despite its popularity, the full-contact sport of rugby has the potential for serious injury. Tackling is the phase of play that produces the highest proportion of injuries in rugby union football (Garraway et al. 1999).

Coughlan et al. (2014) developed main groupings and categories for classifying rugby injury types: bone injuries include fractures and other bone injuries; joint and ligament (non-bone) injuries include dislocations/subluxations, sprains/ligaments, and meniscus and cartilage or disc lesions; muscle and tendon injuries include muscle rupture/tear/strain/cramps, tendon injury/rupture/tendinopathy/bursitis, and hematoma/contusions/bruises; skin injuries include abrasions and lacerations; brain/spinal cord/peripheral nervous system injuries include concussion (with or without loss of consciousness), structural brain injuries, spinal cord compression/transections, and nerve injuries; and other injuries include dental, visceral, and other injuries.

Although rugby athletes can sustain injuries to all body parts, concussions are the most common injury Schulz et al (2004). Earlier studies have focused on the proportion of injuries that occur during tackling but have not provided detailed information about the tackles themselves. There are no published reports on the influence of personal lifestyle, personality, or other player-related factors on rugby injuries (Garraway et al. 1999). The reasons for injury occurrence in tackling may include the playing situations in which the tackle occurs and/or personal factors associated with the players involved. There are no previous studies of the influence of personal lifestyle, personality, or other player related factors on rugby injuries. Garraway et al. (1999) evaluated the factors that influence tackle injuries in rugby union football. In this study, tackle injury was present in 72 episodes. The tackled player was injured in 40 cases, the tackler in 33 cases, and both were injured in one tackle injury episode. And, they presented concussion was 12 % of all tackle injuries.

Additionally, cervical spinal injury (CSI) is often associated with concussions. G J Browne (2006) evaluated 125 male children (mean age, 12.7 years) with CSI and found that 43 % had simultaneously sustained a concussion. There is little information on rugby league injuries in female players. King DA et al. (2010) tabulated women's rugby league injuries requiring medical treatment and associated costs. They found that women incur substantially fewer injuries (5.7 %) than men (94.3 %). Concussion/brain injuries accounted for 3.8 % of the total female moderate to serious injury claims but accounted for 5.4 % of female injury costs. Thus, concussions and brain injuries resulted (Hollis et al. 2009), in the highest mean cost per claim.

4.3 Prevention of Concussion

Sports involving body contact, projectiles, and/or high speeds are associated with an increased risk of head and neck injury. The personal and social costs of severe head and neck injuries are high. Injury prevention must be multifaceted, addressing environmental factors, training, and supervision. There have been few evaluations of injury prevention methods. Research on epidemiological, medical, and human factors need to be combined with studies involving biomechanical and technological approaches to reduce concussion risks in sports (McIntosh and McCrory 2005).

4.3.1 Protective Equipment

Mcintosh et al. (2009) evaluated the use of padded headgear to prevent head injuries in rugby football. In this study, 1,493 participants (10,040 player hours) were in the control group, 1,128 participants (8,170 player hours) were assigned to the standard headgear group, and 1,474 participants (10,650 player hours) were assigned to a modified headgear group. The investigators concluded that the use of padded headgear did not reduce the rate of head injury or concussion. Although individuals may choose to wear padded headgear, the routine or mandatory use of protective headgear for preventing injury cannot be recommended. In fact, low compliance rates are a limitation. Marshall et al (2005) also investigated the injury prevention effectiveness of protective equipment that consisted of padded headgear and mouth guards as used in the rugby union. The risk of concussion was not lessened by the use of padded headgear (RR, 1.13; 95 % confidence interval [CI], 0.40–3.16) or mouth guards (RR, 1.62; 95 % CI, 0.51–5.11). The results of the above studies make it quite clear that current versions of padded headgear and mouth guards offer very little, if any, potential for ameliorating the risk of concussions in rugby. Marshall et al (2005) likewise concluded that while helmets and headgear have been shown to reduce the risk of severe head and facial injury, they do not reduce the rate of concussions.

Pettersen (2002) investigated the attitudes of players and coaches toward the use of protective headgear, particularly with respect to preventing concussion. Although the players tended to believe that the use of rugby headgear might prevent concussions (62 %), the coaches tended to be less convinced (33 %). Despite the players' belief that headgear offers protection against concussions, only a minority reported actually wearing headgear (27 %), and few (24 %) felt that its use should be mandatory.

4.3.2 Rule Change

High-impact contact with another person appears to be the main risk factor responsible for a majority of the concussions sustained by high school athletes. Gessel et al. (2007) found that full- and partial-contact sports had the highest competition-related rates of concussions. The ICC has stated that fair play and respect for opponents are ethical values that should be encouraged in all sports and by all sporting associations (2013). In the rugby football, rules have been changed to decrease the impact forces incurred by players during scrums (Bleakley et al. 2011).

All states within the US have passed laws mandating medical evaluation and subsequent clearance for high school athletes who have experienced concussions before they are allowed to return to play. No athlete at any level (high school, collegiate, or professional) is allowed to return to play on the same day (game and/or practice) as when a concussion has been diagnosed. Similarly, no athlete who has sustained a concussion is allowed to return to play before being cleared by a health-care professional with experience in sports-related concussions.

4.3.3 Education

Educational campaigns such as distribution of the "Heads Up: Concussion in High School Sports" toolkit by the Centers for Disease Control and Prevention and the National Center for Injury Prevention and Control in the USA, have increased awareness of concussion symptoms among coaches, athletes, and parents (Gessel et al. 2007). As the current ability to treat or reduce the effects of a concussive injury are minimal, the education of athletes, colleagues, and the general public is a critical aspect of progress in this field. The development of effective sports-related concussion preventive measures is dependent upon increasing our knowledge of concussion rates, patterns, and risk factors.

Newton et al. (2014) examined the factors that influence the intended use of concussion guidelines among amateur-level coaches and sports trainers in the Australian football and rugby football leagues. The results showed that personal norms and self-efficacy were significant predictors of intention to use concussion guide-lines. Interestingly, the relevance of self-efficacy and intention was stronger among Australian football coaches than among rugby football coaches. Efforts to increase the intended use of concussion guidelines by both coaches and sports trainers should include the targeting of personal norms and self-efficacy. Programs targeting self-efficacy may be particularly effective for Australian football coaches and rugby football coaches. Such coaches were less familiar with concussion guidelines and had less experience using them than did sports trainers.

References

Bleakley C, Tully M, O'Connor S (2011) Epidemiology of adolescent rugby injuries: a systematic review. J Athl Train 46(5):555–565

Browne GJ (2006) Cervical spinal injury in children's community rugby football. Br J Sports Med 40(1):68–71

Colvin AC, Mullen J, Lovell MR, West RV, Collins MW, Groh M (2009) The role of concussion history and gender in recovery from soccer-related concussion. Am J Sports Med 37(9):1699–1704

Coughlan GF, Fullen BM, McCarthy C (2014) A national survey of clubs medical personnel, facilities and protocols in Irish Rugby Union. Ir J Med Sci 183(1):39–45

Garraway WM, Lee AJ, Macleod DA, Telfer JW, Deary IJ, Murray GD (1999) Factors influencing tackle injuries in rugby union football. Br J Sports Med 33(1):37–41

Gessel LM, Fields SK, Collins CL, Dick RW, Comstock RD (2007) Concussions among United States High School and Collegiate Athletes. J Athl Train 42(4):495–503

Hollis SJ, Stevenson MR, McIntosh AS, Shores EA, Collins MW, Taylor CB (2009) Incidence, risk, and protective factors of mild traumatic brain injury in a cohort of Australian nonprofessional male rugby players. Am J Sports Med 37(12):2328–2333

Kaplan KM, Goodwillie A, Strauss EJ, Rosen JE (2008) Rugby injuries a review of concepts and current literature. Bull NYU Hosp Jt Dis 66(2):86–93

King DA, Hume PA, Milburn P, Gianotti S (2010) Women's rugby league injury claims and costs in New Zealand. Br J Sports Med 44(14):1016–1023

Marshall SW, Loomis DP, Waller AE, Chalmers DJ, Bird YN, Quarrie KL, Feehan M (2005) Evaluation of protective equipment for prevention of injuries in rugby union. Int J Epidemiol 34(1):113–118

McCrory P, Meeuwisse WH, Aubry M, Cantu B, Dvorák J, Echemendia RJ, Engebretsen L, Johnston K, Kutcher JS, Raftery M, Sills A, Benson BW, Davis GA, Ellenbogen RG, Guskiewicz K, Herring SA, Iverson GL, Jordan BD, Kissick J, McCrea M, McIntosh AS, Maddocks D, Makdissi M, Purcell L, Putukian M, Schneider K, Tator CH, Turner M (2013) Consensus statement on concussion in sport: the 4th International Conference on Concussion in Sport held in Zurich, November 2012. Br J Sports Med 47(5):250–258

McIntosh AS, McCrory P (2005) Preventing head and neck injury. Br J Sports Med 39(6):314–318

McIntosh AS, McCrory P, Finch CF, Best JP, Chalmers DJ, Wolfe R (2009) Does padded headgear prevent head injury in rugby union football? Med Sci Sports Exerc 41(2):306–313

Newton JD, White PE, Ewing MT, Makdissi M, Davis GA, Donaldson A, Sullivan SJ, Seward H, Finch CF (2014) Intention to use sport concussion guidelines among community-level coaches and sports trainers. J Sci Med Sport 17(5):469–473

Pettersen JA (2002) Does rugby headgear prevent concussion? Attitudes of Canadian players and coaches. Br J Sports Med 36(1):19–22

Schulz MR, Marshall SW, Mueller FO, Yang J, Weaver NL, Kalsbeek WD, Bowling JM (2004) Incidence and risk factors for concussion in high school athletes, North Carolina, 1996–1999. Am J Epidemiol 160(10):937–944

Solomon GS, Ott SD, Lovell MR (2011) Long-term neurocognitive dysfunction in sports: what is the evidence? Clin Sports Med 30(1):165–177

Chapter 5
Prevention of Head and Neck Trauma in Rugby

Mutsuo Yamada

Abstract The incidence of rugby injuries to the head and neck region differs somewhat depending on the age of players and the level of competition. Although catastrophic head and/or neck trauma is rare, such trauma cannot be ignored given the prognostic significance of sequelae such as paralysis and the impact of that trauma on surrounding tissues and structures, caused by an acute subdural hematoma, acute epidural hematoma, traumatic subarachnoid hemorrhage, contusions, or a cervical spinal cord injury. According to previous studies, most head and neck trauma occurs as a result of tackling. The cause of the injury is described as being largely related to tackling skill. In Japan, catastrophic head and/or neck trauma is very often related to tackling. The next most prevalent cause is a scrum in other countries, but in Japan it is a ruck. Head trauma seldom occurs in a scrum in rugby, but neck trauma often occurs. The biggest issue with neck trauma in a scrum is cervical spinal cord injuries. Many injuries related to mauls and rucks are the result of charging into a tight situation and making "head-first contact" with one's head down while not looking where one is going. The Japan Rugby Football Union has implemented safety measures predicated on the tenet "Safe Technique = Good Technique". This means that coaches and players are informed that correct play is safe and results in better performance. Players need to understand what "Safe Technique = Good technique" actually means.

Keywords Prevention • Rugby • CI • Trauma • JRFU

5.1 Introduction

A study of injuries related to rugby football (denoted here simply as rugby) led to the creation of an injury report form by the International Rugby Board (IRB) in 2007. After the form was featured in the British Journal of Sports Medicine (Fuller

M. Yamada (✉)
Faculty of Health & Sports Science and Graduate School of Health & Sports Science, Ryutsu Keizai University, Okinawa, Japan

Medical Committee, Japan Rugby Football Union, Tokyo, Japan
e-mail: muyamada@rku.ac.jp1

K. Kanosue et al. (eds.), *Sports Injuries and Prevention*,
DOI 10.1007/978-4-431-55318-2_5

Table 5.1 The definition catastrophic injury (International Rugby Board)

1. Spinal cord injuries with an ASIA classification at 48 h of A to D
2. Brain injuries with a Glasgow Coma Scale (GCS) at 48 h of ≤12 (i.e. Graded as moderate or severe)
3. Fatalities (incl. spinal cord, traumatic brain, and cardiac-related fatalities)

et al. 2007), reports in accordance with the form, as well as the associated standardization of injury definitions, gradually increased. Similarly, some countries on the IRB began reporting catastrophic injuries (CIs) in a CI Report in 2010. The CI Report only records instances of fatalities due to rugby or the presence of major sequelae such as paralysis. A CI Report is submitted to the IRB in two stages. In the first stage, an injury that is a potential CI is reported within 48 h of its occurrence. Such injuries include "spinal cord injuries with an American Spinal Injury Association (ASIA) classification at 48 h of A to D," "brain injuries with a Glasgow Coma Scale (GCS) at 48 h of ≤12 (i.e. graded as 'moderate' or 'severe')," and "fatalities (including spinal cord, traumatic brain, and cardiac-related fatalities)." A CI Report of such an injury is submitted to the relevant department of a country's rugby association that handles CI Reports. After 12 months, injuries that ultimately have met the definition of a CI according to the IRB are reported. These injuries include "spinal cord injuries with an ASIA classification at 12 months of A to D," "traumatic brain injuries with a Glasgow Outcome Scale (GOS) at 12 months of 1–3," and "fatalities resulting from any rugby match or training activity." In other words, most CI Reports are ultimately finalized 12 months after a CI has occurred (Table 5.1).

The incidence of rugby injuries to the head and neck region, which include minor maxillofacial injuries, differs somewhat depending on the age of players and the level of competition, but such injuries are quite prevalent, second only to injuries to the lower limbs (Kaplan et al. 2008; Brooks et al. 2005; Best et al. 2005; Fuller et al. 2008; Schick et al. 2008; Collins et al. 2008). However, the incidence of catastrophic head and/or neck trauma is rare in comparison to other injuries. Although catastrophic head and/or neck trauma is rare, such trauma cannot be ignored given the prognostic significance of sequelae such as paralysis and the impact of that trauma on surrounding tissues and structures. The current study analyzes the CIs reported to the Japan Rugby Football Union (JRFU) in order to parlay the results of that analysis into safety programs for rugby.

5.2 Current Realities of Head and Neck Trauma in Rugby in Japan

Trauma of the head and neck in rugby is rare in comparison to other injuries such as concussions, bumps on the head, cuts on the head or face, and nasal fractures. However, a CI can occur in the form of an acute subdural hematoma, acute epidural hematoma, traumatic subarachnoid hemorrhage, brain contusions, or a cervical spinal cord injury. According to previous studies, most head and neck trauma occurs as a result of tackling (Collins et al. 2008; MacManus and Cross 2004). The cause of the injury is described as being largely related to tackling skill (Mcintosh et al. 2010; Silver 2002)

In Japan, individuals suffering CIs and matches are most numerous in high school. Unfortunately, the impression that is given is that CIs often occur. Thus, safety measures to prevent CIs are targeted specifically at high school students. Watanabe et al. examined the number of high school and university students who suffered a CI per 100,000 players in Japan from 2008 to 2011, and they found that university students suffered more CIs. A similar trend was noted in a study overseas. Quarrie et al. stated that the number of CIs peaked at ages 22–25 (Quarrie et al. 2002). Focused safety measures are needed for high school students as well as university students (Table 5.2, Figs. 5.1 and 5.2).

Mcintosh et al. analyzed the tackling techniques of elite players and players ages 15–20, and they found that younger players often made a "passive shoulder tackle" whereby they stopped moving forward and absorbed the brunt of their opponent's energy (Mcintosh et al. 2010). As players improved, they started making an "active shoulder

Table 5.2 Total number of matches in 2011. Rugby union game in 2011

Grade	Total
High school (U18)	1,195
University (U19-22)	822
Club team	376

These totals are the number of regular matches for all high school, college, and club teams nationwide in Japan. Evident is the fact that high school students have a large number of regular matches

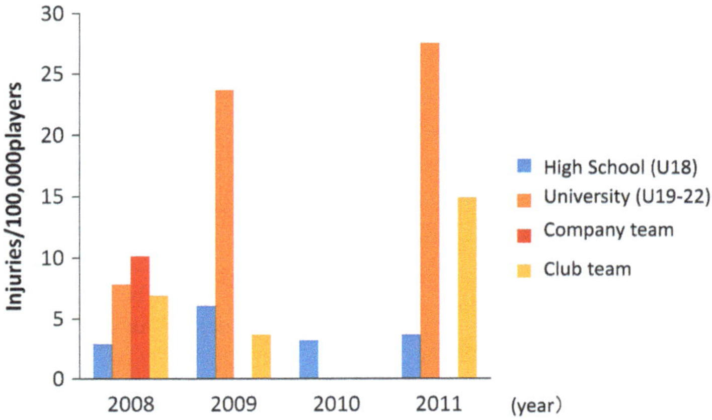

Fig. 5.1 Incidence of cervical spinal cord injuries in Japan (injuries/100,000 players)

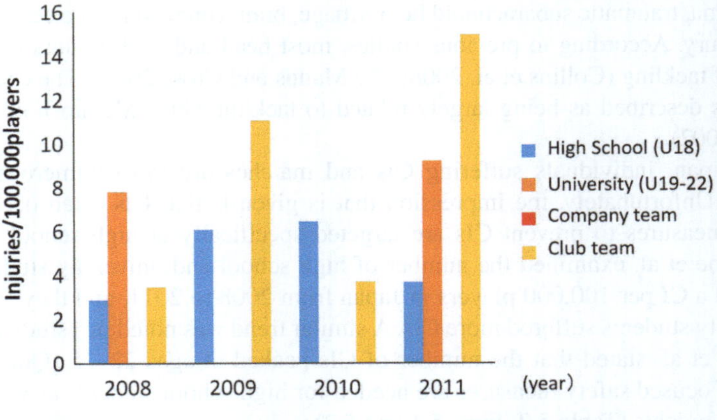

Fig. 5.2 Incidence of severe head trauma in Japan (injuries/100,000 players)

Table 5.3 Study of cervical spinal cord injuries by the IRB

Phases of play	2008–2011	JRFU (%)
Scrum engagement	11	18.2
Scrum collapse on engagement	8	12.5
Scrum collapse not on engagement	3	33.3
Tackler	21	23.8
Ball carrier	10	40.0
Ruck	17	47.1
Maul	3	33.3
Collision	5	0.0
Unknown	3	0.0
Total	81	27.2

The regions studied were England, Scotland, Ireland, France, Italy, Argentina, South Africa, Canada, and Japan. The percentage on the right is the proportion of injuries due to each cause in Japan. The incidence of cervical spinal cord injuries in the 9 regions was 1.27 injuries/100,000 players (Cervical spinal cord injuries related to scrums decreased 32.5 %)

tackle" whereby they oriented their body towards their opponent while moving forward. Mcintosh et al. noted that younger individuals had less developed tackling techniques.

A study examined catastrophic cervical spinal cord injuries suffered by players from ages 12 to 19 (U19) in Scotland, Wales, Ireland, and England, and it noted that tackles were most often responsible for those injuries, followed by scrums (Maclean and Hutchison 2011). In Japan, catastrophic head and/or neck trauma is, when categorized for all age groups, very often related to tackling. The next most prevalent cause is a scrum in other regions, but in Japan it is a ruck. A ruck is a tight scramble for a ball on the ground (Table 5.3).

Thus, instruction in safety measures, which includes teaching correct tackling techniques, is key to preventing head and neck trauma in rugby, regardless of whether the injury is mild or catastrophic (27).

5.3 Causes of Injury

5.3.1 Causes of Injury Related to Tackles

Head and neck trauma in rugby is most often caused by tackling (MacManus and Cross 2004; Silver 2002; Quarrie et al. 2002; Maclean and Hutchison 2011). A key factor is an individual player's tackling skill (Yamada et al. 2004; John et al. 2008). Of the incidents of catastrophic head and/or neck trauma reported to the JRFU, close to 60 % are due to "head-first contact" (Yamada 2011). There are several forms of head-first contact, though the one that most often causes head and/or neck trauma is when a tackler (a player tackling an opponent) makes a tackle with his head down while not looking at the ball carrier (the player with the ball), as shown in Figs. 5.3 and 5.4. When the tackler's head collides with his opponent, the tackler's head moves in the direction the ball carrier was moving. The tackler's head is violently struck by the ball carrier's knees or the front of his/her torso in what is known as a "head-spinning tackle" (Yamada et al. 2004).

In other tackling situations, there is a high risk of head and neck trauma when there are two or more tacklers tackling a single opponent at the same time. This situation puts both the ball carrier and the tacklers at risk. When the tacklers make impact, one of the tacklers may be tackling the ball carrier above his shoulders. Part of the ball carrier's body often comes into contact with the tackler's head and neck region (Fig. 5.5) (Mcintosh et al. 2010; Yamada 2008). When the head is lowered in the event of a passive tackle, as mentioned earlier, head trauma often results.

There are few detailed domestic studies of how ball carriers are injured. There are two possibilities which are both restricted by current game rules. One is a high tackle where the ball carrier is tackled above his shoulders (Fig. 5.11a), and the other is a spear tackle where the player with the ball is driven head-first into the group (Fig. 5.11b) (Fuller et al. 2010). If a high school student lacks skill at being tackled, his head may strike the ground, resulting in a concussion.

Fig. 5.3 A head-down tackle made with the head down while not looking at one's opponent

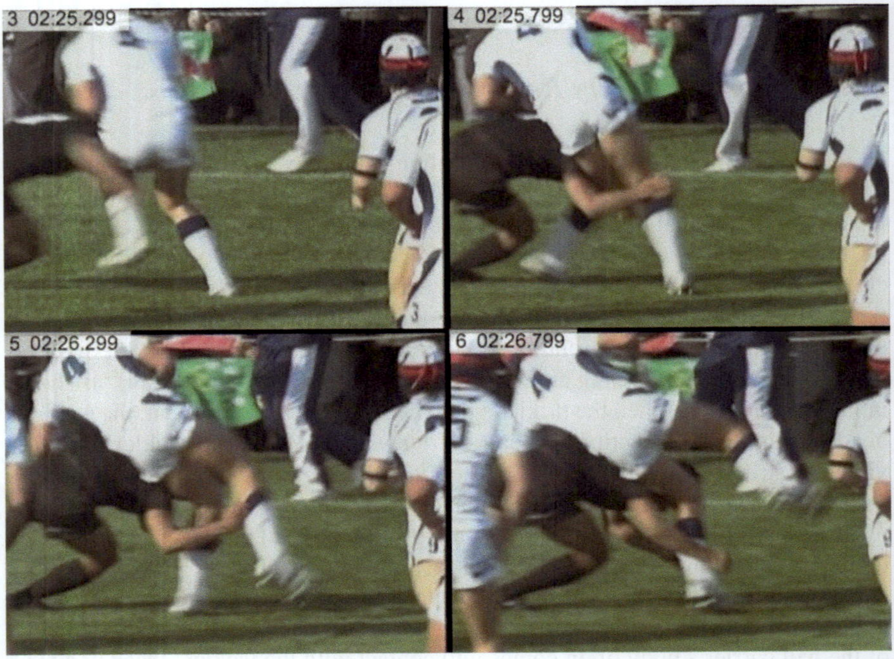

Fig. 5.4 A head-spinning tackle (the head is incorrectly positioned)

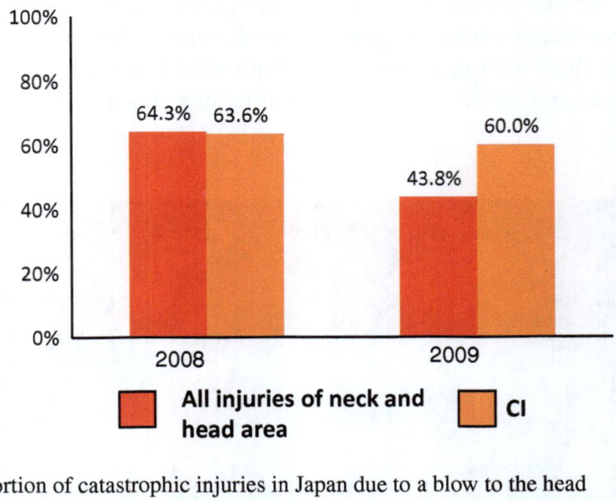

Fig. 5.5 Proportion of catastrophic injuries in Japan due to a blow to the head

5.3.2 Causes of Injury Related to Scrums

Head trauma seldom occurs in a scrum in rugby, but neck trauma often occurs. The biggest issue with neck trauma in a scrum is cervical spinal cord injuries.

Quarrie et al. summarized past statistics, and reported that most cervical spinal cord injuries in a scrum are suffered by players at the front who form a scrum and come into direct contact with their opponents in the front row of the scrum (Quarrie et al. 2002). According to Quarrie et al. cervical spinal cord injuries in a scrum are mostly caused by hyperflexion of the neck (this also includes instances where a player's neck is subjected to rotational stress). Cervical spinal cord injuries occur most often (47.1 %) as accidents during "engagement," which is the moment when a scrum is formed. Those injuries also occur (45.9 %) as a result of "collapse" when a scrum has collapsed (Fig. 5.6). In addition, those injuries can occur (4.7 %) due to "popping," which is when the front row of a scrum intentionally extends their necks to push their opponents upwards (Quarrie et al. 2002; Yamada 2008).

Maclean et al. report that cervical spinal cord injuries in U19 rugby are often due to tackling but that severe injuries, such as complete paralysis, were often due to scrums (Maclean and Hutchison 2011). Two-thirds or more of CIs in scrums occur when the scrum "collapses." Safety measures to deal with the collapse of a scrum are an important topic.

5.3.3 Causes of Injury Related to Tight Play (a Maul or Ruck)

In rugby, a scramble for the ball in a tight situation is known as a maul (the ball is off the ground) or a ruck (the ball is on the ground). Many injuries related to mauls and rucks are the result of charging into a tight situation and making

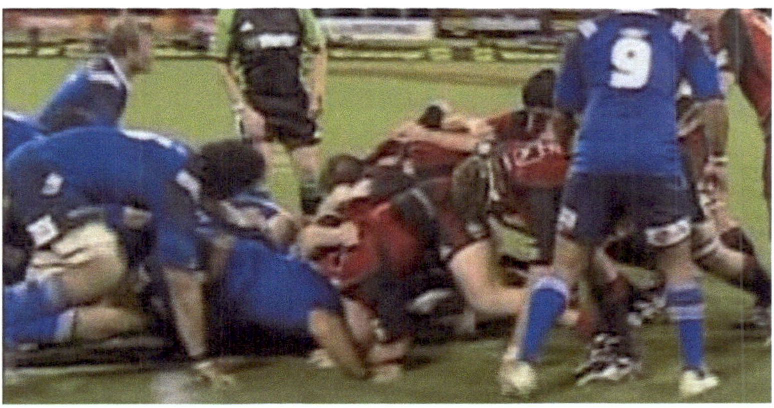

Fig. 5.6 Dangerous conditions in a scrum. An instance where a scrum has collapsed

"head-first contact" with one's head down while not looking where one is going (Fig. 5.5). When a maul collapses, a player's head may strike the ground. When pressed down on by an opponent in a ruck, a player can be pinned with his neck hyperflexed. Statistics assembled from the literature indicate that fewer cervical spinal cord injuries are suffered in a maul or ruck than in a scrum or tackle (Quarrie et al. 2002). Over the past few years, however, the scramble for the ball after a tackle has intensified. Instead of adopting a lower posture than one's opponent in order to scramble for a ball on the ground in a ruck, many players join the ruck with their heads down. As a result, both the head and neck can be injured in a ruck. CIs in Japan are most often caused by tackles, followed by rucks. Safety measures for both tackles and rucks will be needed in the future.

A recent issue with rucks is the "squeeze ball" (or "turtle") (Figs. 5.7, 5.8, and 5.9). This maneuver is illegal in under-18 rugby (U18) in Japan, but it is not prohibited in university. Catastrophic cervical spinal cord injuries due to this maneuver have been reported in university and post university players.

An actual instance of a cervical spinal cord injury due to a squeeze ball will now be described. A player has three points of support from his head and both knees and his neck is flexed, as shown in Figs. 5.7 and 5.8. The player's neck is hyperflexed when he is pushed from behind by a teammate. The player's neck is also subjected to an axial load soon afterwards when an opponent pushes down from above. This results in dislocation of cervical vertebrae and cervical spinal cord injury.

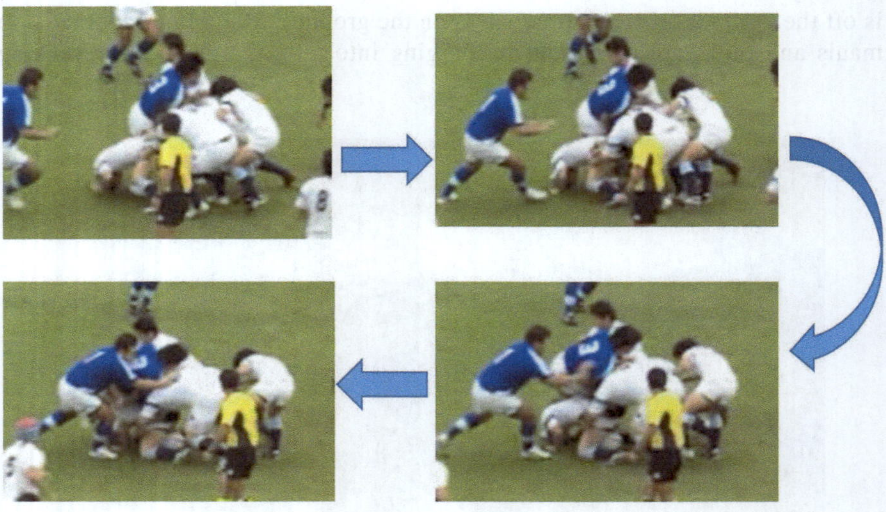

Fig. 5.7 Conditions for a cervical spinal cord injury to occur in a ruck

Fig. 5.8 An explanation of the conditions for a cervical spinal cord injury depicted in Fig. 5.7

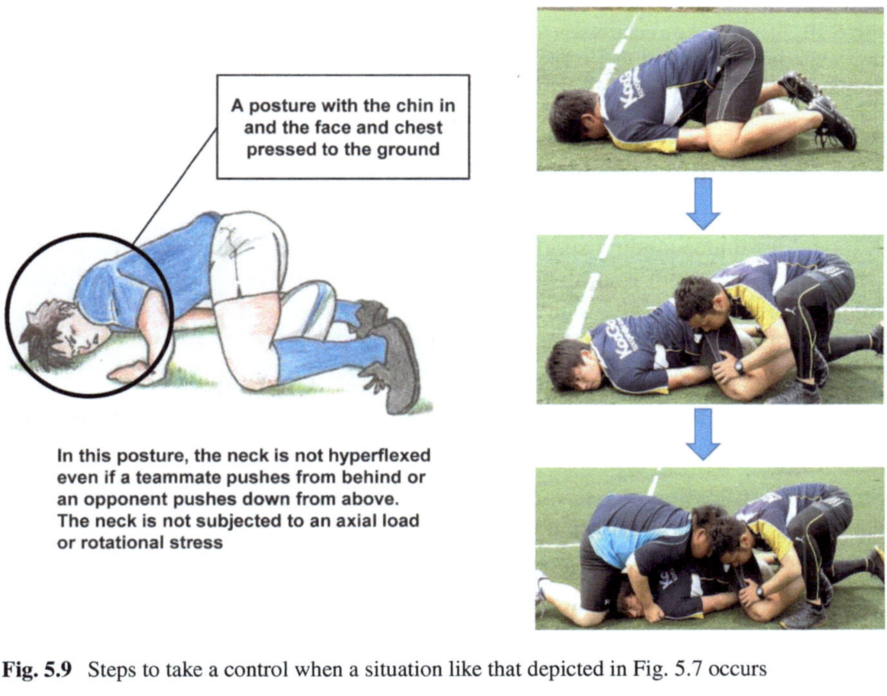

Fig. 5.9 Steps to take a control when a situation like that depicted in Fig. 5.7 occurs

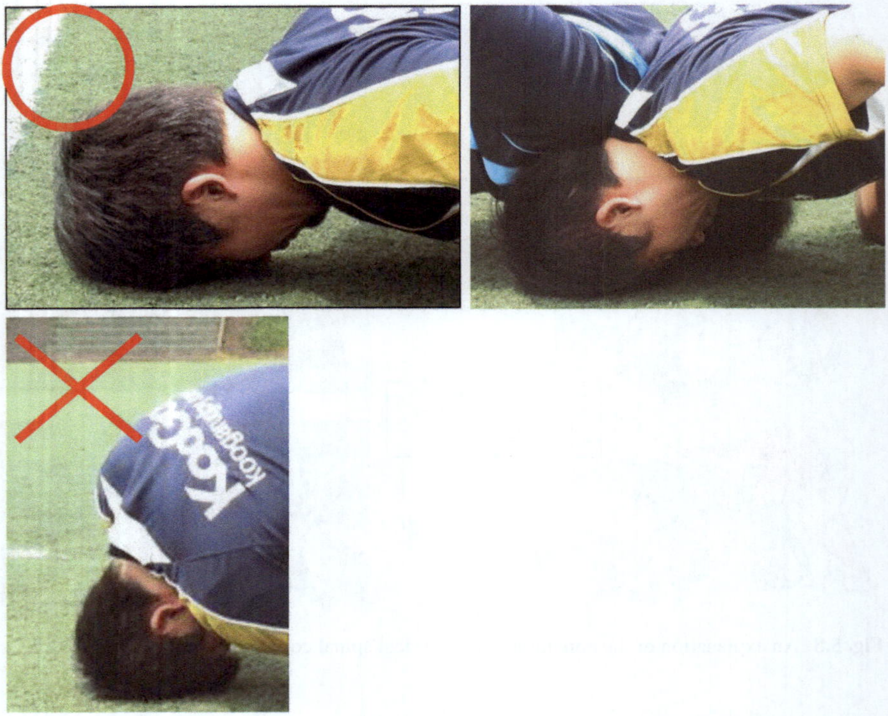

Fig. 5.10 Steps to take a control when a scrum has collapsed. When a scrum has collapsed, a chin-in posture is assumed and the forehead is thrust into the ground. The neck is not subjected to flexion, rotational stress, or an axial load, so danger can be avoided

5.4 Prevention of Head and Neck Trauma

A model for understanding sports trauma is often used to minimize sports injuries. One such model with four steps was described at the World Congress on Sports Injury Prevention (Fukubayashi 2008).

The order of these steps is indicated below.

Step 1: **Ascertaining aspects of trauma (incidence & severity)**
Risky play like a tackle, scrum, or ruck that might lead to a CI is statistically analyzed and identified.

Step 2: **Identification of risk factors (causes and mechanisms of trauma)**
Dangerous aspects of identified play are analyzed using video recordings of actual injuries.

Step 3: **Institution of preventive techniques**
Once causes are identified, players, coaches, and trainers (athletic trainers and strength and conditioning coaches) are instructed on methods for preventing

head and neck trauma. Players, coaches, and trainers also implement measures to prevent such trauma.

Step 4: **The effectiveness of preventive techniques is assessed by studying trauma**
The effectiveness of Step 3 is assessed and whether effective techniques are being accurately communicated is determined. If the techniques are not found to be effective, then the analysis will be performed again. Steps 1 through 3 will be followed once again, techniques will be instituted, and results will be assessed again.

A certain tenet is crucial to prevention, as was mentioned in Step 3. The JRFU has implemented safety measures predicated on the tenet **"Safe Technique = Good Technique."** This means that coaches and players are informed that correct play is safe and results in better performance. As was mentioned earlier, "head-first" tackling involves charging forward with one's head. Such contact is risky and it can lead to poor performance, i.e. failing to bring down one's opponent. In contrast, very safe tackling in line with the basic, accepted technique results in better performance than other, more dangerous forms of tackling. Players need to understand what "Safe Technique Play = Good technique" actually means (Yamada et al. 2004, 2005; Yamada 2008, 2009, 2011).

5.5 Prevention of Head and Neck Trauma Due to Tackles

Preventing head and/or neck trauma in tacklers involves modifying tackling techniques that may cause the type of trauma mentioned earlier. When making a tackle from the front and side, a tackler should keep his opponent in view. The tackler's shoulders and hips should constitute a square, and the tackler orient that square toward the ball carrier. In the sagittal plane, the tackler's body from his head to his torso should be aligned; the tackler must maintain this posture (with no apparent flattening or rounding of the back) without lowering his head. Without diving, the tackler should drive into the ball carrier's base of support with the leg on the same side of the contact shoulder (power leg/too). The tackler should then place his cheek against the ball carrier's buttocks in a position known as going "cheek to cheek." The tackler then should wrap both arms around the ball carrier to bring the ball carrier down. The tackler needs to continue to look ahead and step forward even after making contact with the ball carrier. Once his opponent is down, the tackler should pop up and take possession of the ball. This is basic tackling. When two or more players are tackling the ball carrier, the tacklers should keep the ball carrier in view and communicate with each other to increase their effectiveness and reduce the risk of head and neck trauma.

Important points for ball carriers are not lowering one's head prior to contact and familiarity with how to react when being tackled.

In terms of game law, typical preventive measures are restrictions on high tackles and spear tackles (Fig. 5.11). Moreover, U15 rugby in Japan has a unique rule to protect young people from head and neck trauma in that "head-first contact" is illegal.

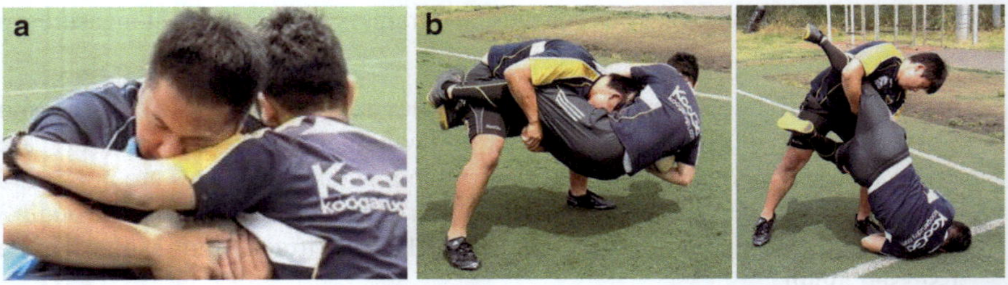

Fig. 5.11 Illegal tackles. (**a**) High tackle. A tackle made above the shoulders of the ball carrier. (**b**) Spear tackle. A tackle in which the tackler lifts the ball carrier up and drives or drops him head-first into the ground

5.6 Prevention of Head and Neck Trauma in a Scrum

For safety, game rules mandate 4 calls for a scrum. A referee issues 4 calls when a scrum is formed, as will now be explained.

- "Crouch" (Assuming the position to form a scrum)
  A player keeps his head and shoulders above his hips and he bends his knees enough to be able to comfortably form a scrum. The player lifts his chin and keeps his body from his head to his torso aligned for safety.
- "Touch" (Ensuring there is a safe distance between one's self and one's opponent)
  Bothprops (Tight head prop and Loose head prop) touch the outside shoulder of their opponents, they measure a safe distance of one arm's length from their opponents, and they return their hands to the original position.
- "Pause" (The moment before a scrum is formed, akin to the initial face off in sumo)
  A player opens his eyes, he looks at his target, and he maintains his posture.
- "Engage" (Forming a scrum)
  Players take hold of one another with their torsos parallel to the ground facing their opponents. Players take hold of their opponents ("binding") while keeping their torsos in line.

This sequence of calls is abbreviated CTPE, which comes from the first letter of each call. This game law was introduced in U19 rugby in Australia in 1986. The incidence of cervical spinal cord injuries decreased markedly, so this law was formally adopted by the IRB in 2007. All of the member countries of the IRB have begun implementing this rule at every level of rugby (Gianotti et al. 2008). However, this procedure alone is not adequate as a safety measure for scrums since scrum safety begins with adoption of the correct posture. Keeping one's shoulders above one's hips (not lowering one's head), keeping one's torso straight from the head down without flattening (flexing) the lumbar spine or rounding the entire spine, and crouching, are crucial. Other crucial actions are not sticking out one's chin, maintaining an interval so that one's fist can just about fit between one's lower jaw and chest, and having the front row adopting a "chin-in" posture and "shrug" as they look up at their opponents (Yamada et al. 2005). Revised game law have had varying effectiveness in different countries (Gianotti et al. 2008). In Japan, the revision was presumably effective given that the incidence of cervical spinal cord injuries due to scrums decreased. To further improve the positive effect on safety of these game law, "Pause" and "Engage" were incorporated into a single call, "Set," in 2012, so the 4 calls involved in forming a scrum were decreased to 3 calls. Since 2013, once the call "Touch" is given the front row players keep their hands on the shoulders of their opponents. This has reduced the force with which packs collide in a scrum by 25 %, as one study revealed. This is how scrums are now formed.

When a scrum collapses, cervical spinal cord injuries occur as a result of a front row player striking the ground with his/her neck flexed. The player is pushed from behind by his teammates, hyperflexing his neck. Front row players must be taught what to do in the event of a collapsing scrum. This instruction should start in practice settings. A front row player can avoid a cervical spinal cord injury by adopting the "chin-in" posture as shown in Fig. 5.10 and by pressing his forehead to the ground.

5.7 Prevention of Head and/or Neck Trauma in Tight Situations (a Maul or Ruck)

The basic posture for a tight play like a maul or ruck is the same as that for a scrum. This means adopting a strong, self-supporting posture that is similar to a scrum posture. The important aspects are to fully bend the knees, crouch, and not lower one's head.

Cervical spinal cord injuries in the event of a squeeze ball, as mentioned earlier, can be avoided by not using one's head as a fulcrum and by placing the region from one's face to one's chest on the ground, as shown in Fig. 5.9.

5.8 Conclusion

The mechanism by which head and neck trauma occurs in rugby, techniques to prevent that trauma, and ways to deal with it have been described here. Rugby is a sport that involves making physical contact at intense speed, so players can suffer accidental trauma. However, the reality is that catastrophic trauma can be minimized through preparation beforehand and learned skills. Relevant parties such as coaches, players (and their families), and medical support staff should have an accurate knowledge of safety measures predicated on the tenets of "Safe Play = Better Performance" and "Safe technique = Good technique." The fervent hope is that such knowledge will reduce the incidence of CIs that occur during rugby practices and matches.

Acknowledgements The author wishes to sincerely thank the Rugby Club of Ryutsu Keizai University, the Safety Committee of the JRFU, J Sports, and Koji Inomata (illustrations) for their help during the preparation of this manuscript.

References

Best JP et al (2005) Rugby World Cup 2003 injury surveillance project. Br J Sports Med 39:812–817

Brooks JHM et al (2005) A prospective study of injuries and training amongst the England 2003 Rugby World Cup squad. Br J Sports Med 39:288–293

Collins CL, Micheli LJ et al (2008) Injuries sustained by high school rugby players in the United States, 2005–2006. Arch Pediatr Adolesc Med 162(1):49–54

Fukubayashi T (2008) The role of prevention in sports trauma and injuries. J Clin Sports Med 25:2–5

Fuller CW, Molloy MG et al (2007) Consensus statement on injury definitions and data collection procedures for studies of injuries in rugby union. Br J Sports Med 41:328–331

Fuller CW et al (2008) International Rugby Board Rugby World Cup 2007 injury surveillance study. Br J Sports Med 42:452–459

Fuller CW, Ashoton T, Brooks JHM et al (2010) Injury risk associated with tackling in rugby union. Br J Sports Med 44:159–167

Gianotti S, Hume PA et al (2008) Interim evaluation of the effect of a new scrum law on neck and back injuries in rugby union. Br J Sports Med 42:427–430

John HM et al (2008) Recent trend in rugby union injuries. Clin Sports Med 27:51–73

Kaplan KM, Andrew G et al (2008) Rugby injuries: a review of concepts and current literature. Bull NYU Hosp Jt Dis 66(2):86–93

Maclean JGB, Hutchison JD (2011) Serious neck injuries in U19 rugby union players: an audit of admissions to spinal injury units in Great Britain and Ireland. Br J Sports Med 23:2–4

MacManus A, Cross DS (2004) Incidence of injury in elite junior Rugby Union: a prospective descriptive study. J Sci Med Sport 7(4):438–445

Mcintosh AS, Savagte TN et al (2010) Tackle characteristics and injury in a cross section of rugby union football. Med Sci Sports Exerc 42(59):977–984

Quarrie KL et al (2002) Rugby union injuries to the cervical spine and spinal cord. Sports Med 32:633–653

Schick DM et al (2008) Injuries during the 2006 Women's Rugby World Cup. Br J Sports Med 42:447–451

Silver JR (2002) The impact of the 21st century on rugby injuries. Spinal Cord 40:552–559

Yamada M (2008) Cervical spinal cord injuries: the mechanism by which they occur, preventing those injuries, and preventing their recurrence. J Clin Sports Med 25:52–62

Yamada M (2009) Cervical spinal cord injuries in rugby. J Clin Sports Med 26(8):1025–1033

Yamada M (2011) Motion analysis with images: contact. J Clin Sports Med 28:167–165

Yamada M et al (2004) Measures to prevent head and/or neck trauma due to tackling. Spine Spinal Cord 17(12):1137–1144

Yamada M et al (2005) Effectiveness of a shrugging motion in rugby. J Ryutsu Keizai Uni Faculty Health Sport Sci 1(1):131–138

Chapter 6
Severe Head & Neck Injury and Its Prevention in Judo

Seiji Miyazaki

Abstract Severe cases of concussion commonly occur in Judo practice. Therefore, it is important to know how to prevent this type of injury in Judo practice as well as how to respond to a concussed individual.

Keywords Concussion • Judo • Prevention • Ukemi (break-fall)

6.1 Introduction

According to the Japan Neurosurgical Society, a concussion is defined as head trauma that includes temporal neurological dysfunction following a head contusion, with the loss of normal brain function being only temporary (A glossary of the Japan Neurosurgical Society 2006). The main symptom of a concussion is the change of mental status or activity level, with or without a loss of consciousness. For diagnostic purposes, a concussion is determined as the condition of loss of consciousness within 6 h. In a report by The Ministry of Education, Culture, Sports, Science, and Technology in Japan, a concussion is defined as a "head trauma that includes temporal neurological dysfunction seen right after the head contusion, but recovers to be back to normal condition before the head contusion." In the report's section on the mechanisms and risk factors for head injury, concussion is a mechanism noted for brain damage. In Japan, this definition tends to confuse medical personnel with their understanding of a concussion. In a comment in *Neurology*, it is stated, "a concussion is a trauma-induced alteration in mental status that may or may not involve loss of consciousness (The Quality Standards Subcommittee, American Academy of Neurology 1997)." The Centers for Disease Control and Prevention states that "a concussion is a type of traumatic brain injury, or TBI, caused by a bump, blow, or jolt to the head that can change the way your brain normally works. Concussions can also occur from a blow to the body that causes the head to move rapidly back and forth. Even a 'ding,' 'getting your bell rung,' or what

S. Miyazaki (✉)
Sports Medicine, Tokai University School of Physical Education, Kanagawa, Japan
e-mail: seijim@tsc.u-tokai.ac.jp

K. Kanosue et al. (eds.), *Sports Injuries and Prevention*,
DOI 10.1007/978-4-431-55318-2_6

seems to be a mild bump or blow to the head can be serious" (Centers for Disease Control and Prevention (CDC) 1997). According to these statements, a concussion is recognized as a traumatic brain injury. Conversely, a concussion is recognized as head trauma with temporal neural dysfunction and plasticity to the neural function; therefore, in Japan, it seems to be recognized as a head trauma regardless of its severity.

6.2 Head and Neck Injuries in Judo Athletes

6.2.1 Concussion

According to injury surveillance data collected in 2008 from 1,443 Judo athletes (1,000 males and 443 females, mean age 17.4 ± 9.1 years), there were 517 athletes (35.8 %) with a medical history of concussion. The total number of concussion cases was 645. Concussions were more common in males (39.0 %) than in females (27.7 %). There were 103/348 (29.6 %) athletes in junior high school with a medical history of concussion, 158/454 (34.8 %) in senior high school, 241/617 (39.1 %) in college, and 14/24 (58.3 %) in amateur sports. As the age of the athletes increased, the athletes' medical history of concussions also increased (Fig. 6.1).

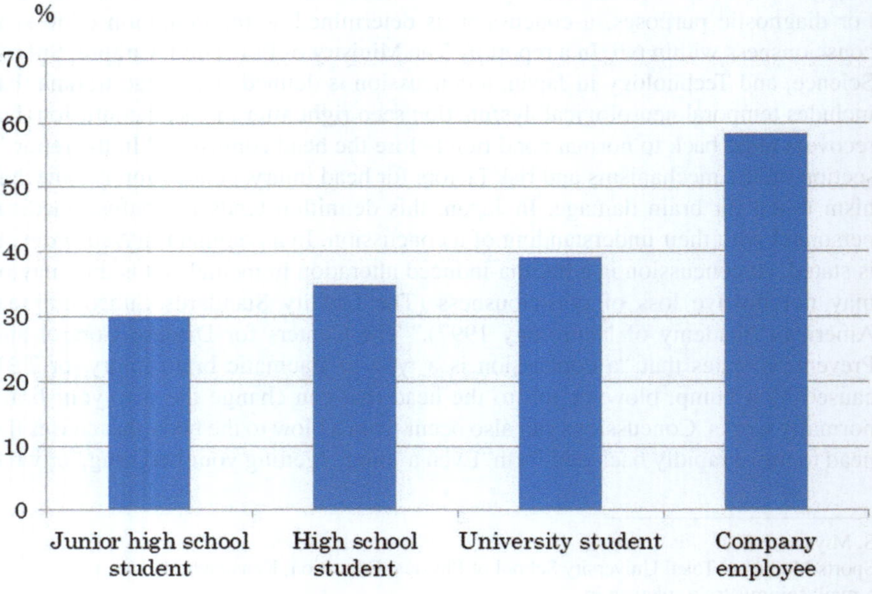

Fig. 6.1 Frequency of concussions

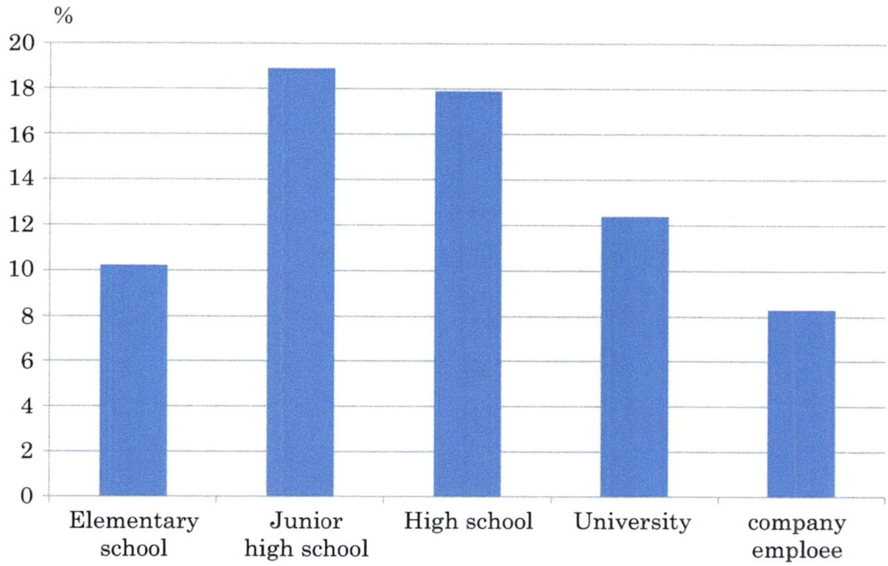

Fig. 6.2 When did the concussion was occur?

In terms of the time when the first concussion occurred, 114/1,114 (10.2 %) athletes were in elementary school, 265/1,400 (18.9 %) were in junior high school, 194/1,084 (17.9 %) were in senior high school, 80/6,410 (12.4 %) were in college, and 2/24 (8.3 %) were in amateur sports (Fig. 6.2). Examination of these data indicates that sustaining a concussion is more frequent in junior and senior high school.

With regard to the cause of concussion, in 645 cases the injury occurred when the Judo athlete's partner executed a Judo technique. The most frequently reported Judo technique was the osotogari (sweeping leg throw) (37.1 %). However, 14.3 % of the athletes reported that they did not remember which Judo technique was executed prior to their concussion. The seoinage (shoulder throw) was also a common Judo technique that caused concussion in 12.4 % of the athletes. Overall, in Judo practice, a concussion occurred more frequently in cases where an athlete was thrown to the mat by the back than when thrown to the mat by the front.

6.2.2 Severe Traumatic Head Injuries

Severe traumatic head injuries (including death or severe sequelae) were investigated using records from insurance claims from a system for disability benefits and consolatory money maintained by the All Japan Judo Federation (2003–2012). Thirty-nine cases of severe traumatic head injuries (35 male and 4 female) were reported. Thirty-five cases were diagnosed as acute subdural hematoma, and 21 of

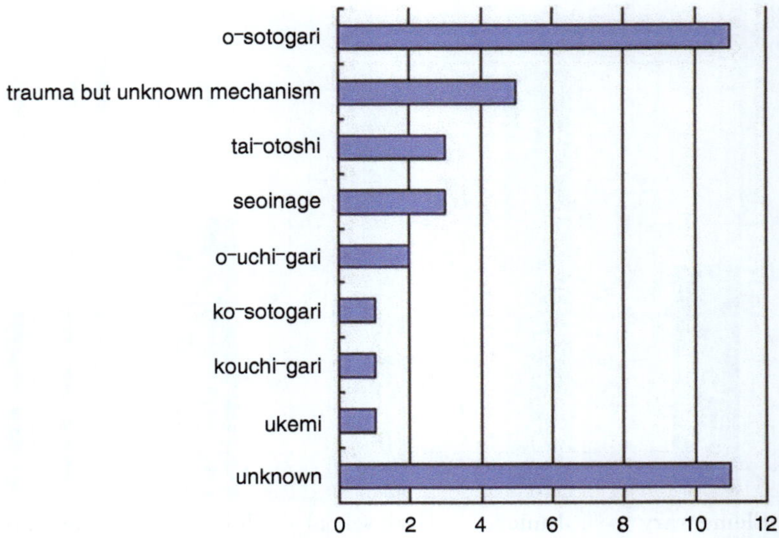

Fig. 6.3 Techniques that lead to severe brain injuries

those resulted in death. Many of the injuries occurred during the first year of junior and senior high school, and most of the injured students were beginners (Fig. 6.3). Based on these 39 cases, the severe traumatic head injuries were caused when an osotogari or seoinage throw was executed. These throws were previously noted as mechanisms of concussion in the study above, which indicates that severe traumatic head injuries and concussions have similar causes.

6.2.3 *Severe Traumatic Neck Injuries*

Severe traumatic neck injuries (including fracture, dislocation, and cervical cord injury) were caused by a fall in which the athlete's forehead hit the mat. Twenty-nine cases of these injuries were reported, and many of the neck injuries were caused by an uchimata (inner thigh throw) or haraigoshi (sweeping hip-to-loin throw), either of which could also cause a concussion. Thus, a direct impact to the forehead can also be one of the risk factors for a concussion (Fig. 6.4).

Any direct or indirect impact to the head that might result in a concussion can cause severe injury. In Judo practice, some form of impact is applied to the head regardless of whether an athlete is thrown to the mat by the front or the back. In summary, if the situations that cause a concussion can be avoided, then severe traumatic neck injuries can also be prevented.

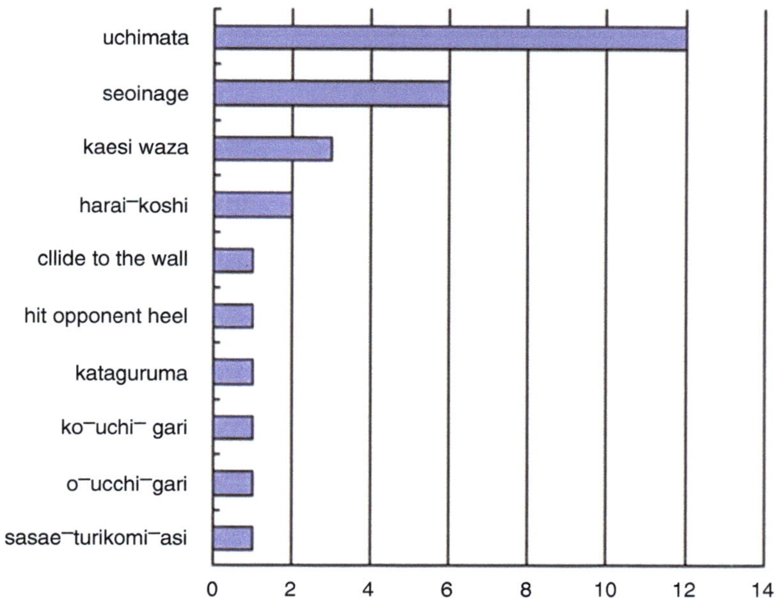

Fig. 6.4 Techniques that lead to severe neck injuries

6.2.4 Judo-Related Injuries During Regular Class and After School Activities

Is the Judo technique itself related to falls in which the athletes' heads or necks hit the mat? According to data reported by the Japan Sport Council, with regard to the number and rate of sports-related injuries by anatomical location of Judo activity, head trauma accounted for 9.8 % of the total injuries (2007–2009). Moreover, in terms of the rate of injury, there was no difference between Judo practiced in a regular class setting or in after school activity clubs. This data suggests that beginners in a regular class are also prone to head trauma (Table 6.1).

6.3 Principle of Judo

Jigoro Kano (a founder of Kodokan Judo) (for details, see Kano Teacher Biography Compiled Association 1977) learned Tenjin Shinyo-ryu (a school of thought) jujutsu and subsequently Kito-ryu jujutsu. Though very knowledgeable in both of these, he also took an interest in other esoteric schools of thought, and devoted his time to researching and borrowing good aspects from the others. Further, along with

Table 6.1 Investigation of athletic injuries in Judo

Parts	Regular class			After school activity		
	Cases	%	Incidence per 100,000	Cases	%	Incidence per 100,000
Head	**2,189**	**9.8**	10.52	**1,366**	3.7	640
Forehead	38	0.2	0.18	107	0.3	50
Eye	479	2.1	2.3	701	1.9	328
Ear	43	0.2	0.21	293	0.8	137
Nose	267	1.2	1.28	240	0.7	112
Mouse	41	0.2	0.2	98	0.3	46
Tooth	187	0.8	0.9	376	1	176
Cheek	24	0.1	0.12	44	0.1	21
Jaw	35	0.2	0.17	57	0.2	27
Neck	**1,656**	**7.4**	7.96	**1,108**	3	519
Shoulder	**2,222**	10	10.68	**5,590**	15.3	2,618
Upper arm	251	1.1	1.21	675	1.8	316
Elbow	898	4	4.32	**2,816**	7.7	1,319
Forearm	548	2.5	2.63	979	2.7	458
Hand	711	3.2	3.42	**1,100**	3	515
Finger	**2,195**	9.8	10.55	**3,426**	9.4	1,604
Chest	704	3.2	3.38	994	2.7	465
Abdomen	35	0.2	0.17	68	0.2	32
Back	125	0.6	0.6	149	0.4	70
Lumber	673	3	3.23	**1,678**	4.6	786
Buttock	67	0.3	0.32	142	0.4	66
Hip	82	0.4	0.39	203	0.6	95
Thigh	272	1.2	1.31	339	0.9	159
Knee	**1,003**	4.5	4.82	**4,534**	12.4	2,123
Leg	587	2.6	2.82	**1,219**	3.3	571
Ankle	**1,750**	7.8	8.41	**3,102**	8.5	1,453
Foot,toe	**5,211**	23.3	25.04	**4,939**	13.5	2,313
Whole body	33	0.1	0.16	190	0.5	89
Total	22,326	100	107.29	36,533	100	17,107

establishing systematized techniques that reflected his own originality and ingenuity, he also made advances in the theoretical realm. From the principle of "flexibility is stronger than stiffness" he developed the concept of "utilizing the power of the mind and body most effectively". Kano defined jujutsu as "offensive or defensive techniques while unarmed or with an accompanying small weapon against an unarmed or armed opponent". Jujutsu was seen as being different from other martial arts (such as swordsmanship, spearmanship, artillery, archery, etc., including self-defense). When today's Judo is performed as a competition, the outcome (win or lose) is determined either by winning "ippon by throw or grappling" or by points given for "yuko" during the match. The definition of the "ippon" is as follows:

(a) When a contestant with control throws the other contestant largely on his back with considerable force and speed.
(b) When a contestant holds the other contestant with osaekomi-waza, and the other contestant is unable to get away for 20 s after the announcement of osaekomi.
(c) When a contestant gives up by tapping twice or more with his hand or foot or says maitta generally as a result of a grappling technique, shime-waza or kansetsu-waza.
(d) When a contestant is incapacitated by the effect of a shime-waza or kansetsu-waza. (International Judo Federation: Refereeing Rules)

In Kodokan Judo, throwing is the most frequently used technique. The purpose of throwing is to win ippon by having the back of opponent against the matting. Under the current rules, the opponents should be thrown by grasping their Judo uniform.

The fundamental principle of Judo is expressed as "seiryoku-zenyo" meaning a maximully-efficient use of body and spirit. Techniques based on this principle follow the "tsukuri" and "kake" theories. "Tsukuri" consists of "kuzushi" (making the posture of the opponent unstable) as well as establishing a self-posture that works best for executing the techniques. "Kake" involves selecting and executing the optimal technique. Although the "kake" and "tsukuri" are technical principles that are in accordance with the fundamental principles of Judo, the person being thrown loses the correct posture of the upper body and head since the Judo uniform is being held.

6.4 Prevention of Traumatic Head Injury in Judo

6.4.1 All Japan Judo Federation

In an effort to prevent traumatic head injury, the All Japan Judo Federation established a special committee to oversee a safety guidance project in 2010. The committee accomplished the following:

- Revision of a booklet on "Safety guidance of Judo (Safety guidance Project Committee 2011)".
- Analysis and investigation of the cause of accidents in Judo.
- Mandatory attendance of a "Safety guidance" seminar for coaches (Training seminar for Judo coaches associated with the transitional arrangements leader qualification system beginning in 2014)
- Convene the National safety guidance personnel committee.
- Perform research on coaching methods specialized for judo athlete safety.

Efforts that followed included revision of the booklet "Safety guidance of Judo (first edition in 2006)". Manuals were prepared that gave instructions for the prevention of head accidents as well as methods for coping with accidents that

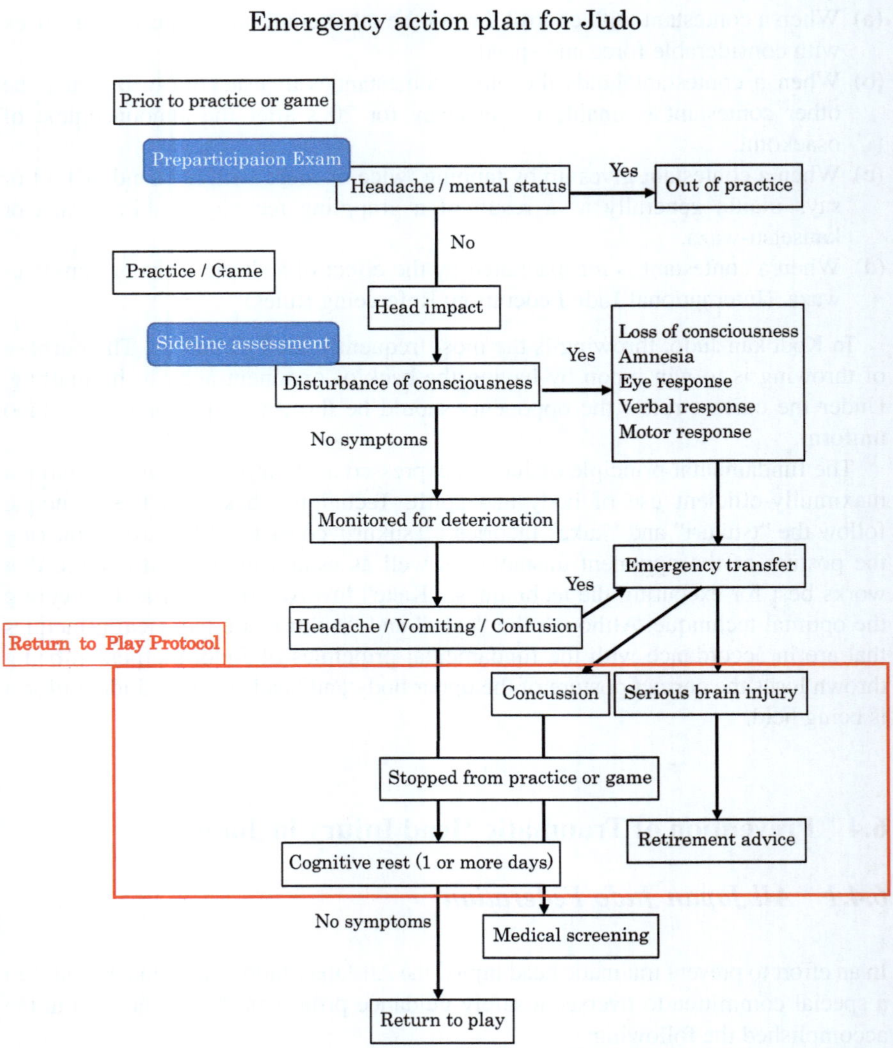

Fig. 6.5 Manual for traumatic head injury in Judo

did occur. Information was provided that analyzed the extent of various incidents (Fig. 6.5). Specific directions were given for actions that must be taken at accident incidence and guidance provided on the general signs which would give a rough indication of the degree recovery. Unfortunately, there is still a lack of awareness among coaches of the potential severity of concussions, because such injuries are relatively frequent and many cases show relatively transient minor symptoms.

6.4.2 *The Ministry of Education in Japan*

Following amendment to the Fundamental Law of Education in 2006, the official curriculum guidelines for junior high school were updated in 2008. To this end, there was an indication of an awareness on the need for increased safety in Judo training. Beginning in 2013 it was deemed important to properly instruct coaches before they taught compulsory martial arts and dance skills in physical education classes. The Ministry of Education, Culture, Sports, Science and Technology established a "Conference for research cooperation on accident prevention in physical education class" in July, 2011. Through discussions in the conference, statements on "Safe and smooth implementation of martial arts classes associated with the new curriculum guidelines", "Thorough safe administration in Judo associated with compulsory of martial arts (Sports and Youth Bureau Counsellor 2012)", and "Toward safe implementation of Judo classes" were released in March, 2012. The included statements raised the following six points for safe administration of Judo classes:

- Prior confirmation of safe conditions in the practice environment
- Precautions for avoiding incidents which would lead to accidents.
- Cooperation with outside coaches; communication and confirmation of guiding principle among coaches.
- Designing of a coaching plan (far-seeing coaching plan through 3 years; reasonable and stepwise coaching based on learning stages and inter-individual differences).
- Specific points to remember for safe Judo coaching (attention to physical condition of students); stepwise coaching based on cognition that most students are beginners; extensive practice of "ukemi" (break-fall) not to and not to be hit on the head; grappling is limited to "osaekomi-waza" (holding techniques), "shime-waza" (choking techniques) and "kansetsu-waza" (locking techniques) must not be coached; appropriate coaching of throwing techniques consistent with proficiency of students only after students have become well-skilled in "ukemi" (break-fall).
- Coping in case of an accident (first aid); proper handling of bruises, sprains, bone fractures, abarticulations; careful handling of neck injuries and with head impacts.

On the basis of the points raised above, seminars for preventative measures in school classes involved with producing Judo coaches are to be held at the level of the province, municipality, and physical education colleges.

6.4.3 *Practice in Prevention of Traumatic Head Injury*

The "Safety guidance of Judo (3rd edition) (Safety guidance Project Committee 2011) advocates, for prevention of accidents, three points to be utilized in practice:

- acquisition of the correct techniques and particularly "ukemi" (break-fall).
- respect for opponent, and reasonable practice levels
- safety check of apparel and training hall

In 1889 Kano stated that: The use of Judo game techniques can kill, hurt, or capture someone if one attempts to do so, and can also protect oneself against any such attempt by an opponent. To summarize, he described Judo as a method to physically take control of and not to be taken control of by an opponent. Therefore, the possible risk of injury has been recognized and "ukemi" has been emphasized as a method to protect oneself. "Ukemi" involves being able to fall down comfortably without injury and suffering regardless of whether falling down oneself or being thrown by an opponent. It serves as the most basic Judo skill and is recognized as extremely important. Further, since it is well recognized that it is difficult to acquire this skill, an example of an appropriate multi-staged coaching method would include:

Being able to utilize "ukemi" in the following situations.

Stage 1. By oneself
Stage 2. By oneself under pressure of opponent
Stage 3. When thrown by opponent
Stage 4. When thrown by opponent with successive (combinational) throwing techniques
Stage 5. When counterattacked by opponent

Whether acquisition of the "ukemi" skill by itself is sufficient to protect oneself remains a subject for discussion. Although the focus of a Judo competition is to throw opponents, excessive focus on this activity during the early practice stages may enhance possible risk of injury. The skills of the throwers should also be considered. When there is a difference in skill level between beginners and experts, using the "kuzushi" (making posture of the opponent unstable) and "osotogari" (a major sweeping leg throw) techniques, for example, results in movement of the body in a posterior direction with the body weight supported equally by both legs. In experts on the other hand, a pull on one side results in relative weight-bearing on the side, body rotation, and consequently in a lateral "ukemi". It is therefore important to consider the skill level of both parties.

6.4.4 Referee Regulations in Judo Contests for Injury Prevention

Currently, refereeing regulations of the International Judo Federation are used in Judo contests for adults and children. Although there are regulations to prevent spinal cord injuries from occurring during the contests, there are no regulations for preventing head injuries like concussions. To prevent the second impact syndrome as well as head and neck injuries, it is important to consider implementing a penalty for techniques that have the potential to cause these types of injuries.

Glossary of Judo Terms

Osotogari	a major sweeping leg throw
Seoinage	a shoulder throw
Uchimata	an inner thigh throw
Haraigoshi	a sweeping hip-to-loin throw
Osaekomi-waza	a hold-down technique (a mat hold)
Shime-waza	a choking technique (a stranglehold)
Kansetsu-waza	a joint locking technique
Tsukuri	a set-up to execute technique
Kake	an execution of technique
Kuzushi	a balance breaking
Ukemi	a break-fall
Ippon	a sign used to indicate a contest decision
Yuko	a point which is awarded in accordance with the Judgment of a technique (effective/moderate advantage)

*Please refer to the link: All Japan Judo Federation (http://www.judo-ch.jp/english/dictionary/)

References

Centers for Disease Control and Prevention (CDC) (1997) Sports-related recurrent brain injuries-United States. Morb Mortal Wkly Rep 46(10):24–227

Kano Teacher Biography Compiled Association (1977) Jigoro Kano (in Japanese), Kodokan, pp 285–308

Safety guidance Project Committee (2011) Safety guidance of Judo (in Japanese), 3rd edn. Foundation All Japan Judo Federation, Tokyo, pp 38–41

Sports and Youth Bureau Counsellor (2012) Thorough safety management of Judo associated with compulsory of martial arts, vol 23. Ministry of Education, Culture, Sports, Science and Technology, p S910

The Japan Neurosurgical Society (2006) A glossary of the Japan Neurosurgical Society, 2nd edn. Nankodo, Tokyo (in Japanese)

The Quality Standards Subcommittee, American Academy of Neurology (1997) Practice parameter: the management of concussion in sports (summary statement). Neurology 48(3):581–585

Chapter 7
Concussion and Severe Head-Neck Injury: An Approach for Their Prevention in Rugby and Judo

Satoshi Tani

Abstract Concussion is a common sequel to head injuries incurred during sports activity. In the last decade, there has been a significant increase in the number of studies dealing many aspects of concussions. Included approaches involve epidemiology, clinical aspects including modern neuroimaging as well as neuropsychiatric evaluation, and prophylactic possibilities for reducing concussion incidence. In this chapter I cover the epidemiology, clinical symptoms, mechanical aspects, experimental studies, computer-based simulation, and pathophysiology of the clinical symptoms of concussion. I also cover the aims of stroke prophylaxis and current prophylactic issues.

Keywords Acute subdural hematoma • Angular acceleration • Chronic traumatic encephalopathy • Concussion

7.1 Introduction

Concussion is a common sequel to head injuries incurred during sports activity. Neither clinical problems nor the pathophysiology of concussions have been clearly delineated in the past. However, over the last decade there has been a significant increase in the literature involving concussions. This has included the epidemiology, clinical aspects including modern neuroimaging, and neuropsychiatric evaluations as well as the prophylactic possibilities for reducing concussions. This increase is partially the result of an increased awareness of the severity and damaging complications that can follow head trauma. In this chapter, I cover the epidemiology, clinical symptoms, mechanical aspects, experimental studies, computer-based simulations, and pathophysiology of clinical symptoms. I also present the aims of prophylaxis, as well as current prophylactic issues.

S. Tani (✉)
Department of Neurosurgery, Jikei University, Tokyo, Japan
e-mail: tani@jikei.ac.jp

K. Kanosue et al. (eds.), *Sports Injuries and Prevention*,
DOI 10.1007/978-4-431-55318-2_7

7.2 Epidemiology of Concussions

The incidence of concussions remain unclear, since concussions can occur in schoolyards, stadiums, and mountainous as well as other areas, and the victim may or may not be hospitalized. Statistical analyses in the epidemiology of concussion have been proposed in various venues, including sports organizations, sports education, social insurance, and hospital based patient registries.

Many publications have addressed concussion incidence in specific sports such as soccer, rugby, and American football. In these sports, approximately 5–10 % of the players experience apparent concussions. A recent 3-year-surveillance program of Japanese junior high and high school students involved in sports reported a prominent occurrence of concussions in Rugby, Judo and gymnastics (Fukubayashi 2013). In addition, a hospital-based surveillance questionnaire in Japan revealed a prominent occurrence of concussions the in winter leisure sports of snowboarding and skiing as well as in the school based sports of baseball, soccer, rugby, and basketball (Tani et al. 2008).

7.3 Clinical Symptoms of Concussions

While the symptoms of concussion can be categorized in a number of ways, there are four major characteristics. Some of these are prominent in acute concussions and others in sub-acute stages. These characteristics are:

1. Loss of consciousness
2. Disturbance of neurocognitive function
   Retrograde amnesia, post-traumatic amnesia, disorientation, confusion, mental irritability, sleepless, and other
3. Imbalance
   Difficulty in tandem stance, and others
4. Physical symptoms
   Headache, nausea, dizziness, double vision, tinnitus, and others

7.4 Clinical Grading of Concussions

Some clinical assessments of concussion severity are primarily based on the length of loss of consciousness, however, recent developments have led to the proposal that clinical severity is related to the length of time that concussion related symptoms remain (McCrory et al. 2009). A difficulty of this approach is that it provides no initial estimate of a concussion's severity.

Objective clinical evaluation of a concussion is an important issue in the athlete's recovery. In an international conference on concussions, a proposal termed SCAT3 invoked a special emphasis on concussed children (McCrory et al. 2013).

7.5 Mechanical Aspects of Concussions

The mechanisms involved in the production of concussions should be carefully evaluated in order to develop prophylactic issues as well as clinical treatments for concussions. A direct impact to the head is not the only way to evoke a concussion as an indirect stress to the brain such that caused by acute neck extension and flexing or head shaking may produce sufficient mechanical stress (e.g. share stress) in the brain parenchyma to result in a concussion.

Rather than translational acceleration, rotational acceleration is mechanically considered to be a more important stress in the production of the diffuse brain injury that is characteristic of a concussion. However, angular velocity and acceleration are critical for primates restrained by the neck. The physical definition of angular velocity is the rate of change of angular displacement and is a vector quantity which specifies the angular speed (rotational speed) of an object and the axis about which the object is rotating. Angular acceleration represents the rate of change of angular velocity (Fig. 7.1).

Angular acceleration produced by an impact can be applied to the head in three planes; sagittal, coronal and axial planes. In a practical sense, a pure impact involving only one of the three planes is feasible only in a controlled experimental situation. Angular acceleration in the coronal plane is most typically associated with diffuse brain injuries in primates (Gennarelli et al. 1982). On the other hand, acceleration forces (translational, rotational, or angular) in the sagittal plane may be associated with life-threating acute subdural hematoma: bleeding from disrupted bridging veins that course along the midline structures of the brain.

About 80 % of the knock-outs (KO) in professional boxing bouts involve concussions; hook shots to the head which result in an angular rotation of the neck produce the most severe type of KO (Tani et al. 2001).

Fig. 7.1 Schema of angular acceleration to the head

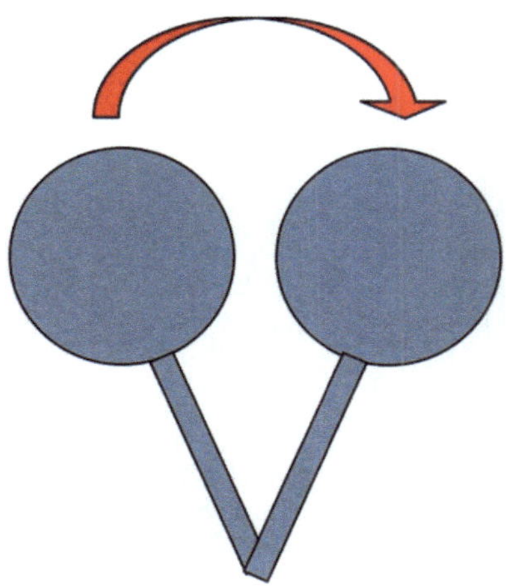

Magnitude of acceleration (angular velocity and acceleration) is another important factor regarding concussion; more than 3,500 rad/s^2 of angular acceleration is reported to produce a concussion (Shewchenko et al. 2005).

7.6 Experimental Studies of Concussion

In 1946 Pudenz and Shelden (1946) reported superficial brain deformation and shift occurred following a head impact to apes. Over thirty years later, Gennarelli et al. (1982) studied the mechanical and pathophysiological aspects of primate head injury. They reported that lateral rotational impacts to the head (coronal acceleration) were the most significant in the production of diffuse brain injury. Little academic research has been done on primates since Gennarelli's report, although primate species present an obvious homology with humans.

7.7 Simulation of Concussions

As primate based experiments have declined, computer-based physical simulation studies have simultaneously been increasing, both in frequency and in sophistication. The author and co-workers developed a 2D simulation study in which it was

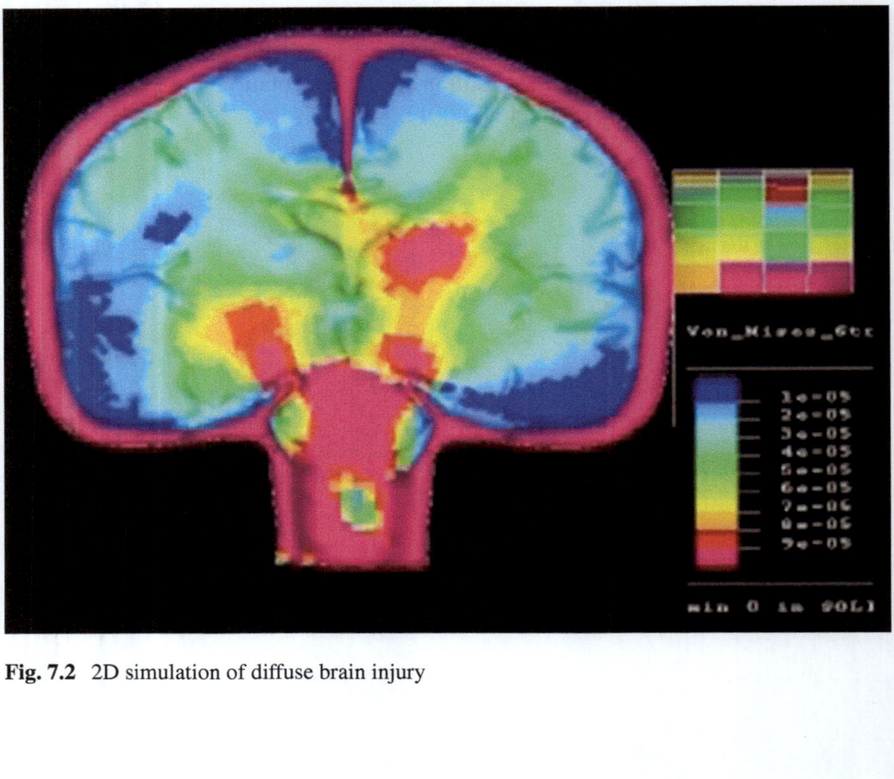

Fig. 7.2 2D simulation of diffuse brain injury

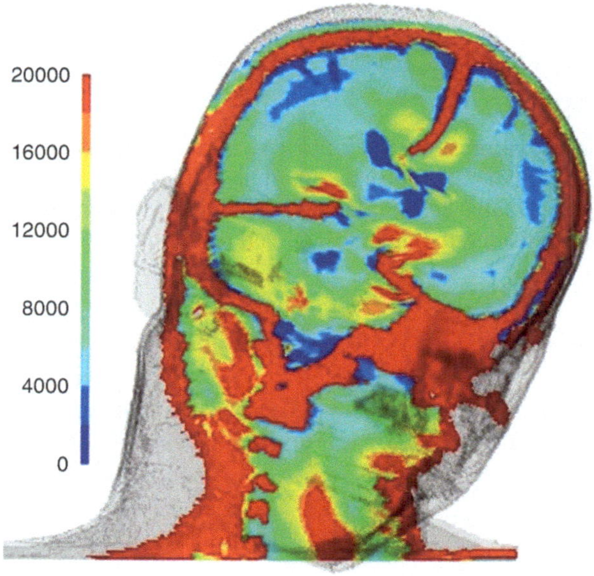

Fig. 7.3 3D voxel simulation of diffuse brain injury

possible to demonstrate diffuse brain injury lesions similar to those seen in clinicopathological settings. An impactor was applied to the right temporal side of the model at the level of the third ventricle, at a point which was at the same level as the center of gravity at a speed of 2 m/s. The model moved freely after the impact without any fixation. Major mechanical stresses on the brain parenchyma were observed in the midbrain, basal ganglia and corpus callosum underneath the falx cerebri (Fig. 7.2). Recent developments of the 3D voxel method enable it to provide a more precise representation of shear stress in the brain parenchyma after various impacts to the head in coronal and/or sagittal planes under different strain conditions (Fig. 7.3).

Further development of the methods involved in simulation studies are expected to further elucidate the mechanical aspects of concussions.

7.8 Pathophysiology of Clinical Symptoms

Pathophysiological reactions and consequences of mechanical stress in brain parenchyma have been proposed as follows:

1. Abrupt neuronal depolarization
2. Release of excitatory neurotransmitters

3. Ionic shifts
4. Changes in glucose metabolism
5. Altered cerebral blood flow
6. Impaired axonal function

These pathological reactions may provoke various lesions and their associated loss of fiber connections in neuronal structures, and may result in transient brain dysfunction similar to that seen under various clinical syndromes. These pathological reactions are considered to be transient and reversible.

7.9 Aims of Prophylaxis for Concussions

Recent advances in sports medicine have reached the conclusion that concussions should be minimized. The main reason for prophylaxis of concussions is to avoid the following associated conditions:

1. Second impact syndrome (SIS)
2. Post-concussion syndrome (PCS)
3. Chronic traumatic encephalopathy (CTE)

SIS is an acute and fatal diffuse brain swelling that occurs when an athlete who has sustained a concussion receives a second head injury before the initial symptoms have fully cleared (Cantu 1992). However, in retrospect this proposed unique pathological entity appears to be an over-diagnosis after the re-evaluation of a number of cases (McCrory and Berkovic 1998). Careful estimation of SIS should be carried out along with correct diagnostic criteria. The clinical significance of SIS now needs to be reinterpreted, as post-injury cognitive slowing and delayed reaction time can have a negative effect on an athlete's ability to play safely and effectively (Giza et al. 2014). Repeated head injuries followed by the suspicion of concussions produce a potentially fatal situation which should be recognized and avoided at all costs.

Most of the symptoms related to concussions disappear spontaneously within 10 days. However, persistent symptoms such as headache, dizziness, tinnitus, emotional changes, and insomnia are noted in some patients. There is some degree of confusion in the nomenclature and definition of PCS. A diagnosis of PCS can be made when the persistent symptoms have not exceeded 3 months. Even if no apparent pathophysiology has been documented, repeated neuroimaging (such as MRI) may still be useful in order to rule out small traumatic lesions including thin subdural hematomas and minor hemorrhagic lesions in the brain parenchyma.

It is likely that concussion-related symptoms place an athlete at greater risk for recurrence of a concussion. The concept of accumulation of concussive injuries has been recognized (Kelly and Rosenberg 1997) and accepted because of the significant occurrence of impaired neurocognitive function in athletes who have sustained repetitive concussions. As a result, deteriorated neurocognitive function and/or positive extrapyramidal or cerebellar signs are included as CTE. CTE has become a medico-socially important issue in sports, not only professional boxing, but also in American football and other contact sports such as soccer and rugby.

7.10 Prophylaxis of Concussions

There is no such thing as a tough brain. It is apparent that we should prevent CTE and the ensuing acute catastrophic complication of acute subdural hematoma. It is important for prophylactic measures to prevent or reduce concussions to be put in place, but so far no definite measures have been established. Some proposals and candidate issues for preventing concussions follow:

1. Headgear
   While headgear is considered to be protective against concussions, and although it does prevent or reduce scalp contusions or skull fractures, available information indicates that it provides no reduction in the rate of diffuse brain injury. However, an aversive effect of wearing headgear such as aggressive play or head checking should be recognized. Also, the raised mass volume of the head may produce a detrimental increment in received angular acceleration.
2. Mouth piece
   The brain receives no significant protective effect against concussions from the wearing of a mouse piece (Giza et al. 2014). However, their use is quite popular and likely decreases facial and dental injuries in contact sports.
3. Regulation changes
   It is well known that a number of regulation changes in American football were designed to prevent fatal injuries due to acute subdural hematoma. Such regulations have also resulted in a proportionally reduced frequency of concussions as well. We have to make clear the risk factors for the promotion of concussions: Unskilled beginning players, training in hot and humid conditions, exhausted players, and illegal or foul plays. Based on the delineated risk factors, regulation changes even for leisure sports should be established and revised as new information accrues.
4. Other potential measures
   Proposals that develop from information of simulation studies may play an important role in the prevention of concussions and other catastrophic brain injuries. Such studies may also clarify other influences and external stresses on the brain parenchyma due, among other things, to endogenous brain conditions such as severe dehydration, congenital anomalies, or lowered neck tonus.

References

Cantu RC (1992) Cerebral concussion in sport. Management and prevention. Sports Med 14:64–74

Fukubayashi T (2013) Surveillance system in Japanese sports-related injury. Net prints. http://www.japan-sports.or.jp/Portals/0/data0/publish/pdf/H2401.pdf. Accessed on 22 Dec 2013

Gennarelli TA, Thibault LE, Adams JH, Graham DI, Thompson CJ, Marcincin RP (1982) Diffuse axonal injury and traumatic coma in the primate. Ann Neurol 12:564–74

Giza CC, Kutcher JS, Ashwal S Summary of evidence-based guideline update: evaluation and management of concussion in sports. Report of the guideline development subcommittee of the

American Academy of Neurology. Net print. http://www.neurology.org/content/early/2013/03/15/WNL.0b013e31828d57dd.full.pdf. Accessed on 29 Jan 2014
Kelly JP, Rosenberg JH (1997) Diagnosis and management of concussion in sports. Neurology 48:575–580
McCrory PR, Berkovic SF (1998) Second impact syndrome. Neurology 50:677–683
McCrory P, Meeuwisse W, Johnston K, Dvorak J, Aubry M, Molloy M, Cantu B (2009) Consensus Statement on Concussion in Sport: the 3rd International Conference on Concussion in Sport held in Zurich, November 2008. Br J Sports Med 43(Suppl 1):i76–90. doi:10.1136/bjsm.2009.058248
McCrory P, Meeuwisse WH, Aubry M, Cantu B, Dvořák J, Echemendia RB, EngebretsenL JK, Kutcher JS, Raftery M, Sills A, Benson BW, Davis GA, Ellenbogen RG, Guskiewicz K, Herring SA, Iverson GL, Jordan BD, Kissick J, McCrea M, McIntosh AS, Maddocks D, Makdissi M, Purcell L, Putukian M, Schneider K, Tator CH, Turner M (2013) Consensus statement on concussion in sport: the 4th international conference on concussion in sport held in Zurich, November 2012. Br J Sports Med 47:250–258. doi:10.1136/bjsports-2013-092313
Pudenz RH, Shelden CH (1946) The lucite calvarium—a method for direct observation of the brain. J Neurosurg 3:487–505
Shewchenko N, Withnall C, Keown M, Gittens R, Dvorak J (2005) Heading in football. Part 1: development of biomechanical methods to investigate head response. Br J Sports Med 39(Suppl 1):i10–i25
Tani S, Ohashi G, Ohtsuki G, Murakami S, Abe T (2001) Efficacy of punches in professional boxing bouts. Jpn J Clin Sports Med 9:413–416
Tani S, Kawamata T, Ogino M, Nariai T, Fukuda O, Mori T (2008) Head injury in sports. Clin Sports Med 25:309–312

Part III
ACL Injury: Injury Mechanism, Prevention Programs, and Their Usefulness

Chapter 8
Video Analysis of ACL Injuries in Sports

Tron Krosshaug and Roald Bahr

Abstract A complete understanding of the causes for injury requires a description of the mechanism of injury. Video analysis is the only method where we can objectively describe the motion patterns that are involved in a real injury situation. A new model-based image-matching (MBIM) method for 3D motion reconstruction from uncalibrated camera sequences was developed and validated against a marker-based 3D motion analysis system for running and side-step cutting. Flexion/extension joint angles agreed well in both hip and knee, but some underestimation from the video analysis was seen in knee flexion and hip abduction. The internal/external rotation angles varied the most. Next, the model was successfully applied on real injury situations outside the lab. A four-camera basketball video, a three-camera handball video and a one-camera video of an alpine skier was analysed using the MBIM method. Although it was demonstrated that important information for understanding the mechanism of injury could be extracted, some limitations exist. Poor picture quality, occluded body parts (due to clothing or other players), lack of visible landmarks or having only one camera angle may give sub-optimal matchings. It is therefore crucial to obtain high-resolution video footage, as this will open for improved understanding of injury mechanisms.

Keywords Video analysis • Biomechanics • ACL • Injury mechanism

8.1 Injury Causation in Sports

Sports injury prevention can be described as a "step-by-step process" (van Mechelen et al. 1992) where information on the causes is systematically collected and utilized to develop preventive measures. An important part of this process is the investigation of internal and external factors that characterizes the athlete's risk of sustaining an injury (Fig. 8.1). Internal risk factors such as e.g. insufficient muscle strength and increased joint laxity may predispose an athlete for injury, whereas additional

T. Krosshaug (✉) • R. Bahr
Oslo Sports Trauma Research Center, Norwegian School of Sport Sciences, Oslo, Norway
e-mail: tron.krosshaug@nih.no

K. Kanosue et al. (eds.), *Sports Injuries and Prevention*,
DOI 10.1007/978-4-431-55318-2_8

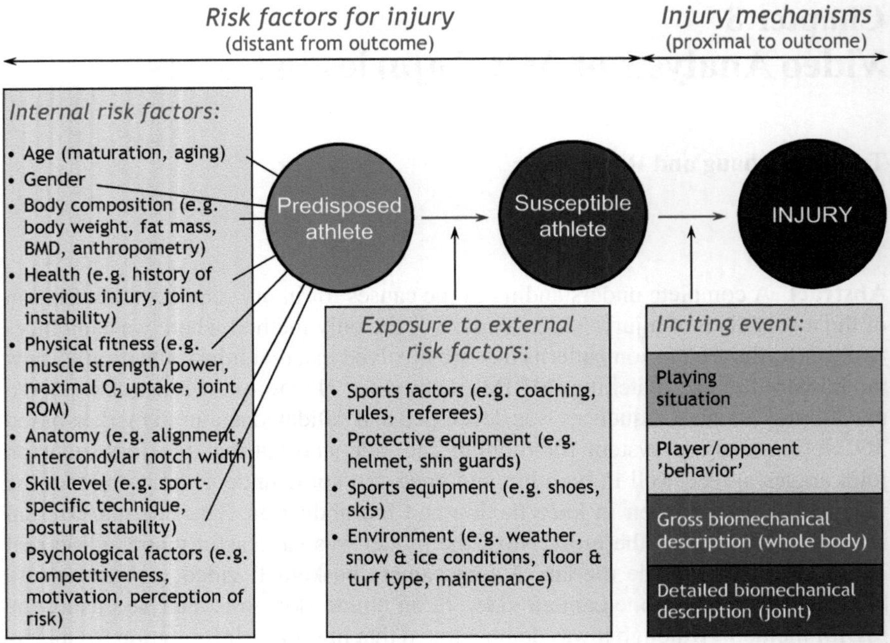

Fig. 8.1 A comprehensive model for injury causation (Bahr R, Krosshaug T (2005) Understanding the injury mechanisms – a key component to prevent injuries in sport. *Br J Sports Med* 39:324–329)

exposure to external risk factors such as high shoe-floor friction and high match intensity may make the athlete susceptible to injury (Meeuwisse 1994). However, a complete understanding of the causes for injury requires a description of the mechanism of injury. According to Bahr & Krosshaug, the mechanism of injury should describe the playing situation, player/opponent behaviour as well as gross and detailed biomechanics leading up to the moment of injury (Bahr and Krosshaug 2005).

8.2 Anterior Cruciate Ligament Injuries

Studies from handball football and other sports has shown that non-contact ACL injuries can be prevented through training programs focusing on knee control, balance, technique and strength (Renstrom et al. 2008). There exist some solid knowledge on the risk factors, e.g. that females are at higher risk compared to men, that match situations involves higher risk than training and that higher surface friction will increase the likelihood of sustaining an injury (Renstrom et al. 2008). However, there is still a lack of knowledge on all the numerous potential risk factors that have been suggested in the literature. The same is true for the injury mechanisms where different injury scenarios are heavily debated in the scientific community over the later years (Yu and Garrett 2007; Quatman and Hewett 2009). What seems to be clear is that most of the ACL injuries in team sports like handball, basketball and football are non-contact injuries occurring in cutting and landing movements. However, there exist several hypotheses for what joint loading that are causing the injury. The "Quadriceps drawer" theory suggests that the quadriceps muscles, through the patellar ligament, will pull the tibia forward relative to the femur. This effect is highest at low knee flexion angles. Valgus bending, internal rotation and vertical compression are other suggested mechanisms that have been suggested to load the ACL, but it is still unknown how important the different loading mechanisms are. This lack of understanding limits our ability to design optimal preventive measures (Krosshaug et al. 2005).

8.3 Research Approaches to Investigate Injury Mechanisms

So how can we then investigate injury mechanisms? From a systematic literature review eight different research approaches were identified. These methods were athlete interview, video analysis, motion analysis in the lab, in-vivo measurements of force/strain in the tissue, cadaver studies, mathematical modeling and simulation, clinical studies (MRI, CT) and in rare cases – accidents that may occur during biomechanical experiments (Krosshaug et al. 2005).

8.4 The Importance of Video Analysis

With exception of the last category (accidents during research experiments), video analysis is the only method where we can objectively describe the motion patterns that are involved in a real injury situation. Such kinematical estimates from real injury situations are essential for several reasons: In addition to the information that can be directly interpreted, these data are also necessary to validate studies that attempt to simulate the injury situation (e.g. cadaver studies, mathematical modeling, motion analysis in the lab and in-vivo studies). Such simulations must produce similar motions to those which are observed in real injury situations in order to be valid. If not, we may risk simulating situations that rarely or never occurs in real life. Additionally, video analysis may be used to describe other important aspects of the injury situation, not only biomechanical information, but information related to the game situation or player/opponent behavior as well. Such information can be used to design preventive measures. Video analysis has therefore been utilized in different sports to study various injury types. Andersen and co-workers (Andersen et al. 2003, 2004a, b) customized methods, previously used in game analysis, to study ankle sprains and head injuries in football. McIntosh and co-workers used video analysis to quantify head velocity during impacts in Australian Rules Football (McIntosh et al. 2000). These studies did, however, not attempted to quantify joint motion.

8.5 Video Analysis of ACL Injuries in Ball/Team Sports

Several video analysis studies have been conducted to investigate ACL injury mechanisms in ball/team sports (Olsen et al. 2004; Boden et al. 2000; Cochrane et al. 2007). These studies concluded that the knee is relatively straight at the moment of injury. Furthermore, is has been reported that the knee was either internally or externally rotated and in many cases in combination with valgus or valgus collapse. Some of these studies have also attempted to quantify joint motion (Olsen et al. 2004; Cochrane et al. 2007).

Unfortunately, there are several obvious limitations in these studies. They are based on relatively few cases and the video quality was in many cases poor, making it difficult to assess joint motion and the time point of injury (Olsen et al. 2004). However, the main problem with previous studies was the fact that only simple visual inspection, e.g. merely "guesstimating", was used to assess the joint motion.

8.6 The Validity of Simple, Visual Inspection

To investigate the accuracy of visual inspection, a systematic study was carried out where three test subjects performed sport specific movements (Krosshaug et al. 2007a). The movements were filmed with three ordinary consumer cameras; front, side and rear. The test subjects had 33 reflective markers attached to anatomical landmarks. Eight infrared, high-speed cameras (ProReflex, Qualisys Inc., Gothenburg, Sweden.) recorded the marker paths, enabling accurate measurement of joint angles in three dimensions.

The study revealed that experienced researchers were unable to produce reliable estimates of joint angles. Knee flexion angles were underestimated by 19° on average (Fig. 8.2). This implies that we must be cautious when interpreting results from studies using this approach. In other words, we can not trust that the injuries occur with a straight leg. As can be seen from Fig. 8.2, the experienced researchers estimated the knee flexion angle to be 30° when it in fact was 60°!

8.7 Development of a Model-Based Image-Matching (MBIM) Method for 3D Motion Reconstruction from Uncalibrated Camera Sequences

As a result of the poor results from visual inspection, there was a need for a new, more accurate and less subjective method. To make possible a more detailed biomechanical analysis, it was desirable to find a method that could produce continuous estimates of joint kinematics. A literature review showed that such methods had already been developed (Halvorsen 2002; Trewartha et al. 2001). However these

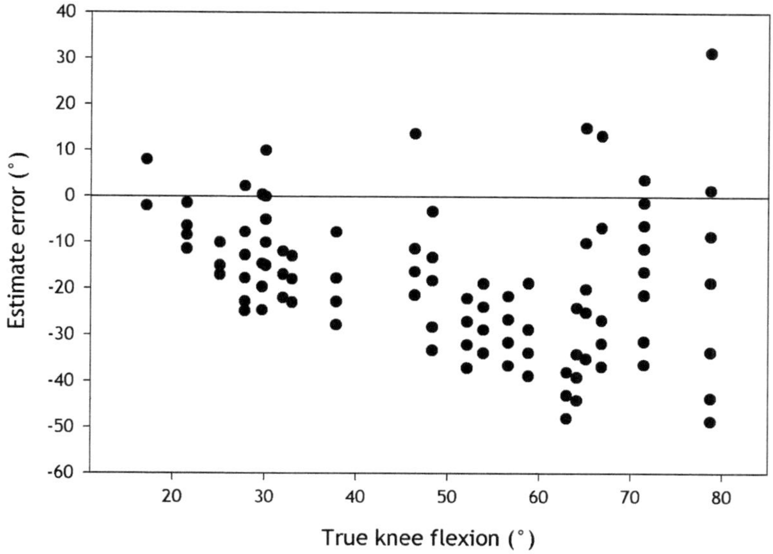

Fig. 8.2 Error estimates of knee joint angles as a function of true joint angle, as measured with a marker-based motion capture system (Krosshaug T, Nakamae A, Boden B, Engebretsen L, Smith G, Slauterbeck J, Hewett TE, Bahr R (2007) Estimating 3D joint kinematics from video sequences of running and cutting maneuvers – assessing the accuracy of simple visual inspection. *Gait Posture* 26:378–385)

methods were designed with automatic motion reconstruction in mind rather than accuracy. Additionally these methods could not be used for analyzing injury videos since they required e.g. fixed cameras, clothes with color codes, one-colored background or other simplifications in order to work. For this reason it was necessary to develop a new method for 3D reconstruction of injury situations. A solution to the problem was to manually configure a computerized skeleton model to imitate the injured person by superimposing it onto the video. This method is based on commercially available software Poser® 4 Pro Pack (Curious Labs, Inc., Santa Cruz, California, USA). Poser has the possibility to import video from up to four video cameras.

In order to reconstruction the injury situation in the computer, it is necessary to build a model of the surroundings (Fig. 8.3), as well as a customized skeleton model of the person of interest (Fig. 8.4). The model of the surroundings can for instance utilize the standard measurements of e.g. a handball court, including exact positioning of floor markings, commercial boards, goal posts etc.

The skeleton model consists of 21 segments (forefoot, rear foot, tibia, femur, pelvis, lower spine, upper spine/chest, neck, head, shoulder girdle, humerus, ulna, hand). All joints have three degrees of freedom with the exception of the sternoclavicular, elbow and hand joint which are assigned with two degrees of freedom. In addition, the pelvis has three rotational and three translational degrees of freedom.

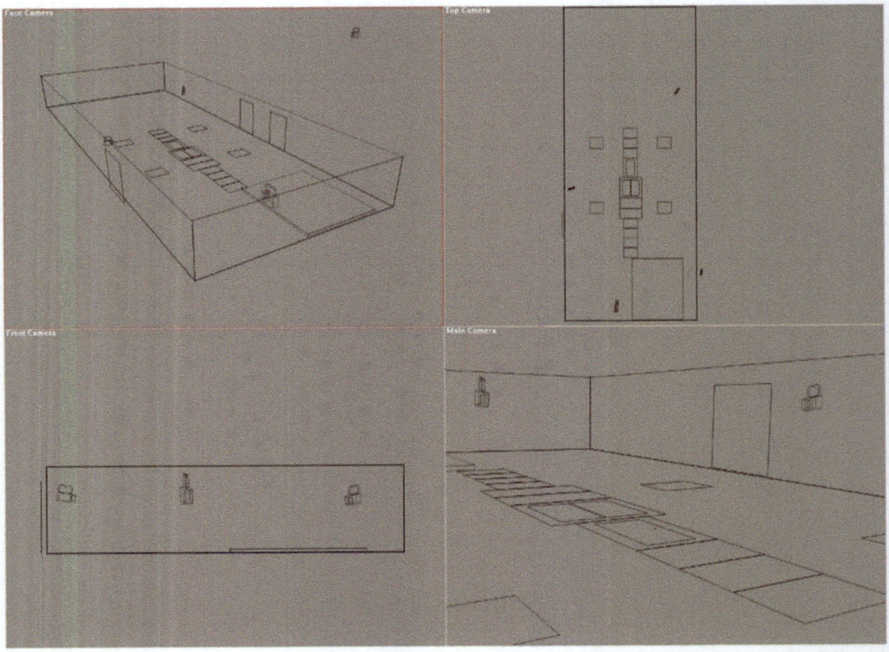

Fig. 8.3 Customizing the model of the surroundings. A computer model of the surroundings is built based on exact field measurements

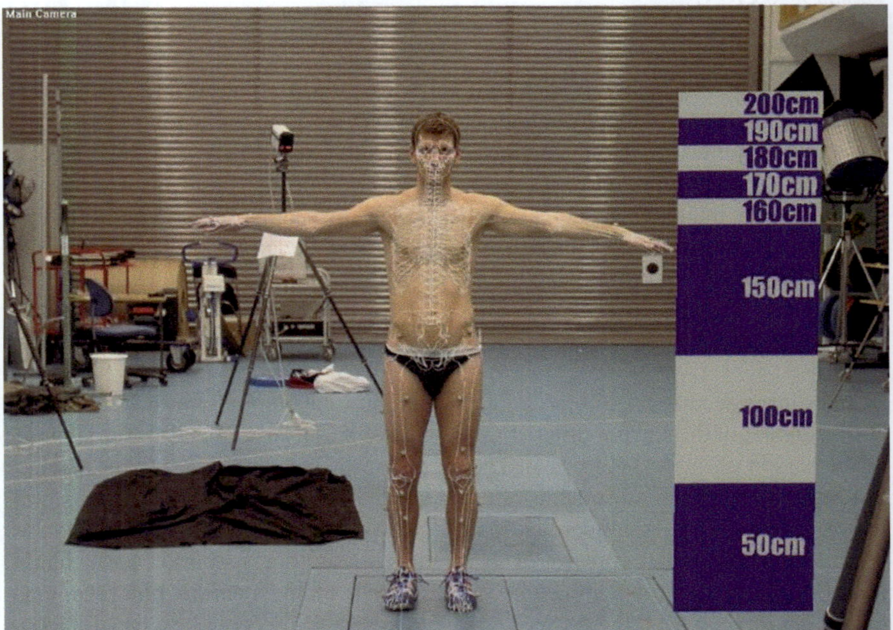

Fig. 8.4 Customizing the skeleton model. The model dimensions are adjusted to the exact measurements of the subject. If these measurements are not available, scaling of the segments will be part of the matching process

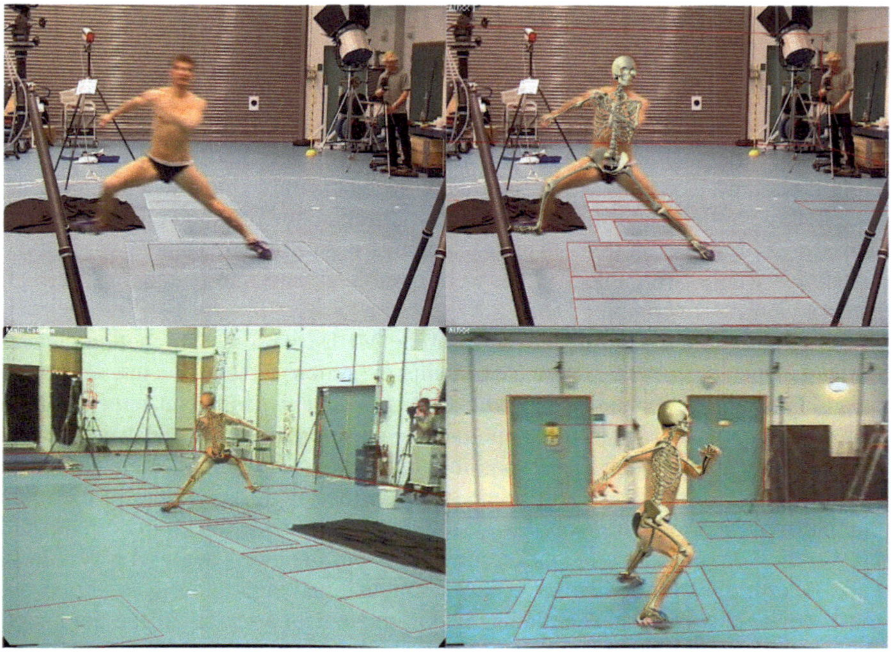

Fig. 8.5 Matching of the laboratory and skeleton model in three camera angles. The picture on the *upper left panel* shows the original camera picture. The poser cameras are located in the approximate same position as the real cameras and can be seen in the *upper right panel* (one camera) and the *bottom left panel* (two cameras) (Krosshaug T, Bahr R (2005) A model-based image-matching technique for three-dimensional reconstruction of human motion from uncalibrated video sequences. *J Biomech* 38:919–929)

In total, this gives 57 degrees of freedom for the skeleton model. All segments are individually customized to the body dimensions of the person by scaling along the three axes of the segment. In the cases where this information is not available, scaling of the segments will be part of the manual matching process.

The next step of the matching process is to adjust the camera parameters so that the surrounding model (here – the lab) matches the video sequence (Fig. 8.5). This is done by adjusting each of the seven camera parameters; rotation about three axes, translation along three axes as well as the focal length (zoom). Next, the skeleton model is matched. Unfortunately, manual matching is very time consuming since it is necessary to adjust all the 57 degrees of freedom for the skeleton model plus the 7 degrees of freedom for the camera motion for each frame. However, Poser includes a spline interpolation feature that substantially reduces the matching time. Depending on the number of camera view, length of the video sequence and skill of the analyst, one full-body matching can take between 3 and 6 weeks.

8.8 Validation of the MBIM Method

To validate the method, we compared its results with marker-based, three-dimensional motion analysis (Krosshaug and Bahr 2005). One test subject with 33 reflective markers attached to anatomical body landmarks performed running and cutting motions. The trials were recorded with eight infrared high-speed cameras as well as three ordinary consumer camcorders. This resulted in a total of seven matchings for each of the motions (Three matchings based on one camera view, three matchings where two camera views were combined, and one matching where all camera views were combined (Fig. 8.6).

To be able to compare the results from the MBIM with the 3D motion capture measurements it was necessary to have similar definition of the joint axes. These calculations were done in the software Matlab. Also signal filtering, joint centers, inertia parameters, joint angles and other important parameters were calculated using Matlab.

The study showed that it was possible to reconstruct the movement in all the matchings. No differences were found between the jogging motion and the more complex cutting motion. Flexion/extension joint angles agreed well in both hip and knee (Fig. 8.7a, b), but some underestimation from the video analysis was seen in knee flexion and hip abduction. The internal/external rotation angles varied the most, however considering the inherent skin movement artefacts of the gold standard, we did not exactly know how accurate the MBIM estimates actually were.

Analyses of the segment movements revealed that the hip angle error originated from matching the pelvis. This was not a surprise due to the pelvis shape which makes it more difficult to assess it's orientation compared with e.g. the thigh and shank.

The velocity of the center of mass was clearly better in the cases where two perpendicular cameras were available (Fig. 8.7c). In the case where only camera angles from front and rear are available, it was difficult to achieve good estimates of anterior-posterior velocity. A side view camera will then be necessary.

8.9 Application on ACL Injury Situations

The next logical step was to test whether the method could be applied to real injury situations outside the lab (Krosshaug et al. 2007b). A four-camera basketball video, a three-camera handball video and a one-camera video of an alpine skier was analysed using the MBIM method. The handball injury occurred as the player performed a right-left cutting maneuver (Fig. 8.7). As can be seen from Fig. 8.8, the model of the handball court could be matched to the background video image in all three camera views. The visual match was very good and all joints as well as hands, feet and head were positioned stably on top of the real player throughout the matching sequence. However, the visual match appeared to be somewhat better for the lower extremity than the upper extremity.

The horizontal velocity at initial foot contact was estimated to be 3.6 m/s, and the cutting angle was approximately 67°. At foot strike the hip was flexed 19°, abducted 26° and externally rotated 16°. The knee was flexed 11°, whereas the rotation and

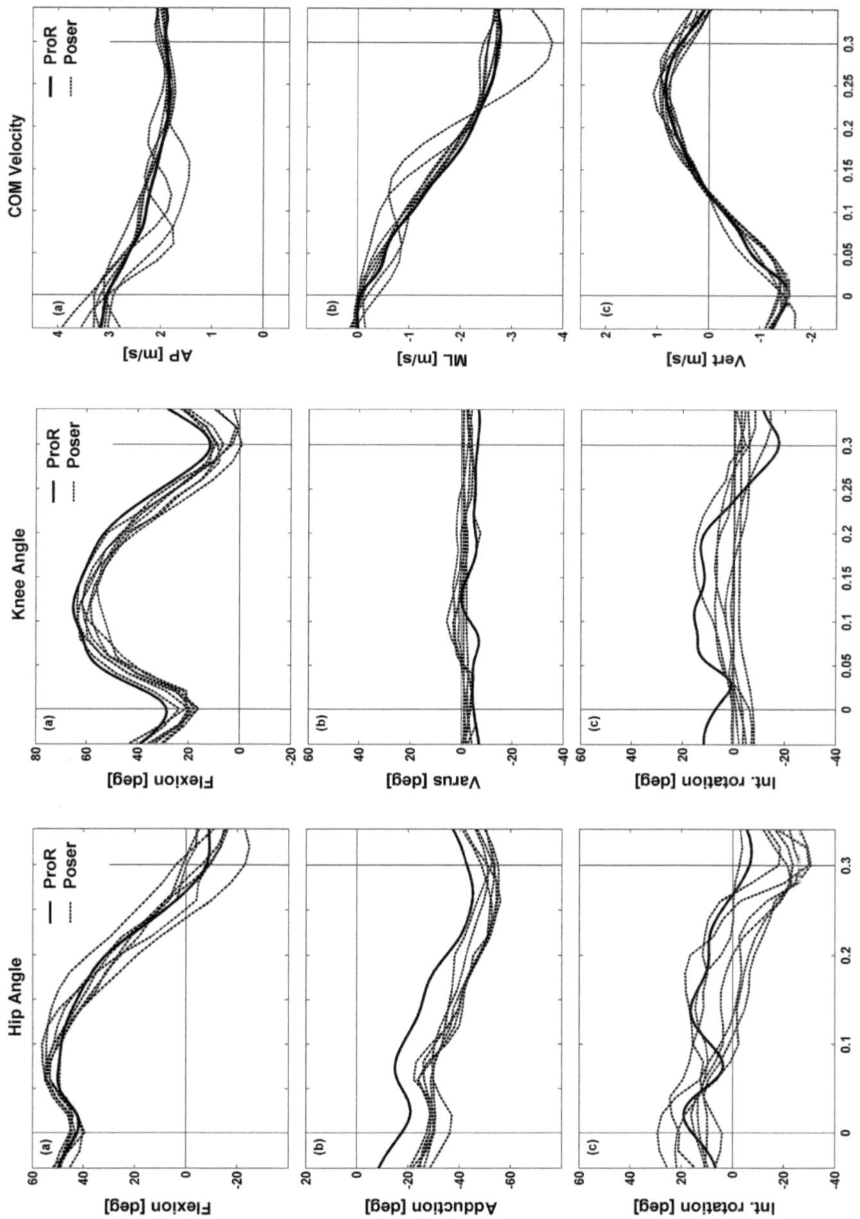

Fig. 8.6 Hip joint angles (**a**), knee joint angles (**b**) and velocity of the center of mass (**c**) in the cutting maneuver as measured with the motion analysis system (ProR – *solid line*) and the MBIM method (Poser – *dotted lines*). Each dotted line represents one of the seven matchings. The vertical lines represent initial contact (time = 0.0 s) of the foot and toe-off (time = 0.3 s) (Krosshaug T, Bahr R (2005) A model-based image-matching technique for three-dimensional reconstruction of human motion from uncalibrated video sequences. *J Biomech* 38:919–929)

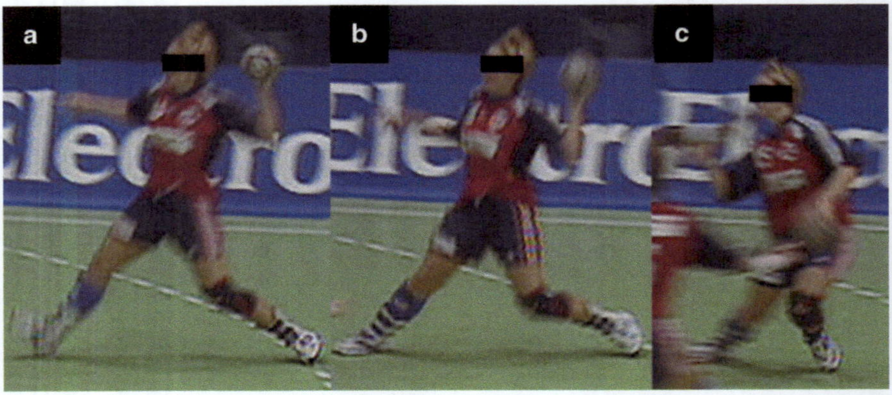

Fig. 8.7 Image sequence of a handball ACL injury. The images are taken from initial contact (**a**), 40 ms later (**b**) and 100 ms later (**c**) (Krosshaug T, Slauterbeck JR, Engebretsen L, Bahr R (2007) Biomechanical analysis of anterior cruciate ligament injury mechanisms: three-dimensional motion reconstruction from video sequences. *Scand J Med Sci Sports* 17:508–519)

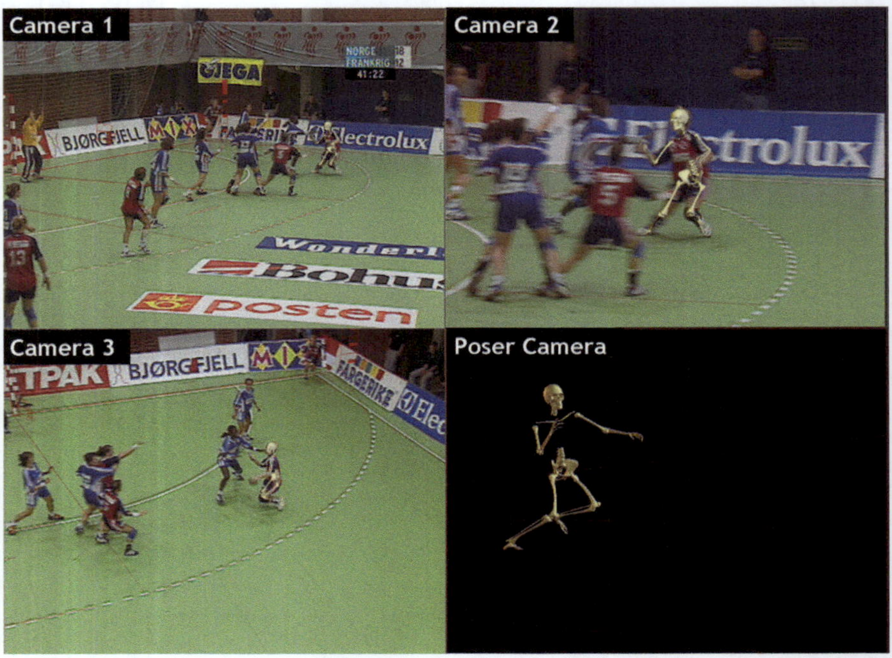

Fig. 8.8 Matching of the handball injury situation, based on three camera views. The down, *right panel* (Poser camera) shows the skeleton model from an alternative viewing angle (Krosshaug T, Slauterbeck JR, Engebretsen L, Bahr R (2007) Biomechanical analysis of anterior cruciate ligament injury mechanisms: three-dimensional motion reconstruction from video sequences. *Scand J Med Sci Sports* 17:508–519)

varus-valgus angles were neutral. During the first 40 ms after initial contact the hip flexion remained constant whereas the knee flexion was increased to 31°. At the same time a considerable valgus angle development of 15° was seen. A similar valgus development was also seen in the basketball injury.

8.10 Limitations and Possibilities

Although a match may seem to be good there is no guarantee that axial rotations of the segments are correct. Since our ability to assess axial rotations depends on how well we are able to see e.g. surface contours and landmarks, good video quality is essential for accuracy. Poor picture quality, occluded body parts (due to clothing or other players), lack of visible landmarks or having only one camera angle may give sub-optimal matchings.

To increase the reliability of video analyses it will be necessary to analyse several situations. Visually, it appears that several of the non-contact ACL injury situations are strikingly similar. For instance, valgus collapse has been frequently reported in several studies (Olsen et al. 2004; Krosshaug et al. 2007c). Therefore, if several matchings show similar abnormal motion patterns it will strengthen our trust in the findings.

An important question is when the injury occurs. This may potentially be investigated by studying the development of abnormal joint configurations or assessing sudden changes in joint angle time histories. Furthermore, it is important to find at what knee flexion angles these injuries occur, as this will have implications for the injury mechanism, e.g. to determine if quadriceps drawer may contribute to ACL loading.

Fortunately, technological developments will make possible even better quality video analyses in the future. The development of high-definition video and high camera coverage at sports events will provide us with far more detailed images that what has been the case so far. There are, in fact examples of videos where one can assess tibial translation relative to the femur in an ACL injury situation (Koga et al. 2011). Analysis of such high-resolution video footage opens for improved understanding of injury mechanisms based on video recordings.

Acknowledgements The Oslo Sports Trauma Research Center has been established at the Norwegian School of Sport Sciences through generous grants from the Norwegian Eastern Health Corporate, the Royal Norwegian Ministry of Culture, the Norwegian Olympic Committee & Confederation of Sport, Norsk Tipping AS, and Pfizer AS.

References

Andersen TE, Larsen O, Tenga A et al (2003) Football incident analysis: a new video based method to describe injury mechanisms in professional football. Br J Sports Med 37:226–232

Andersen TE, Floerenes TW, Arnason A et al (2004a) Video analysis of the mechanisms for ankle injuries in football. Am J Sports Med 32:69S–79S

Andersen TE, Arnason A, Engebretsen L et al (2004b) Mechanisms of head injuries in elite football. Br J Sports Med 38:690–696

Bahr R, Krosshaug T (2005) Understanding the injury mechanisms – a key component to prevent injuries in sport. Br J Sports Med 39:324–329

Boden BP, Dean GS, Feagin JA Jr et al (2000) Mechanisms of anterior cruciate ligament injury. Orthopedics 23:573–578

Cochrane JL, Lloyd DG, Buttfield A et al (2007) Characteristics of anterior cruciate ligament injuries in Australian football. J Sci Med Sport 10:96–104

Halvorsen, K (2002) Model-based methods in motion capture. Department of Information Technology, University of Uppsala, Sweden. Ref Type: Thesis/dissertation

Koga H, Bahr R, Myklebust G et al (2011) Estimating anterior tibial translation from model-based image-matching of a noncontact anterior cruciate ligament injury in professional football: a case report. Clin J Sport Med 21:271–274

Krosshaug T, Bahr R (2005) A model-based image-matching technique for three-dimensional reconstruction of human motion from uncalibrated video sequences. J Biomech 38:919–929

Krosshaug T, Andersen TE, Olsen OE et al (2005) Research approaches to describe the mechanisms of injuries in sports: limitations and possibilities. Br J Sports Med 39:330–339

Krosshaug T, Nakamae A, Boden B et al (2007a) Estimating 3D joint kinematics from video sequences of running and cutting maneuvers – assessing the accuracy of simple visual inspection. Gait Posture 26:378–385

Krosshaug T, Slauterbeck JR, Engebretsen L et al (2007b) Biomechanical analysis of anterior cruciate ligament injury mechanisms: three-dimensional motion reconstruction from video sequences. Scand J Med Sci Sports 17:508–519

Krosshaug T, Nakamae A, Boden BP et al (2007c) Mechanisms of anterior cruciate ligament injury in basketball: video analysis of 39 cases. Am J Sports Med 35:359–367

McIntosh AS, McCrory P, Comerford J (2000) The dynamics of concussive head impacts in rugby and Australian rules football. Med Sci Sports Exerc 32:1980–1984

Meeuwisse WH (1994) Athletic injury etiology: distinguishing between interaction and confounding. Clin J Sport Med 4:171–175

Olsen OE, Myklebust G, Engebretsen L et al (2004) Injury mechanisms for anterior cruciate ligament injuries in team handball: a systematic video analysis. Am J Sports Med 32:1002–1012

Quatman CE, Hewett TE (2009) The ACL injury controversy: is "valgus collapse" a sex-specific mechanism? Br J Sports Med 43(5):328–335. http://www.ncbi.nlm.nih.gov/pubmed/19372087

Renstrom P, Ljungqvist A, Arendt E et al (2008) Non-contact ACL injuries in female athletes: an International Olympic Committee current concepts statement. Br J Sports Med 42:394–412

Trewartha G, Yeadon MR, Knight JP (2001) The sensitivity of video tracking performance to the simulated environment, the human body model and the image matching algorithm. In: Proceedings of the International Symposium on Computer Simulation in Biomechanics, July 5–6, Milano, Italy, pp 187–192

van Mechelen W, Hlobil H, Kemper HC (1992) Incidence, severity, aetiology and prevention of sports injuries. A review of concepts. Sports Med 14:82–99

Yu B, Garrett WE (2007) Mechanisms of non-contact ACL injuries. Br J Sports Med 41(Suppl 1):i47–i51

Chapter 9
Video Analysis of ACL Injury Mechanisms Using a Model-Based Image-Matching Technique

Hideyuki Koga, Takeshi Muneta, Roald Bahr, Lars Engebretsen, and Tron Krosshaug

Abstract Model-based image-matching (MBIM) technique has enabled detailed video analysis of injury situations that previously had been limited to simple visual inspection. We have analyzed anterior cruciate ligament (ACL) injury situations from ten analogue and one HD video sequences using the MBIM technique. The knee kinematical patterns were remarkably consistent, with immediate valgus and internal rotation (IR) motion occurring within 40 ms after initial contact (IC), and then external rotation was observed. Peak vertical ground reaction force (GRF) occurred at 40 ms after IC. Based on these results, it is likely that the ACL injury occurred approximately 40 ms after IC. In the one HD video available, 9 mm of abrupt anterior tibial translation at the time of injury was also detected. On the other hand, the hip kinematics were constant with an abducted, flexed and IR position during 40 ms after IC. Based on these results and other previous studies, we propose a new hypothesis for ACL injury mechanisms that valgus loading and lateral compression generate IR motion and anterior translation of the tibia, due to the joint geometry, result in ACL rupture. Moreover, it seems that the hip is relatively 'locked' at IC, cannot absorb energy from GRF and thus the knee is exposed to a larger force, which leads to ACL injury. These results suggest that prevention programs should focus on acquiring a good cutting and landing technique promoting knee flexion and avoiding knee valgus and foot internal rotation, and with greater hip flexion to absorb energy from GRF. Moreover, the fact that the ACL injury occurs within 40 ms after IC suggests that "feed-forward" strategies before landing may be critical, as reflex-based "feed-back" strategies are too slow to prevent ACL injuries.

Keywords Anterior cruciate ligament • Video analysis • Injury mechanism

H. Koga (✉) • T. Muneta
Department of Joint Surgery and Sports Medicine, Tokyo Medical and Dental University, Tokyo, Japan
e-mail: koga.orj@tmd.ac.jp

R. Bahr • L. Engebretsen • T. Krosshaug
Oslo Sports Trauma Research Center, Norwegian School of Sport Sciences, Oslo, Norway

K. Kanosue et al. (eds.), *Sports Injuries and Prevention*,
DOI 10.1007/978-4-431-55318-2_9

9.1 Introduction

Anterior cruciate ligament (ACL) injuries commonly occur during sports activities, and the number has increased in recent years. Development of ACL reconstruction procedures has enabled athletes to return to sports and favorable short-term results have been achieved; yet it takes a relatively long period for most athletes to get back to sports activities. It has also been reported that ACL reconstruction cannot prevent progression of osteoarthritis (Oiestad et al. 2009). Therefore, the importance of ACL injury prevention has been recognized, and establishment of effective injury prevention programs is needed. As of today, various ACL injury prevention programs have been developed successfully (Caraffa et al. 1996; Gilchrist et al. 2008; Mandelbaum et al. 2005; Myklebust et al. 2003; Olsen et al. 2005); however, it is not well understood how the different elements in these multicomponent programs play particular roles in preventing the injury. We know that ACL injuries predominantly result from a non-contact mechanism and occur during cutting or one-leg landing maneuvers (Boden et al. 2000; Krosshaug et al. 2007a; Olsen et al. 2004). Nevertheless, to develop more targeted injury preventive programs, a more detailed description of the mechanism(s) of non-contact ACL injuries is needed.

9.1.1 Previously Proposed ACL Injury Mechanisms

Several theories have been proposed regarding the mechanisms for non-contact ACL injury; however, it is still a matter of controversy, with the main opponents favoring either sagittal or non-sagittal plane knee joint loading. DeMorat et al. proposed that aggressive quadriceps loading was responsible, based on a cadaver study which demonstrated that aggressive quadriceps loading could take the ACL to failure (DeMorat et al. 2004; Yu and Garrett 2007). In contrast, Mclean et al., based on a mathematical simulation model, argued that sagittal plane loading alone could not produce such injuries (McLean et al. 2004, 2005). A prospective cohort study among female athletes, showing that increased dynamic valgus and high valgus loads increased injury risk, lead Hewett et al. to suggest valgus loading as an important component (Hewett et al. 2005; Quatman and Hewett 2009). Some video analyses also showed that valgus collapse seemed to be the main mechanism among female athletes (Krosshaug et al. 2007a; Olsen et al. 2004). However, cadaver studies and mathematical simulation have shown that pure valgus motion would not produce ACL injuries without tearing the medial collateral ligament (MCL) first (Mazzocca et al. 2003; Shin et al. 2009).

Nevertheless, other simulation studies have suggested that valgus loading would substantially increase ACL force in situations where an anterior tibial shear force is applied (Withrow et al. 2006). Based on MRI findings, Speer et al. reported that bone bruises of the lateral femoral condyle or posterolateral portion of tibial plateau occurred in more than 80 % of acute ACL non-contact injuries. They concluded that valgus in combination with internal rotation and/or anterior tibial translation occurred at the time of ACL injuries (Speer et al. 1992). Furthermore, it has been shown that valgus loading induces a coupled motion of valgus and internal tibial rotation (Matsumoto 1990; Matsumoto et al. 2001).

Although both cadaver studies and MRI studies have suggested that internal rotation is present in ACL injury situations, video analyses have suggested that valgus in combination with external rotation is the most frequent motion pattern (Olsen et al. 2004; Ebstrup and Bojsen-Moller 2000).

9.1.2 Research Approaches to Injury Mechanisms

As mentioned above, several different approaches have been used to investigate ACL injury mechanisms, including athlete interviews, clinical studies, laboratory motion analysis, video analysis, cadaver studies, and mathematical simulations (Krosshaug et al. 2005). Among these, video analysis of injury tapes is the only method available to extract kinematic data from actual injury situations. However, video analyses have so far been limited to simple visual inspection (Boden et al. 2000; Olsen et al. 2004; Cochrane et al. 2007), and the accuracy has been shown to be poor, even among experienced researchers (Krosshaug et al. 2007b). In addition, simple visual inspection is not sufficient to extract time course data for joint angles, velocities and accelerations; therefore, it is difficult to determine the point of ACL rupture.

9.1.3 Development of Model-Based Image-Matching Technique

Model-based image-matching (MBIM) has been developed as an alternative to simple visual inspection, in order to extract joint kinematics from video recordings using one or more uncalibrated cameras (Krosshaug and Bahr 2005). The detailed procedures of the MBIM technique have been described in the literature, as well as in the previous section of this chapter. The idea underpinning this technique is that matching a skeleton model to the background video sequences provides an estimate of the actual three-dimentional body kinematics using the commercially available program Poser® and Poser® Pro Pack (Curious Labs Inc., Santa Cruz, California, USA). This technique has been validated in non-injury situations in a laboratory environment (Krosshaug and Bahr 2005) and also has been found to be feasible for use in actual ACL injury situations (Krosshaug et al. 2007c). In the following we propose a new description of the mechanism for non-contact ACL injury based on analyses using the MBIM technique. In addition, suggestions for injury programs based on the proposed mechanism are introduced.

9.2 Biomechanics in Non-contact ACL Injury

Ten ACL injury situations from women's team handball (n=7) and basketball (n=3), recorded with at least two analogue cameras during TV broadcasts, were analyzed using MBIM technique (Fig. 9.1). All the players were handling the ball in the injury situation; seven were in possession of the ball at the time of injury, two had shot and one had passed the ball. In six cases, there was player-to-player contact with an opponent at the time of injury, all of them to the torso being pushed or held. There was no direct contact to the knee. The injury situations could be classified into two groups; seven cases occurred when cutting and three during one-leg landings.

The knee kinematical patterns were remarkably consistent among the ten cases (Fig. 9.2). The knee was relatively straight, with a flexion angle of 23° (range, 11–30°), at initial contact (IC) and had increased by 24° (95 % CI, 19–29°, $p<0.001$) 40 ms later. The knee abduction angle was neutral, 0° (range, −2° to 3°) at IC, but had increased by 12° (95 % CI, 10–13°, $p<0.001$) 40 ms later. As for knee rotation angle, the knee was externally rotated 5° (range, −5° to 12°) at IC,

Fig. 9.1 An example of a video matched in Poser, three-camera basketball injury situation 50 ms after IC. The *two top panels* and *left bottom panel* show the customized skeleton model and the basketball court model superimposed on and matched with the background video image from three cameras with different angles. The *right bottom two panels* show the skeleton model from a side view created in Poser

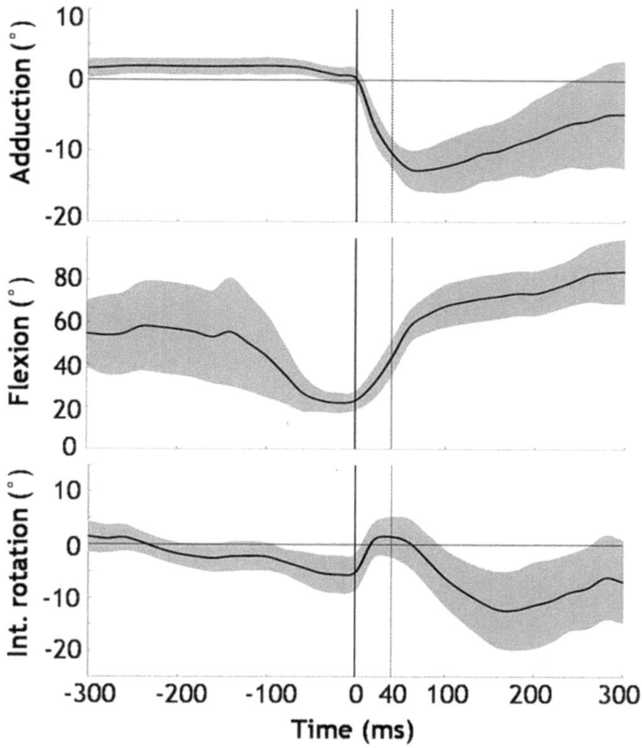

Fig. 9.2 Time sequences of the mean knee angles (°) (*black line*) of the 10 cases with 95 % confidence intervals (CI) (*grey area*). Time 0 indicates IC and the dotted vertical line indicates the time point 40 ms after IC

but abruptly rotated internally by 8° (95 % CI, 2–14°, $p=0.037$) during the first 40 ms. From 40 ms to 300 ms after IC, however, we observed an external rotation of 17° (95 % CI, 13–22°, $p<0.001$). In addition, the estimated peak vertical ground reaction force (GRF) was 3.2 times body weight (95 % CI, 2.7–3.7), and occurred at 40 ms (range, 0–83) after IC. On the other hand, the hip kinematics was relatively constant at a 20° abducted, 50° flexed and 30° IR position during 40 ms after IC (Fig. 9.3).

However, a limitation of the above mentioned analysis was how accurate the joint kinematics and timing of peak GRF could be estimated from the relatively low frame rate (50 or 60 Hz) and low quality images (768×576 pixels) in analog video sequences, and therefore we were unable to assess the anterior translation of the tibia. However, a noncontact ACL injury situation in a male footballer was available which had been recorded using four high-definition (HD, 1080i) cameras, including two high-speed recordings (100 and 300 Hz). In this case, the 26-year old male elite football player suffered a noncontact ACL injury to his right knee during a national

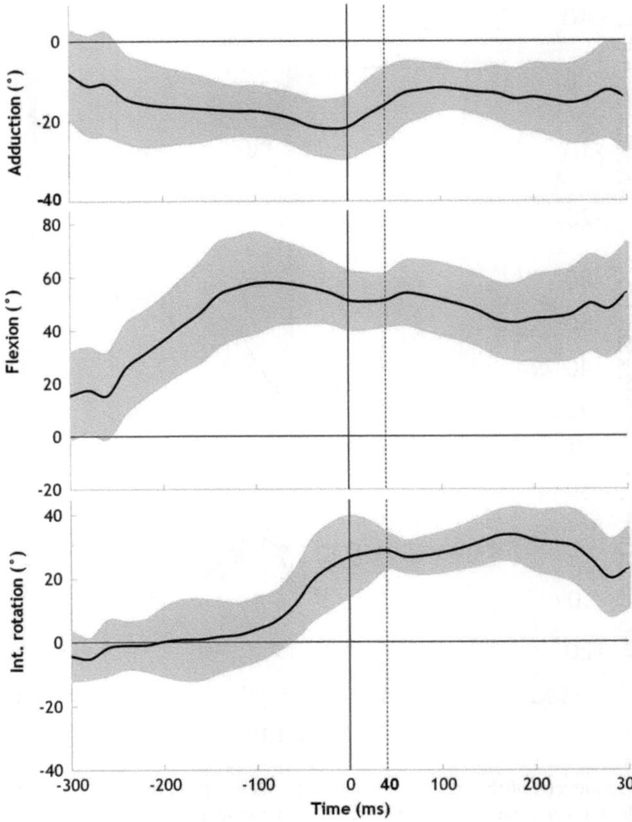

Fig. 9.3 Time sequences of the mean hip angles (°) (*black line*) of the 10 cases with 95 % CI (*grey area*). Time 0 indicates IC and the dotted vertical line indicates the time point 40 ms after IC

team match, when he tried to stop after having passed the ball with his right leg. This case was analyzed using the MBIM technique to describe the more detailed joint kinematics, including tibial translations (Fig. 9.4). Knee kinematics in this case were strikingly consistent with the previous analyses of the ten cases (Fig. 9.5). The knee was flexed 35° at IC, with initial extension (26° of flexion) until 20 ms after IC, after which flexion angle continued to increase. The knee abduction angle was neutral at IC, but had increased by 21° 30 ms later. The knee was externally rotated 11° at IC, but abruptly rotated internally by 21° during the first 30 ms, then changed its direction to external rotation after this. In addition, anterior tibial translation was able to be detected; it started to occur at 20 ms after IC, where the knee was the most extended, and by 30 ms after IC approximately 9 mm of anterior translation had occurred. The translations plateaued by 150 ms, and then shifted back to a reduced position between 200 ms and 240 ms after IC.

Fig. 9.4 A soccer injury situation recorded using HD cameras. Each panel shows the customized skeleton model and the football pitch model superimposed on and matched with the background video image from each camera. Overview camera and rear camera had an effective frame rate after being deinterlaced of 50 Hz, frontal camera 100 Hz and side camera 300 Hz

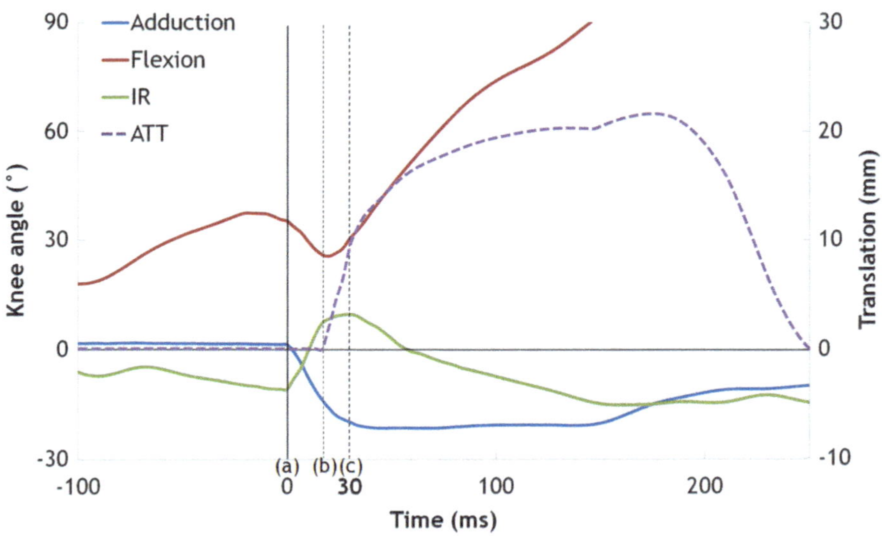

Fig. 9.5 Time sequences of knee joint angles (left axis) and anterior tibial translation (right axis) in the soccer case. Time 0 (*a*) indicates IC and the *dotted vertical lines* (*b*) and (*c*) indicate the time point 20 and 30 ms after IC, respectively

9.3 Timing of Non-contact ACL Injury

It has not been possible to determine the exact timing of ACL injury from video analysis based on simple visual inspection (Boden et al. 2000; Krosshaug et al. 2007a; Olsen et al. 2004). However, this may be possible by using the MBIM technique, by assessing abnormal joint configurations, sudden changes in joint angular motion and timing of GRFs. The extracted knee kinematics during ACL injuries using the MBIM technique showed that sudden increase of valgus and internal rotation angle occurred within the first 40 ms after IC. These periods also correspond to the average peak vertical GRF in these cases. Moreover, in the case recorded using HD cameras, abrupt anterior tibial translation reached 9 mm in 30 ms after IC, which corresponds to the maximum anterior translation in intact knees (Jakob et al. 1987; Meyer and Haut 2008). Based on these results, together with the previous studies showing that the ACL was strained shortly (approximately 40 ms) after IC in simulated landing (Withrow et al. 2006; Shin et al. 2007), it seems likely that the injury occurs within 40 ms for the majority of these cases.

9.4 Mechanism for Non-contact ACL Injury

As already mentioned, valgus collapse in combination with external rotation (i.e. knee in, toe out) has frequently been identified as an ACL injury mechanism based on simple visual inspection of injury video tapes. However, it has been discussed as to whether these kinematics actually represent the cause for ACL injuries or simply are a result of the ACL being torn (Olsen et al. 2004; Ebstrup and Bojsen-Moller 2000). Our results using the MBIM technique showed that immediate valgus motion occurred within 40 ms after IC. The abrupt internal rotation also occurred during the first 40 ms after IC, then external rotation was observed, which seems to have occurred after the ACL was torn. In addition, anterior tibial translation started a little after IC, and increased abruptly until when the injury might have occurred. The discrepancy between the previous studies and our results could be that the abrupt internal rotation and anterior tibial translation observed using the MBIM technique analysis are likely not easily detected from visual inspection alone; the external rotation that occurs afterwards is more pronounced and therefore easier to observe. The internal-to-external rotation sequence with anterior tibial translation has also been reported previously. In a recent cadaver study, the application of pure compressive loads led to anterior tibial translation and internal tibial rotation of up to 8°, followed by a sudden external rotation of 12° (Meyer and Haut 2008). The combination of internal tibial rotation and anterior tibial translation is probably caused by the joint surface geometry. The concave geometry of the medial tibia facet combined with the slightly convex lateral tibia facet may cause the lateral femoral condyle to slip back. This may also explain why ACL-injured patients tend to have greater posterior lateral tibial plateau slopes than uninjured controls (Brandon et al. 2006; Stijak et al. 2008; Hashemi et al. 2010).

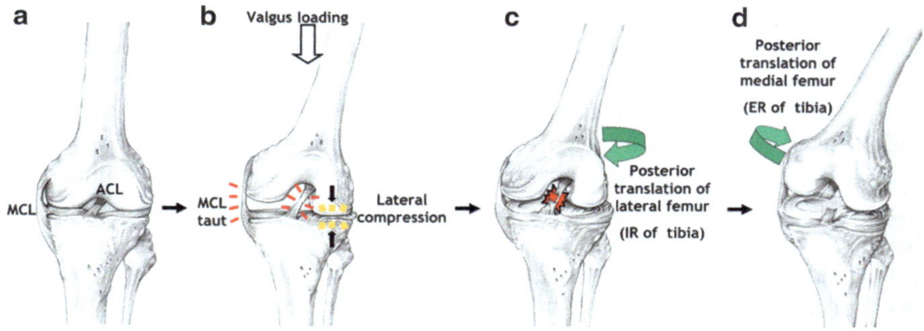

Fig. 9.6 The proposed non-contact ACL injury mechanism. (**a**) An unloaded knee. (**b**) When valgus loading is applied, the MCL becomes taut and lateral compression occurs. (**c**) This compressive load causes a lateral femoral posterior displacement, probably due to the posterior slope of lateral tibial plateau, and the tibia translates anteriorly and rotates internally, resulting in ACL rupture. (**d**) After the ACL is torn, the primary restraint to anterior translation of the tibia is gone. This causes the medial femoral condyle to also be displaced posteriorly, resulting in external rotation of the tibia

Combining the results obtained using the MBIM technique with previous findings, the following hypothesis for the mechanism of non-contact ACL injury is proposed (Fig. 9.6) (Oiestad et al. 2009): when valgus loading is applied, the MCL becomes taut and lateral compression occurs (Caraffa et al. 1996). This compressive load causes a lateral femoral posterior displacement, probably due to the posterior slope of lateral tibial plateau, and the tibia translates anteriorly and rotates internally, resulting in ACL rupture (Gilchrist et al. 2008). After the ACL is torn, the primary restraint to anterior translation of the tibia is gone. This causes the medial femoral condyle to also be displaced posteriorly, resulting in external rotation of the tibia. This external rotation may be exacerbated by the typical movement pattern when athletes plant and cut, where the foot typically rotates externally relative to the trunk.

9.5 The Role of the Hip in Preventing ACL Injury

The lower extremities act as a kinetic chain during dynamic tasks and the control of hip motion substantially affects the knee motion. Researchers have studied the relationships between hip biomechanics and ACL injury. As for hip biomechanics being a risk factor for ACL injury, Decker et al. (2003) reported that, in drop landing, energy absorption at the hip joint, and hip flexion angles at IC were less in females than in males. Schmitz et al. (2007) reported that, in single-leg landing, energy absorption at the hip and total hip flexion displacement were smaller in females, whereas peak vertical GRF was larger in females. Yu et al. (2006) also reported that hip flexion angular velocity at IC was negatively correlated with peak vertical GRF in a stop-jump task. When it comes to ACL injury mechanisms, Heshemi et al.

(2007) reported that, in a cadaver study, a restricted flexion of the hip at 20° combined with low quadriceps and hamstrings force levels in simulated single-leg landing were found to be conducive to ACL injury. A video analysis has shown that ACL injured subjects' hip flexion and abduction angles were constant during 100 ms after IC, whereas uninjured control subjects' hip flexion increased by 15° in cutting/landing maneuvers (Boden et al. 2009). Our study using MBIM technique also showed that hip kinematics was constant during 40 ms after IC in an abducted, flexed and internally-rotated position, which seems to play a significant role in the mechanism of ACL injury. In this regard, Hashemi et al. (2011) have proposed a mechanism called "hip extension, knee flexion paradox", i.e. that a mismatch between hip and knee flexion in landing is the cause of ACL injury. In normal conditions, both the knee and the hip flex together in landings, whereas in unbalanced landings, the knee is forced to flex while the hip is forced to extend, and the tibia will undergo anterior translation, which will increase the risk of ACL injury.

There are some possible causes of hip/knee mismatch (Oiestad et al. 2009): in the sagittal plane, an upright or backward-leaning trunk position at IC makes the center of mass posterior to knee, and increased GRF may encourage more knee flexion than hip flexion, and relatively act to extend hip (Caraffa et al. 1996). In other planes insufficient hip abductor/external rotator strength or activation would lead to adducted/internally-rotated position of the hip, causing knee valgus (Gilchrist et al. 2008). Large hip internal rotation at IC seen in our video analysis could also be an explanation; ACL injured patients could have limited range of motion in internal rotation, and the hip joint may be locked at a large internally-rotated position. As a matter of fact, hip dysplasia has also been reported to be a risk factor of ACL injury (Yamazaki et al. 2011).

For these reasons, it seems that the hip joint is relatively locked at IC, cannot absorb energy from GRF and the knee joint is thus exposed to larger forces, which may lead to ACL injury. Therefore, it is important that prevention efforts should focus not only on the knee joint, but also on the hip joint.

9.6 ACL Injury Prevention Based on the Proposed Mechanisms

Based on the mechanisms clarified using MBIM technique, prevention strategy for ACL injury can be proposed as follows (Oiestad et al. 2009): as the kinematics when ACL injury is happening is knee valgus and internal rotation with the hip being locked, it is important to acquire a good cutting and landing technique with knee flexion, avoiding knee valgus and foot internal rotation, and with hip flexion to absorb energy from GRF, avoiding hip internal rotation (Caraffa et al. 1996). As ACL injuries occur approximately 40 ms after IC, it is likely that a "feed-back" strategy, i.e. ACL prevention program focusing on training after landing cannot prevent ACL injury; it takes at least 150–200 ms to react after landing at risk. Prevention efforts should focus on a "feed-forward" strategy before landing, i.e. training muscular activation and neural control during the pre-landing phase.

References

Boden BP, Dean GS, Feagin JA Jr, Garrett WE Jr (2000) Mechanisms of anterior cruciate ligament injury. Orthopedics 23:573–578

Boden BP, Torg JS, Knowles SB, Hewett TE (2009) Video analysis of anterior cruciate ligament injury: abnormalities in hip and ankle kinematics. Am J Sports Med 37:252–259

Brandon ML, Haynes PT, Bonamo JR, Flynn MI, Barrett GR, Sherman MF (2006) The association between posterior-inferior tibial slope and anterior cruciate ligament insufficiency. Arthroscopy 22:894–899

Caraffa A, Cerulli G, Projetti M, Aisa G, Rizzo A (1996) Prevention of anterior cruciate ligament injuries in soccer. A prospective controlled study of proprioceptive training. Knee Surg Sports Traumatol Arthrosc 4:19–21

Cochrane JL, Lloyd DG, Buttfield A, Seward H, McGivern J (2007) Characteristics of anterior cruciate ligament injuries in Australian football. J Sci Med Sport 10:96–104

Decker MJ, Torry MR, Wyland DJ, Sterett WI, Richard SJ (2003) Gender differences in lower extremity kinematics, kinetics and energy absorption during landing. Clin Biomech (Bristol, Avon) 18:662–9

DeMorat G, Weinhold P, Blackburn T, Chudik S, Garrett W (2004) Aggressive quadriceps loading can induce noncontact anterior cruciate ligament injury. Am J Sports Med 32:477–483

Ebstrup JF, Bojsen-Moller F (2000) Anterior cruciate ligament injury in indoor ball games. Scand J Med Sci Sports 10:114–116

Gilchrist J, Mandelbaum BR, Melancon H et al (2008) A randomized controlled trial to prevent noncontact anterior cruciate ligament injury in female collegiate soccer players. Am J Sports Med 36:1476–1483

Hashemi J, Chandrashekar N, Jang T, Karpat F, Oseto M, Ekwaro-Osire S (2007) An alternative mechanism of non-contact anterior cruciate ligament injury during jump-landing: in-vitro simulation. Exp Mech 47:347–354

Hashemi J, Chandrashekar N, Mansouri H et al (2010) Shallow medial tibial plateau and steep medial and lateral tibial slopes: new risk factors for anterior cruciate ligament injuries. Am J Sports Med 38:54–62

Hashemi J, Breighner R, Chandrashekar N et al (2011) Hip extension, knee flexion paradox: a new mechanism for non-contact ACL injury. J Biomech 44:577–585

Hewett TE, Myer GD, Ford KR et al (2005) Biomechanical measures of neuromuscular control and valgus loading of the knee predict anterior cruciate ligament injury risk in female athletes: a prospective study. Am J Sports Med 33:492–501

Jakob RP, Staubli HU, Deland JT (1987) Grading the pivot shift. Objective tests with implications for treatment. J Bone Joint Surg Br 69:294–299

Krosshaug T, Bahr R (2005) A model-based image-matching technique for three-dimensional reconstruction of human motion from uncalibrated video sequences. J Biomech 38:919–929

Krosshaug T, Andersen TE, Olsen OE, Myklebust G, Bahr R (2005) Research approaches to describe the mechanisms of injuries in sport: limitations and possibilities. Br J Sports Med 39:330–339

Krosshaug T, Nakamae A, Boden BP et al (2007a) Mechanisms of anterior cruciate ligament injury in basketball: video analysis of 39 cases. Am J Sports Med 35:359–367

Krosshaug T, Nakamae A, Boden B et al (2007b) Estimating 3D joint kinematics from video sequences of running and cutting maneuvers–assessing the accuracy of simple visual inspection. Gait Posture 26:378–385

Krosshaug T, Slauterbeck JR, Engebretsen L, Bahr R (2007c) Biomechanical analysis of anterior cruciate ligament injury mechanisms: three-dimensional motion reconstruction from video sequences. Scand J Med Sci Sports 17:508–519

Mandelbaum BR, Silvers HJ, Watanabe DS et al (2005) Effectiveness of a neuromuscular and proprioceptive training program in preventing anterior cruciate ligament injuries in female athletes: 2-year follow-up. Am J Sports Med 33:1003–1010

Matsumoto H (1990) Mechanism of the pivot shift. J Bone Joint Surg Br 72:816–821
Matsumoto H, Suda Y, Otani T, Niki Y, Seedhom BB, Fujikawa K (2001) Roles of the anterior cruciate ligament and the medial collateral ligament in preventing valgus instability. J Orthop Sci 6:28–32
Mazzocca AD, Nissen CW, Geary M, Adams DJ (2003) Valgus medial collateral ligament rupture causes concomitant loading and damage of the anterior cruciate ligament. J Knee Surg 16:148–151
McLean SG, Huang X, Su A, Van Den Bogert AJ (2004) Sagittal plane biomechanics cannot injure the ACL during sidestep cutting. Clin Biomech (Bristol, Avon) 19:828–38
McLean SG, Andrish JT, van den Bogert AJ (2005) Aggressive quadriceps loading can induce noncontact anterior cruciate ligament injury. Am J Sports Med 33:1106; author reply –7
Meyer EG, Haut RC (2008) Anterior cruciate ligament injury induced by internal tibial torsion or tibiofemoral compression. J Biomech 41:3377–3383
Myklebust G, Engebretsen L, Braekken IH, Skjolberg A, Olsen OE, Bahr R (2003) Prevention of anterior cruciate ligament injuries in female team handball players: a prospective intervention study over three seasons. Clin J Sport Med 13:71–78
Oiestad BE, Engebretsen L, Storheim K, Risberg MA (2009) Knee osteoarthritis after anterior cruciate ligament injury: a systematic review. Am J Sports Med 37:1434–1443
Olsen OE, Myklebust G, Engebretsen L, Bahr R (2004) Injury mechanisms for anterior cruciate ligament injuries in team handball: a systematic video analysis. Am J Sports Med 32:1002–1012
Olsen OE, Myklebust G, Engebretsen L, Holme I, Bahr R (2005) Exercises to prevent lower limb injuries in youth sports: cluster randomised controlled trial. BMJ 330:449
Quatman CE, Hewett TE (2009) The anterior cruciate ligament injury controversy: is "valgus collapse" a sex-specific mechanism? Br J Sports Med 43:328–335
Schmitz RJ, Kulas AS, Perrin DH, Riemann BL, Shultz SJ (2007) Sex differences in lower extremity biomechanics during single leg landings. Clin Biomech (Bristol, Avon) 22:681–8
Shin CS, Chaudhari AM, Andriacchi TP (2007) The influence of deceleration forces on ACL strain during single-leg landing: a simulation study. J Biomech 40:1145–1152
Shin CS, Chaudhari AM, Andriacchi TP (2009) The effect of isolated valgus moments on ACL strain during single-leg landing: a simulation study. J Biomech 42:280–285
Speer KP, Spritzer CE, Bassett FH 3rd, Feagin JA Jr, Garrett WE Jr (1992) Osseous injury associated with acute tears of the anterior cruciate ligament. Am J Sports Med 20:382–389
Stijak L, Herzog RF, Schai P (2008) Is there an influence of the tibial slope of the lateral condyle on the ACL lesion? A case-control study. Knee Surg Sports Traumatol Arthrosc 16:112–117
Withrow TJ, Huston LJ, Wojtys EM, Ashton-Miller JA (2006) The effect of an impulsive knee valgus moment on in vitro relative ACL strain during a simulated jump landing. Clin Biomech (Bristol, Avon) 21:977–83
Yamazaki J, Muneta T, Ju YJ, Morito T, Okuwaki T, Sekiya I (2011) Hip acetabular dysplasia and joint laxity of female anterior cruciate ligament-injured patients. Am J Sports Med 39:410–414
Yu B, Garrett WE (2007) Mechanisms of non-contact ACL injuries. Br J Sports Med 41(Suppl 1):i47–i51
Yu B, Lin CF, Garrett WE (2006) Lower extremity biomechanics during the landing of a stop-jump task. Clin Biomech (Bristol, Avon) 21:297–305

Chapter 10
Injury Rate of Soccer Players and the Efficacy of the FIFA 11 + Program

Yasuaki Saho

Abstract Soccer is the most popular sport in the world. There are more than 265 million soccer players worldwide (in 2006, http://www.fifa.com/worldfootball/bigcount/). In particular, youth and female players are increasing. In the 2008 Beijing Olympics and the 2012 London Olympics, soccer was categorized among the higher injury risk sports (Junge et al., Am J Sports Med 37(11):2165–2172, 2009; Engebretsen et al., Br J Sports Med 47(7):407–414, 2013). Injuries affect team performance negatively and exact high medical and social costs. In soccer, about 70 % of all injuries occur in the lower extremities (Dvorak et al., Br J Sports Med 45(8):626–630, 2011; Theron et al., Clin J Sport Med 23(5):379–383, 2013). The most commonly injured body parts are the thigh, ankle, lower leg and knee, and the most common types of acute injuries are contusions, ligament sprains, and muscle strains (Dvorak et al., Br J Sports Med 45(8):626–630, 2011; Theron et al., Clin J Sport Med 23(5):379–383, 2013; Dvorak et al., Br J Sports Med 41(9):578–581, 2007). Therefore a soccer-specific injury prevention plan was needed. The Federation Internationale de Football Association (FIFA) Medical and Research Center (F-MARC) developed "The 11" and then the "FIFA 11 +", soccer specific injury prevention programs designed to reduce soccer injuries and promote soccer as a health-enhancing leisure activity. In this chapter, the epidemiology of soccer injuries and the efficacy of the "The 11" and "FIFA 11 +" are covered.

Keywords Soccer • Injury prevention • Epidemiology

Y. Saho (✉)
Department of Sports and Medical Science, Faculty of Medical Technology, Teikyo University, Tokyo, Japan
e-mail: y-saho@main.teikyo-u.ac.jp

K. Kanosue et al. (eds.), *Sports Injuries and Prevention*,
DOI 10.1007/978-4-431-55318-2_10

10.1 Characteristics of Soccer Injury

10.1.1 Injury Rate of Soccer Players

Recently, the establishment of an injury surveillance system was suggested (Fuller et al. 2006). This system was created and has been used in soccer (football) tournaments and the Olympic games (Junge et al. 2004a, b, c, 2009; Dvorak et al. 2007, 2010; Junge and Dvorak 2007, 2013; Engebretsen et al. 2010). Since this system's inception, the use of this system has received much attention and has created a greater awareness about soccer injuries. Information on injuries has been augmented on many fronts including the injured body part, type of injury, diagnosis, severity, circumstance (non-contact, contact or foul play), and the nature of the exposures in both soccer training and matches. The system's guidelines recommend that injury incidence be expressed as the number of injuries per 1,000 athlete exposure hours.

The epidemiology of soccer injuries has been investigated in several studies.

Soccer is one of the highest injury risk sports. In the 2008 Olympic Games, Junge et al. (2009) reported injuries by all sports. The results of this study indicated that the incidence of injuries was highest in soccer, taekwondo, hockey, handball, and boxing. The results were almost the same for the 2012 Olympic Games (Engebretsen et al. 2013). Hootman et al. (2007) reported on the epidemiology of collegiate injuries over a 16 year period. The injury rate was highest for men's football, followed by men's wrestling, men's soccer and then women's soccer (soccer was highest of the women's sports).

The characteristics of soccer injuries are relatively similar across various studies, although there are some differences. The majority of soccer injuries affect the lower extremities, and account for about 70 % of all injuries (Junge et al. 2006; Junge and Dvorak 2007, 2013; Theron et al. 2013). The most frequently injured body parts are the thigh, ankle, lower leg and knee. The most frequent types of injuries are contusions, ligament sprains, and muscle strains (Junge et al. 2006; Junge and Dvorak 2007, 2013; Ekstrand et al. 2011; Walden et al. 2005). The most frequent diagnoses were thigh and lower leg contusion, ankle sprain, and thigh muscle strain. The most common injury mechanism involved contact with another player, and almost half of the contact injuries were caused foul play in a match (Junge and Dvorak 2007, 2013). Injury incidence is higher in matches than in training (Ekstrand et al. 2011; Walden et al. 2005; Hagglund et al. 2009).

In the FIFA World Cup matches, the over-all injury rate was 2.7 injuries per match or 81.0 injuries per 1,000 match hours in 2002 (Korea/Japan Venue), 2.3 or 68.7 in 2006 (German Venue), and 2.0 or 61.1 in 2010 (South African Venue) (Dvorak et al. 2007, 2010; Junge et al. 2004a). The time-loss injury rate was 1.7 injuries per match or 50.7 injuries per 1,000 match hours in 2002 (Korea/Japan Venue), 1.5 or 49.5 in 2006 (German Venue), and 1.3 or 40.1 in 2010 (South African Venue). A third of the injuries involved no absence, while a third to a half of the injuries involved a 1–3 day player loss during the tournament.

In a different international tournament that was monitored, the incidence of time-loss injuries was 41.6 per 1,000 match hours (Hagglund et al. 2009). The incidence of time-loss injuries is lower for matches between professional clubs than for matches of international tournaments. The rate of time-loss injuries for the 11 top clubs in several European countries was 30.5 per 1,000 match hours (Walden et al. 2005). Ekstrand et al. (2011) examined the 23 of the European professional football teams over 7 seasons. The incidence of time-loss injuries was 27.5 per 1,000 match hours. The incident rate differed between countries (Walden et al. 2005). Thus, several factors such as region and match style (international tournament or professional club match) influence injury risk.

Risk of injuries is also influenced by the score. Ryynanen et al. (2013) reported that players on winning teams had a higher risk of suffering injury than did players on a drawing or losing team in the FIFA World Cup. They surmised that the reason for the above results was that the winning team adopted a more defensive strategy and tried to simply maintain ball possession in order to control the game. The losing team, however, needed to be more aggressive in order to steal the ball. This lead to a more risky situation for the winning team, as they had to endure aggressive attacks.

The relationship between match results and injury incidence is somewhat confused. Ekstrand et al. (2004) found a higher injury incidence for matches lost as compared to matches won or drawn in by a national soccer team. A study by Bengtsson et al. (2013) reported similar results for professional clubs in that injury occurrence in a match that was lost or drawn was higher than for matches that were won. On the other hand, Hagglund et al. (2009) reported that there was no difference in the injury incidence rate for matches that were lost, won or drawn.

10.1.2 Injury Rate of Junior Soccer Players

There are about 22 million youth soccer players in the world (Big count 2006). The number of youth soccer players increased by 7 % during the period from 2000 to 2006.

Fauda et al. (2013) summarized soccer injuries in children and adolescent players. In their report, training injury incidence was nearly constant in adolescent players, and ranged from 1 to 5 injuries per 1,000 training hours. Match injury incidence tended to increase with age and ranged from 15 to 20 injuries per 1,000 match hours.

For adolescents, the likelihood of injury varies according to form level, age, and sex. Petersen et al. (2000) and Junge et al. (2002) report that low level youth players have more than twice the injury incidence level as compared with high level youth players. Overall, the incidence of injury in youth football increases with age (Schmidt-Olsen et al. 1991).

Thus older soccer players have a higher injury incidence as compared to younger soccer players (Peterson et al. 2000; Schmidt-Olsen et al. 1991; Inklaar et al. 1996), and the incidence of injury increases with each year/grade of school (Malina et al. 2006).

A study by Hagglund et al. (2009) on the epidemiology of injuries in European soccer showed a tendency toward increased match injury incidence with increased age in male tournaments.

The injury rate of youth soccer players is different from that of professional players. Schmidt-Olsen et al. (1991) suggest that adolescent players incur fewer injuries than senior players because of better flexibility, as well as having less weight and moving slower during collisions.

The study also reported that 14 % of the adolescent had back pain trouble.

10.1.3 Injury Rate of the Female Soccer Players

Injury rates for female soccer players have been reported in several studies. Junge and Dvorak (2013) reported the incidence of injuries during international top-level tournaments from 1998 to 2012. In women, the average incidence of time-loss injury was 1.0 per match in the FIFA World Cups, 1.0 per match in the Olympic Games, and 1.0 per match in the FIFA U19/U20 World Cup and 0.7 per match in the FIFA U17 World Cup. The incidence of time-loss injuries was similar for males and females in the same type of tournament, except for the FIFA World Cup. The average number of injuries increased from 1.28 in the FIFA Women's World cup 1999 to 2.34 in 2007 and from 2.13 in the Olympic Games of 2000 to 2.88 in 2008. The authors felt that the increase in injury incidence was due to an increasingly physical style of play. This interpretation was supported by data which indicated that while the non-contact injury incidence remained constant, contact injuries showed an increased incidence. This is a further indication that injury incidence is affected by playing style and the type of tournament.

Injury incidence in adolescent female players varied from 8.3 to 23.3 injuries per 1,000 match hours and from 1.1 to 4.6 injuries in training hours in a prospective study (Emery et al. 2005; Le Gall et al. 2008; Soderman et al. 2001; Faude et al. 2005; Steffen et al. 2007).

For female players, skilled players have a greater risk of injury. Soligard et al. (2010a) reported that highly skilled female soccer players (players with a high level of technical and tactical training and good physiological abilities) had a higher incidence of injury than lower skilled players. However, as mentioned above, a previous study suggested that low-level male players had a higher risk of injury. More studies are needed to examine the reason for this discrepancy.

Injury incidence in adolescent female players is different from that of male players (Le Gall et al. 2008). Gall documented injuries sustained by elite French female soccer players between 15 and 19 years of age, over the course of 8 seasons. The risk of injuries was lower in the U19 categories than in the U15 categories and overall, injury rate decreased with age. A study of Emery et al. (2005) supported these results that incidence of time-loss injuries in U18 was lower compared to that in U16.

10.2 Efficacy of the FIFA 11 + Program

10.2.1 FIFA "The 11" and "11 +"

In 2003 The Federation Internationale de Football Association (FIFA) Medical and Research Center (F-MARC) developed "The 11", a soccer-specific injury prevention program. This program has ten exercises and a section on the spirit of fair play. Junge et al. (2011)noted the effectiveness of this program for reducing the incidence of soccer injuries (11.5 % fewer match injuries and 25.3 % fewer training injuries) and its cost effectiveness. However, some studies questioned the value of this program on the physical performance of the players (Steffen et al. 2008a; Kilding et al. 2008). Problems included a low rate of compliance and a negative impression by some of the players toward the program (Kilding et al. 2008; Steffen et al. 2008b). Since then, F-MARC developed a more comprehensive program, the "11 +". The 11+ is a warm-up program which takes about 20 min to complete. The 11+ has 3 parts and involves 15 exercises. This program involves a progression through three levels. A video of the exercises and an accompanying manual for the 11+ can be downloaded without cost at the FIFA 11+ website (http://f-marc.com/11plus/). A summary of the 11+ program is mentioned below:

Part 1 involves slow speed running exercises combined with active stretching and controlled partner contact. This part has six exercises; straight ahead, hip out, hip in circling the partner, jumping with shoulder contact, and quick forward and backward sprints.

Part 2 involves strength, plyometrics, and balance exercises. This part has six sessions of exercise activities; the bench, sideways bench hamstrings, single leg balance, squats, and jumping. These sessions have three levels of variation and progression.

Part 3 involves running exercises at moderate to high speeds combined with planting and cutting movements. This part has three exercises; across the pitch, bounding, and plant and cut.

A key point of this program is that the use of proper technique is emphasized during all of the exercises; keeping the player's upper body straight, hips, knees and feet straight, and not letting the knees buckle inward (http://issuu.com/vongrebelmotion/docs/11plus_workbook_english/1?e=0).

10.2.2 Efficacy of the FIFA 11+ on Injury Risk

Previous studies have supported the efficacy of the 11 +. Soligard et al. (2008) reported that the 11+ reduced the risk of all injuries by 32 % if it was performed at least twice per week by adolescent female soccer players. This study also suggested that the 11+ reduced overuse injuries by 53 % and severe injuries by 45 %. Grooms et al. (2013)reported on the effect of the 11 + for male collegiate soccer players. The intervention (11 +) group, as compared to the control group, had a 72 % reduction of lower extremity injury risk as well as less days lost to this type of injury. Longo et al. (2012) investigated the effect of the 11 + on elite male basketball players. Over a 9 month period, the 11+ group had a lower overall injury rate as well as a lower rate for training injuries, lower extremity injuries, acute injuries, severe injuries, trunk injuries, leg injuries, and hip and groin injuries. Soligard et al. (2010b) reported that compliance with the 11+ reduced injury risk and that players with a high compliance rate (more than 1.5 session per week), had significantly lower injury risks than players with a lower compliance rate. F-MARC has recommended that players perform the 11+ at least twice a week to get the optimum benefit from this program. In an unpublished study, we investigated the effect of the 11 + on injury rate in Japanese male adolescent players. Six Japanese U-18 (age: 16–18) and U-15 (age: 13–15) soccer teams (totaling 986 players) were followed for three seasons (2010, 2011 and 2012). In the 2010 season, the players completed the usual warm-up as a control season, whereas for the 2011 and 2012 intervention seasons, the players performed the 11+ more than twice per week throughout the entire season. The injury rates in the 2010 control season were 3.4 per 1,000 training hours and 8.9 per 1,000 match hours while those in 2012 intervention season were 1.5 per 1,000 training hours and 4.4 per 1,000 match hours. The injury rate was significantly lower in the intervention season than in the control season. In summary, use of the 11+ reduces injury risk for male and female soccer players. And the 11 + also reduces injury risk for basketball players. This was also true for adolescents, although there was some degree of discrepancy between several articles

10.2.3 *Efficacy of the FIFA 11+ on Physical Performance*

The efficacy of the 11+ has also been examined for its effect on physical performance as well as injury risk. Bizzini et al. (2013a) investigated post-exercise effects of the 11+ on physiological and performance responses. After the warm-up, there were pre-post differences in the 20-m sprint, vertical jump, stiffness, agility, balance (star excursion balance test), lactate, VO2, and core temperature. They concluded that the 11+ is an appropriate warm-up, inducing performance improvements for soccer players that are comparable with those obtained with other warm-up routines. Impellizzeri et al. (2013) investigated the long-term training effects of the 11+. They tested some variables before and after a 9 week intervention period and compared the intervention and control groups. The results showed that performing the 11+ for 9 weeks improved neuromuscular control, time-to-stabilization, and knee flexor strength. These factors are likely key mechanisms for explaining the injury prevention effects of the 11+. However, the intervention produced no meaningful effects on balance ability, vertical jump, sprint, or agility between the groups. Reis et al. (2013) demonstrated a performance enhancement in youth futsal players from performing the 11+ twice a week for 12 weeks. The 11+ group increased the quadriceps concentric and hamstrings concentric and eccentric peak torque. The intervention group also exhibited an improved jumping ability, sprint time, and agility time. Steffen et al. (2013a) noted that the 11+ improved functional balance in adolescent female football players. The preceding articles indicate that performing the 11+ can improve some aspects of physical performance. Though the original "11" had certain problems involving low compliance and a negative player impression, the 11+ program seems to have solved these problems.

Steffen et al. (2013b) showed that a preseason coaching workshop was the most effective delivery method for producing compliance with the 11+ recommendations instead of unsupervised website viewing and additional physiotherapist supervision to teach team. FIFA has initiated a worldwide dissemination of 11+ (Bizzini et al. 2013b). Not all teams, and particularly adolescent teams, have staff who are familiar with the physical training and/or injury prevention of amateur teams. The 11+ is a simple warm-up program that needs no special equipment, is supported by scientific studies, and can be easily accessed on the (http://f-marc.com/11plus/home/). In summary, the FIFA 11+ is a structured warm-up program which has the benefit of both preventing injuries and improving physical performance.

References

Bengtsson H, Ekstrand J, Walden M, Hagglund M (2013) Match injury rates in professional soccer vary with match result, match venue, and type of competition. Am J Sports Med 41(7):1505–1510

Bizzini M, Impellizzeri FM, Dvorak J, Bortolan L, Schena F, Modena R, Junge A (2013a) Physiological and performance responses to the "FIFA 11+" (part 1): is it an appropriate warm-up? J Sports Sci 31(13):1481–1490

Bizzini M, Junge A, Dvorak J (2013b) Implementation of the FIFA 11+ football warm up program: how to approach and convince the Football associations to invest in prevention. Br J Sports Med 47(12):803–806

Dvorak J, Junge A, Grimm K, Kirkendall D (2007) Medical report from the 2006 FIFA World Cup Germany. Br J Sports Med 41(9):578–581; discussion 581

Dvorak J, Junge A, Derman W, Schwellnus M (2011) Injuries illnesses of football players during the FIFA World Cup 2010. Br J Sports Med 45(8):626–630

Ekstrand J, Walden M, Hagglund M (2004) Risk for injury when playing in a national football team. Scand J Med Sci Sports 14(1):34–38

Ekstrand J, Hagglund M, Walden M (2011) Injury incidence and injury patterns in professional football: the UEFA injury study. Br J Sports Med 45(7):553–558

Emery CA, Meeuwisse WH, Hartmann SE (2005) Evaluation of risk factors for injury in adolescent soccer: implementation and validation of an injury surveillance system. Am J Sports Med 33(12):1882–1891

Engebretsen L, Steffen K, Alonso JM, Aubry M, Dvorak J, Junge A, Meeuwisse W, Mountjoy M, Renstrom P, Wilkinson M (2010) Sports injuries and illnesses during the Winter Olympic Games 2010. Br J Sports Med 44(11):772–780

Engebretsen L, Soligard T, Steffen K, Alonso JM, Aubry M, Budgett R, Dvorak J, Jegathesan M, Meeuwisse WH, Mountjoy M et al (2013) Sports injuries and illnesses during the London Summer Olympic Games 2012. Br J Sports Med 47(7):407–14

Faude O, Junge A, Kindermann W, Dvorak J (2005) Injuries in female soccer players: a prospective study in the German national league. Am J Sports Med 33(11):1694–1700

Faude O, Rossler R, Junge A (2013) Football injuries in children and adolescent players: are there clues for prevention? Sports Med 43(9):819–837

Fuller CW, Ekstrand J, Junge A, Andersen TE, Bahr R, Dvorak J, Hagglund M, McCrory P, Meeuwisse WH (2006) Consensus statement on injury definitions and data collection procedures in studies of football (soccer) injuries. Clin J Sport Med 16(2):97–106

Grooms DR, Palmer T, Onate JA, Myer GD, Grindstaff T (2013) Soccer-specific warm-up and lower extremity injury rates in collegiate male soccer players. J Athl Train 48(6):782–789

Hagglund M, Walden M, Ekstrand J (2009) UEFA injury study – an injury audit of European Championships 2006 to 2008. Br J Sports Med 43(7):483–489

Hootman JM, Dick R, Agel J (2007) Epidemiology of collegiate injuries for 15 sports: summary and recommendations for injury prevention initiatives. J Athl Train 42(2):311–319

Impellizzeri FM, Bizzini M, Dvorak J, Pellegrini B, Schena F, Junge A (2013) Physiological and performance responses to the FIFA 11+ (part 2): a randomised controlled trial on the training effects. J Sports Sci 31(13):1491–1502

Inklaar H, Bol E, Schmikli SL, Mosterd WL (1996) Injuries in male soccer players: team risk analysis. Int J Sports Med 17(3):229–234

Junge A, Dvorak J (2007) Injuries in female football players in top-level international tournaments. Br J Sports Med 41(Suppl 1):i3–i7

Junge A, Dvorak J (2013) Injury surveillance in the World Football Tournaments 1998–2012. Br J Sports Med 47(12):782–788

Junge A, Rosch D, Peterson L, Graf-Baumann T, Dvorak J (2002) Prevention of soccer injuries: a prospective intervention study in youth amateur players. Am J Sports Med 30(5):652–659

Junge A, Dvorak J, Graf-Baumann T (2004a) Football injuries during the World Cup 2002. Am J Sports Med 32(1 Suppl):23S–27S

Junge A, Dvorak J, Graf-Baumann T, Peterson L (2004b) Football injuries during FIFA tournaments and the Olympic Games, 1998–2001: development and implementation of an injury-reporting system. Am J Sports Med 32(1 Suppl):80S–89S

Junge A, Langevoort G, Pipe A, Peytavin A, Wong F, Mountjoy M, Beltrami G, Terrell R, Holzgraefe M, Charles R et al (2006) Injuries in team sport tournaments during the 2004 Olympic Games. Am J Sports Med 34(4):565–76

Junge A, Engebretsen L, Mountjoy ML, Alonso JM, Renstrom PA, Aubry MJ, Dvorak J (2009) Sports injuries during the Summer Olympic Games 2008. Am J Sports Med 37(11):2165–2172

Junge A, Lamprecht M, Stamm H, Hasler H, Bizzini M, Tschopp M, Reuter H, Wyss H, Chilvers C, Dvorak J (2011) Countrywide campaign to prevent soccer injuries in Swiss amateur players. Am J Sports Med 39(1):57–63

Kilding AE, Tunstall H, Kuzmic D (2008) Suitability of FIFA's "the 11" training programme for young football players – impact on physical performance. J Sports Sci Med 7(3):320–326

Le Gall F, Carling C, Reilly T (2008) Injuries in young elite female soccer players: an 8-season prospective study. Am J Sports Med 36(2):276–284

Longo UG, Loppini M, Berton A, Marinozzi A, Maffulli N, Denaro V (2012) The FIFA 11+ program is effective in preventing injuries in elite male basketball players: a cluster randomized controlled trial. Am J Sports Med 40(5):996–1005

Malina RM, Morano PJ, Barron M, Miller SJ, Cumming SP, Kontos AP (2006) Incidence and player risk factors for injury in youth football. Clin J Sport Med 16(3):214–222

Peterson L, Junge A, Chomiak J, Graf-Baumann T, Dvorak J (2000) Incidence of football injuries and complaints in different age groups and skill-level groups. Am J Sports Med 28(5 Suppl):S51–S57

Reis I, Rebelo A, Krustrup P, Brito J (2013) Performance enhancement effects of Federation Internationale de Football Association's "The 11+" injury prevention training program in youth futsal players. Clin J Sport Med 23(4):318–320

Ryynanen J, Junge A, Dvorak J, Peterson L, Karlsson J, Borjesson M (2013) The effect of changes in the score on injury incidence during three FIFA World Cups. Br J Sports Med 47(15):960–964

Schmidt-Olsen S, Jorgensen U, Kaalund S, Sorensen J (1991) Injuries among young soccer players. Am J Sports Med 19(3):273–275

Soderman K, Adolphson J, Lorentzon R, Alfredson H (2001) Injuries in adolescent female players in European football: a prospective study over one outdoor soccer season. Scand J Med Sci Sports 11(5):299–304

Soligard T, Myklebust G, Steffen K, Holme I, Silvers H, Bizzini M, Junge A, Dvorak J, Bahr R, Andersen TE (2008) Comprehensive warm-up programme to prevent injuries in young female footballers: cluster randomised controlled trial. BMJ 337:a2469

Soligard T, Grindem H, Bahr R, Andersen TE (2010a) Are skilled players at greater risk of injury in female youth football? Br J Sports Med 44(15):1118–1123

Soligard T, Nilstad A, Steffen K, Myklebust G, Holme I, Dvorak J, Bahr R, Andersen TE (2010b) Compliance with a comprehensive warm-up programme to prevent injuries in youth football. Br J Sports Med 44(11):787–793

Steffen K, Andersen TE, Bahr R (2007) Risk of injury on artificial turf and natural grass in young female football players. Br J Sports Med 41(Suppl 1):i33–i37

Steffen K, Bakka HM, Myklebust G, Bahr R (2008a) Performance aspects of an injury prevention program: a ten-week intervention in adolescent female football players. Scand J Med Sci Sports 18(5):596–604

Steffen K, Myklebust G, Olsen OE, Holme I, Bahr R (2008b) Preventing injuries in female youth football – a cluster-randomized controlled trial. Scand J Med Sci Sports 18(5):605–614

Steffen K, Emery CA, Romiti M, Kang J, Bizzini M, Dvorak J, Finch CF, Meeuwisse WH (2013a) High adherence to a neuromuscular injury prevention programme (FIFA 11+) improves functional balance and reduces injury risk in Canadian youth female football players: a cluster randomised trial. Br J Sports Med 47(12):794–802

Steffen K, Meeuwisse WH, Romiti M, Kang J, McKay C, Bizzini M, Dvorak J, Finch C, Myklebust G, Emery CA (2013b) Evaluation of how different implementation strategies of an injury prevention programme (FIFA 11+) impact team adherence and injury risk in Canadian female youth football players: a cluster-randomised trial. Br J Sports Med 47(8):480–487

Theron N, Schwellnus M, Derman W, Dvorak J (2013) Illness and injuries in elite football players – a prospective cohort study during the FIFA Confederations Cup 2009. Clin J Sport Med 23(5):379–383

Walden M, Hagglund M, Ekstrand J (2005) UEFA Champions League study: a prospective study of injuries in professional football during the 2001-2002 season. Br J Sports Med 39(8):542–546

Schmidt-Olsen S, Jørgensen U, Kaalund S, Sørensen J (1991) Injuries among young soccer players. Am J Sports Med 19(3):273–275
Söderman K, Adolphson J, Lorentzon R, Alfredson H (2001) Injuries in adolescent female players in European football: a prospective study over one outdoor soccer season. Scand J Med Sci Sports 11(5):299–304
Soligard T, Myklebust G, Steffen K, Holme I, Silvers H, Bizzini M, Junge A, Dvorak J, Bahr R, Andersen TE (2008) Comprehensive warm-up programme to prevent injuries in young female footballers: cluster randomised controlled trial. BMJ 337:a2469
Soligard T, Grindem H, Bahr R, Andersen TE (2010) Are skilled players at greater risk of injury in female youth football? Br J Sports Med 44(15):1118–1123
Soligard T, Nilstad A, Steffen K, Myklebust G, Holme I, Dvorak J, Bahr R, Andersen TE (2010) Compliance with a comprehensive warm-up programme to prevent injuries in youth football. Br J Sports Med 44(11):787–793
Steffen K, Andersen TE, Bahr R (2007) Risk of injury on artificial turf and natural grass in young female football players. Br J Sports Med 41(Suppl 1):i33–i37
Steffen K, Bakka HM, Myklebust G, Bahr R (2008a) Performance aspects of an injury prevention program: a ten-week intervention in adolescent female football players. Scand J Med Sci Sports 18(5):596–604
Steffen K, Myklebust G, Olsen OE, Holme I, Bahr R (2008b) Preventing injuries in female youth football – a cluster-randomized controlled trial. Scand J Med Sci Sports 18(5):605–614
Steffen K, Emery CA, Romiti M, Kang J, Bizzini M, Dvorak J, Finch CF, Meeuwisse WH (2013) High adherence to a neuromuscular injury prevention programme (FIFA 11+) improves functional balance and reduces injury risk in Canadian youth female football players: a cluster randomised trial. Br J Sports Med 47(12):794–802
Steffen K, Meeuwisse WH, Romiti M, Kang J, McKay C, Bizzini M, Dvorak J, Finch C, Myklebust G, Emery CA (2013) Evaluation of how different implementation strategies of an injury prevention programme (FIFA 11+) impact team adherence and injury risk in Canadian female youth football players: a cluster-randomised trial. Br J Sports Med 47(8):480–487
Theron N, Schwellnus M, Derman W, Dvorak J (2013) Illness and injuries in elite football players – a prospective cohort study during the FIFA Confederations Cup 2009. Clin J Sport Med 23(5):379–383
Waldén M, Hägglund M, Ekstrand J (2005) UEFA Champions League study: a prospective study of injuries in professional football during the 2001–2002 season. Br J Sports Med 39(8):542–546

Chapter 11
Biomechanical Studies on ACL Injury Risk Factor During Cutting; Utilizing the Point Cluster Technique

Yasuharu Nagano, Hirofumi Ida, Hideyuki Ishii, and Toru Fukubayashi

Abstract The purpose of this paper is to clarify gender specific differences in strategies adopted by athletes during a cutting maneuver, based on results of recent research. The experimental task selected for detection of the risk of an injury to the anterior cruciate ligament (ACL) during a cutting maneuver should also involve trunk control. An example of such a task is shuttle run cutting. Trunk control during this task can then be analyzed. Important factors for understanding the strategy adopted by female athletes during a cutting maneuver involve gender differences in trunk and knee kinematics and kinetics, and the athlete's reaction to instruction. Both posture and reaction to instruction in female athletes during the cutting maneuver differ from those of their male counterparts. To decrease the risk of ACL injury, trunk position during the cutting task should be screened for forward and lateral inclination. A forward inclination of the trunk in female athletes can be achieved by improving hip and trunk muscle activity.

Keywords Gender difference • Biomechanics • Performance • Injury • Anterior Cruciate ligament injury

11.1 Introduction

The cutting maneuver is an essential sports action that is repeatedly performed in fast moving games. In a soccer game, the players may perform the equivalent of 726 ± 203 turns (Bloomfield et al. 2007). The quality of the cutting maneuver is an important aspect of a player's skill and often has an influence on the results of a game.

Y. Nagano (✉)
Japan Women's College of Physical Education, Tokyo, Japan
e-mail: nagano323@gmail.com

H. Ida
Jobu University, Gunma, Japan

H. Ishii
Rikkyo University, Saitama, Japan

T. Fukubayashi
Faculty of Sport Sciences, Waseda University, Saitama, Japan

K. Kanosue et al. (eds.), *Sports Injuries and Prevention*,
DOI 10.1007/978-4-431-55318-2_11

The cutting maneuver can also be a cause of injury. An anterior cruciate ligament (ACL) injury typically occurs during cutting. ACL injuries occur at a higher rate in female athletes (Agel et al. 2005), and in sports these injuries often occur in non-contact situations such as landing and cutting (Boden et al. 2000). Female athletes are at an increased risk for ACL injuries in sports requiring rapid deceleration during cutting, pivoting, landing, and changing in direction. Females incur injuries at a rate three to five times higher than that of male athletes (Agel et al. 2005). These epidemiological research results suggest that gender differences in the cutting maneuver are responsible for the high ACL injury rate in female athletes.

The relationship between the characteristics of sports tasks (e.g., landing, stopping, and cutting) and ACL injury have been examined. In the landing maneuver, certain characteristics such as increased dynamic valgus and high abduction loads are predictors of ACL injury risk in female athletes (Hewett et al. 2005). Dynamic valgus is also observed at the time of ACL injury (Hewett et al. 2009; Boden et al. 2009). Female athletes whose normal performance involves the above characteristics have an increased risk for ACL injury. Meanwhile, although gender differences in the cutting maneuver have been examined, clear predictors of ACL injury risk have not been identified. We propose that the manner in which female athletes perform a cutting maneuver that leads to injury differs from a cutting response that is commonly seen in sports activities or in experimental conditions. Female athletes often exhibit a different spontaneous response than male athletes, and this increases their risk for ACL injury. Therefore, the purpose of this paper is to use the results of recent research to clarify gender differences in the strategy adopted by females during a cutting maneuver.

11.2 Review of Previous Studies

Several studies have reported gender differences based on a motion analysis of the cutting maneuver. Most of these studies have reported that the angle of knee flexion during the cutting maneuver is the same in male and female athletes (Ford et al. 2005; Landry et al. 2007a, b; McLean et al. 1999; Pollard et al. 2004; Sigward and Powers 2006; Brown et al. 2009), although, some studies have reported that the angle is smaller in female athletes (Malinzak et al. 2001; McLean et al. 2005a). The angle of knee abduction (valgus) was reported to be either greater in female athletes (Ford et al. 2005), or the same as that in male athletes (McLean et al. 1999; Pollard et al. 2004, 2006). The angle of internal rotation at the knee was reported to be smaller in female athletes (Brown et al. 2009; McLean et al. 2004), or the same as that in male athletes (Pollard et al. 2004; McLean et al. 2005a). Results reported on the knee flexion moment were inconclusive (Landry et al. 2007a, b; Pollard et al. 2004, 2006). Female athletes were also reported to sustain a greater knee load during frontal knee moments (Landry et al. 2007a, b; Sigward and Powers 2006; McLean et al. 2005b). Although several studies have been conducted on motion analysis of the cutting maneuver, the results have been inconclusive.

One of the reasons for disagreement in the results of these studies is the wide variety of experimental cutting tasks selected for analysis. Cutting maneuvers that were studied included sidestep cutting (Landry et al. 2007b; Sigward and Powers 2006, 2007), cross step cutting (Landry et al. 2007a), cutting with a defensive opponent (McLean et al. 2004), or cutting after landing (Brown et al. 2009). We think it is very important that the experimental task selected should involve physical loads similar to those that occur at the time of an injury, so the movement against these physical loads can be observed. Recent studies reveal that at the time of ACL injury, there is an increased lateral motion of the trunk (Hewett et al. 2009), decreased forward inclination of the trunk (Sheehan et al. 2012), and extreme posterior position of the center of mass relative to the base of support (Sheehan et al. 2012). Therefore, the experimental task should also involve the use of trunk control by the subject. In the shuttle run cutting maneuver (Nagano et al. 2011a), participants run straight ahead for 5 m, plant their cutting foot on the force plate perpendicular to their initial direction of motion, and then change direction to move 180° relative to their initial direction of motion. This includes not only deceleration but also turning, stopping, and acceleration. Adequate knee and trunk control are necessary because of the change in direction required during the task. Other reasons for the discrepancies between studies may have included the use of different kinematics calculation methods. Some calculation methods result in solutions with a substantial variance from actual knee kinematic values. We calculated knee kinematics using the Point Cluster Technique (Andriacchi et al. 1998), which provides a minimally invasive estimation of the in vivo motion of the knee. The statistical findings of gender differences during the shuttle run cutting maneuver using the Point Cluster Technique are mentioned below.

11.3 Gender Differences in Kinematics of Shuttle Run Cutting

The results of a study that evaluated gender differences in the kinematics of shuttle run cutting (Nagano et al. 2011a) showed that knee flexion during this activity was greater in male than in female athletes. In addition, both the forward and lateral inclination of the trunk toward the cutting limb were significantly greater in male than in female athletes (Fig. 11.1). No differences in knee abduction and knee rotation were observed (Nagano et al. 2011a). These results indicate that the motion pattern in female athletes, with the trunk erect and the knee in slight flexion, is similar to the position reported to be associated with ACL injury (Boden et al. 2000; Sheehan et al. 2012). However, the pattern in female athletes was did not differ from that in male athletes in terms of knee abduction, and also demonstrated lateral trunk inclination to the opposite side as compared with male athletes, which makes the pattern of movement significantly different from the position reported to be associated with ACL injury (Boden et al. 2009). It is, therefore, suggested that not all the characteristics of the female athlete's posture during shuttle run cutting are similar to the position associated with ACL injury.

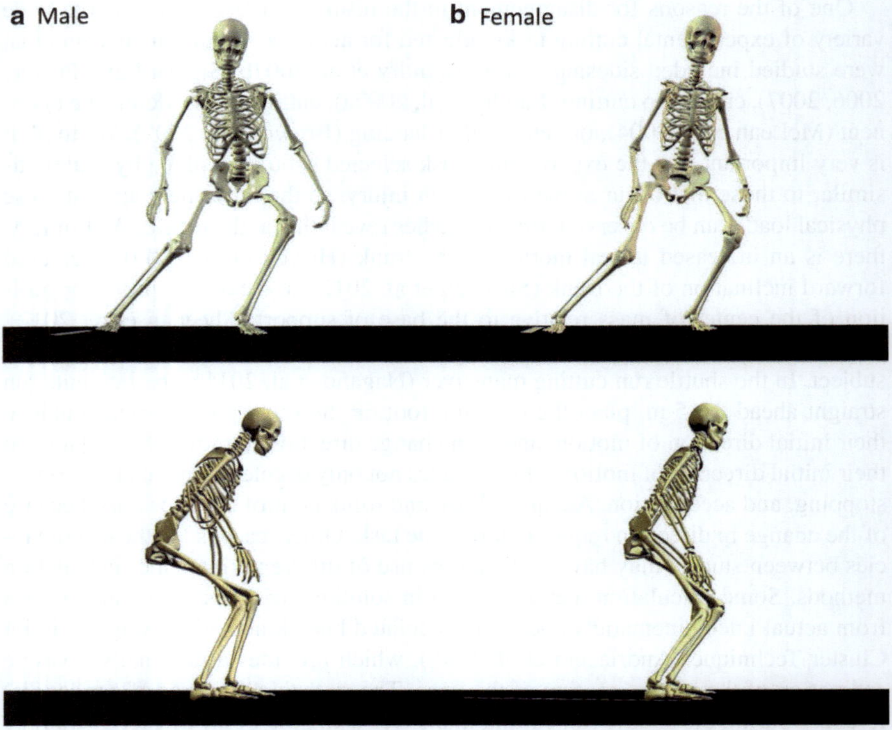

Fig. 11.1 Gender differences in posture during shuttle run cutting; (**a**) male and (**b**) female. Female athletes show a smaller knee flexion angle, smaller forward trunk angles, and greater lateral trunk angles with the inclination opposite to the cutting limb

11.4 Gender Differences in the Capacity of Cutting Kinematics

Gender differences are seen in athletic performance in relation to both reaction and change.. A previous study (Nagano et al. 2011a), noted that an increase in the forward inclination of the trunk may decrease the risk of ACL injury. We examined the gender differences in lower limb kinematics and kinetics when athletes were instructed to perform the shuttle run cutting maneuver with a forward inclination of the trunk (Nagano et al. 2011b). In response to this instruction, female athletes demonstrated an increase in knee and hip flexion angle (Fig. 11.2), a decrease in hip internal rotation angle (Fig. 11.2), and an increase in hip internal extension moment (Nagano et al. 2011b) (Fig. 11.4). On the other hand, female athletes demonstrated an increase in lateral inclination of the trunk toward the cutting limb (Fig. 11.3) instead of a forward inclination of the trunk, and also a tendency to increase the knee external abduction moment (Nagano et al. 2011b) (Fig. 11.5). These results

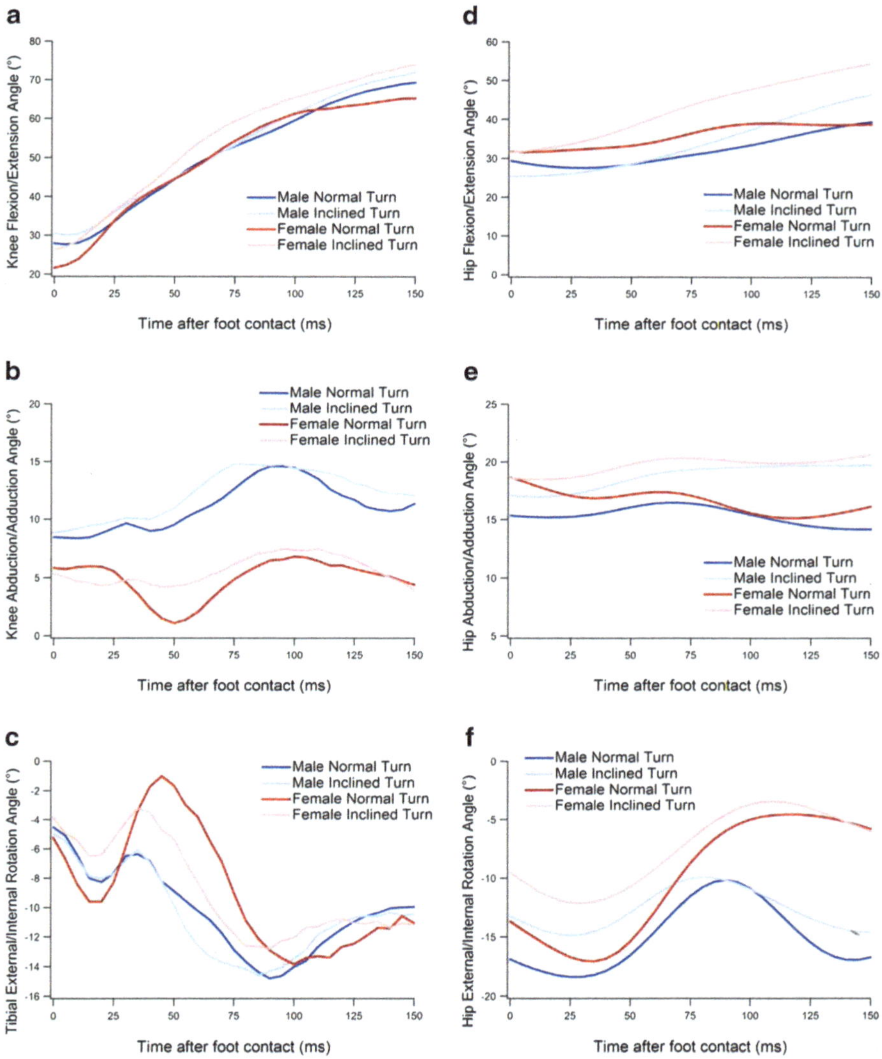

Fig. 11.2 Angular displacement of knee and hip with and without instruction to increase forward inclination of the trunk. Data are presented for (**a**) knee flexion/extension, (**b**) knee abduction/adduction, (**c**) external/internal tibial rotation, (**d**) hip flexion/extension, (**e**) knee abduction/adduction, and (**f**) hip external/internal rotation

suggested that the instruction to increase forward inclination of the trunk alone disturbed the female athlete's trunk control and increased the knee load, thereby increasing the risk of ACL injury.

A detailed knowledge of gender differences for trunk and knee kinematics and kinetics allowed us to better understand the strategy adopted by female athletes during

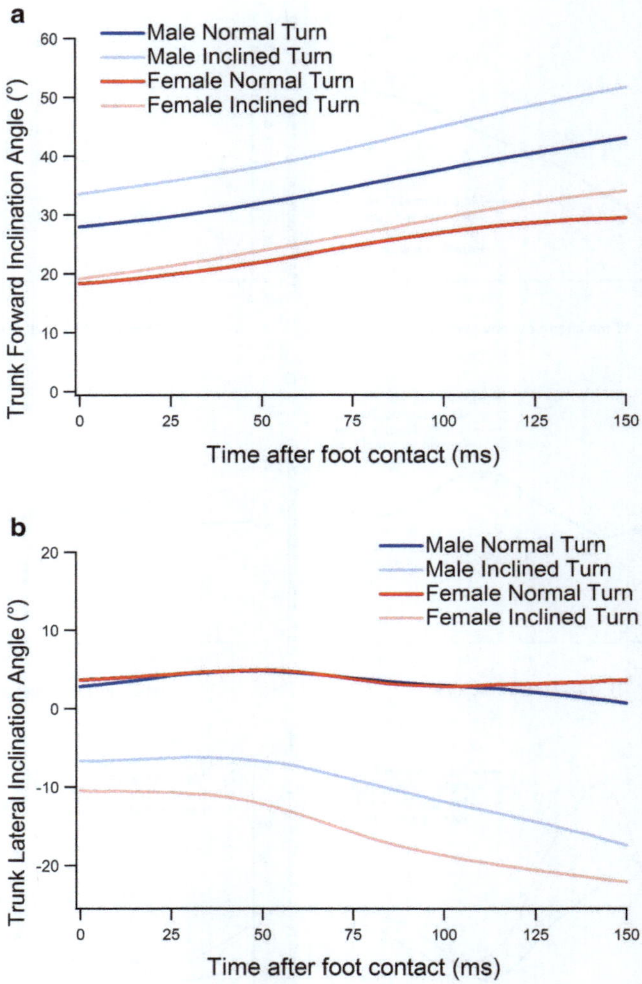

Fig. 11.3 Angular displacement of forward and lateral inclination of the trunk with and without instruction to increase forward inclination of the trunk. Data are presented for (**a**) trunk forward inclination and (**b**) trunk lateral inclination

the cutting maneuver. In a controlled experimental environment, female athletes perform the cutting maneuver with an erect trunk, laterally inclined opposite to the cutting limb, and the knee in slight flexion to avoid an increase in external knee abduction load. This position requires low muscle support, and is similar to the Trendelenburg sign in which the limb is supported by the iliotibial tract. When instruction or other disturbing forces increase the load on the cutting limb, female athletes are unable to control the need to depend on the trunk. This leads to lateral

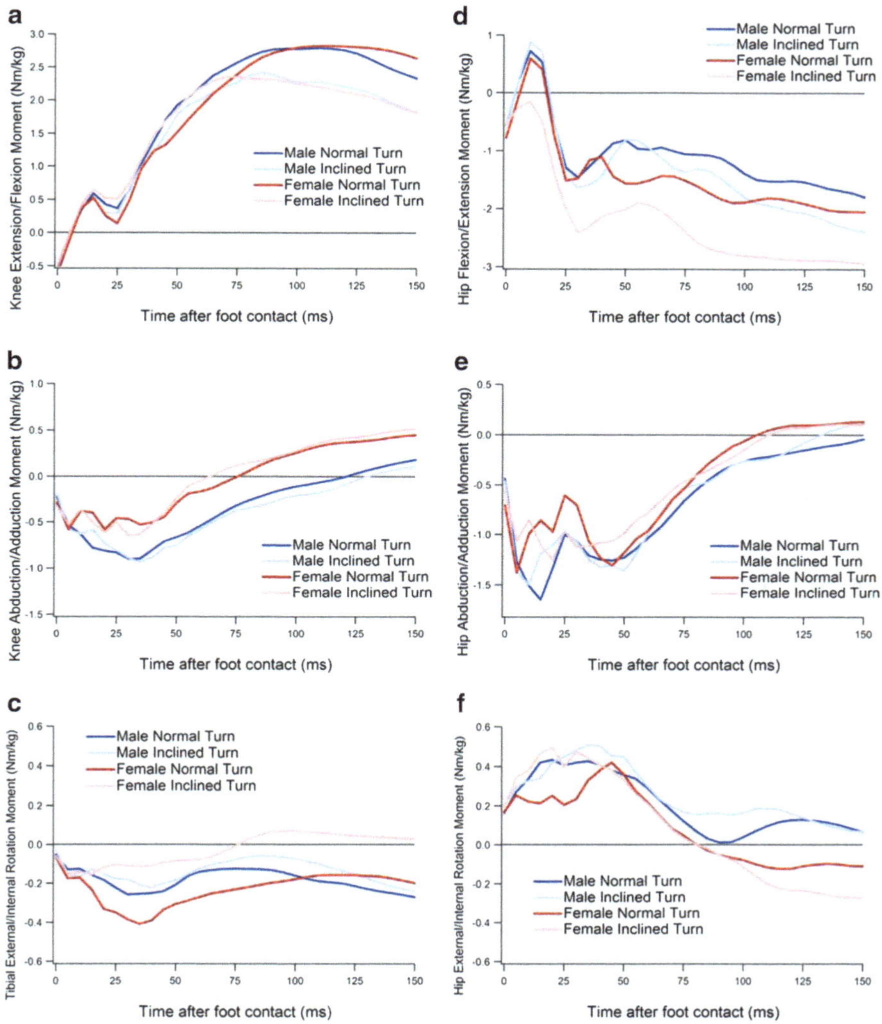

Fig. 11.4 Joint internal moments of knee and hip with and without instruction to increase forward inclination of the trunk. Data are presented for (**a**) knee flexion/extension, (**b**) knee abduction/adduction, (**c**) external/internal tibial rotation, (**d**) hip flexion/extension, (**e**) knee abduction/adduction, and (**f**) hip external/internal rotation

inclination of the trunk toward the cutting limb, and an increase in the knee external abduction moment. This response may be caused by insufficient trunk or muscle activity; moreover, it may also be the mechanism by which ACL injury is caused (Fig. 11.6). To decrease the risk of ACL injury, we should screen the trunk position during the cutting task and improve hip and trunk muscle activity of female athletes to encourage forward inclination of the trunk.

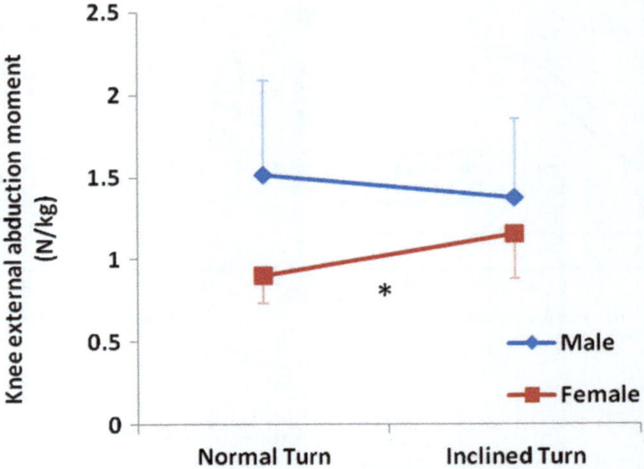

Fig. 11.5 Change in the knee external abduction moment with and without instruction to increase forward inclination of the trunk. As a result of instruction to increase forward inclination of the trunk, there was a significant interaction effect (*: $p < 0.05$). Female athletes showed a greater tendency to increase the knee external abduction moment

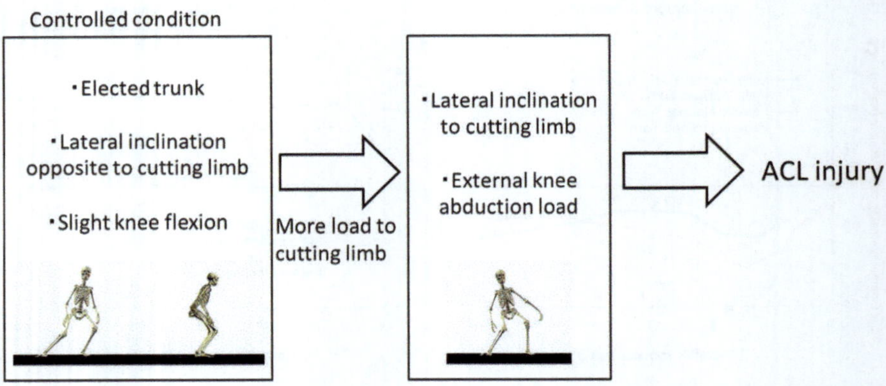

Fig. 11.6 Schematic representation of the strategy adopted by female athletes during the cutting maneuver

11.5 Conclusion

The purpose of this paper is to utilize recent research advances to clarify gender differences in the strategy adopted by athletes during the cutting maneuver. Experimental tasks that detect the risk of ACL injury should involve the use of trunk control, such as shuttle run cutting; trunk control should be included in the overall analysis. A greater understanding of the gender differences in trunk and knee kinematics and kinetics, as well as the reaction to instruction will allow us to better understand the strategy of female athletes during the cutting maneuver. To decrease the risk of ACL injury, trunk position during the cutting task should be screened for forward and lateral inclination, and improved hip and trunk muscle activity employed to encourage forward inclination of the trunk.

Acknowledgements This research was supported by JSPS KAKENHI Grant Number 23700786. The authors thank Sayumi Iwamoto, Toyo University, for providing an opportunity to perform the measurements.

Conflict of Interest Statement No author of this manuscript has any conflict of interest.

References

Agel J, Arendt EA, Bershadsky B (2005) Anterior cruciate ligament injury in national collegiate athletic association basketball and soccer: a 13-year review. Am J Sports Med 33:524–530

Andriacchi TP, Alexander EJ, Toney MK, Dyrby C, Sum J (1998) A point cluster method for in vivo motion analysis: applied to a study of knee kinematics. J Biomech Eng 120:743–749

Bloomfield J, Polman R, O'Donoghue P (2007) Physical demands of different positions in FA Premier League soccer. J Sports Sci Med 6:63–70

Boden BP, Dean GS, Feagin JA Jr, Garrett WE Jr (2000) Mechanisms of anterior cruciate ligament injury. Orthopedics 23:573–578

Boden BP, Torg JS, Knowles SB, Hewett TE (2009) Video analysis of anterior cruciate ligament injury: abnormalities in hip and ankle kinematics. Am J Sports Med 37:252–259

Brown TN, Palmieri-Smith RM, McLean SG (2009) Sex and limb differences in hip and knee kinematics and kinetics during anticipated and unanticipated jump landings: implications for anterior cruciate ligament injury. Br J Sports Med 43:1049–1056

Ford KR, Myer GD, Toms HE, Hewett TE (2005) Gender differences in the kinematics of unanticipated cutting in young athletes. Med Sci Sports Exerc 37:124–129

Hewett TE, Myer GD, Ford KR, Heidt RS Jr, Colosimo AJ et al (2005) Biomechanical measures of neuromuscular control and valgus loading of the knee predict anterior cruciate ligament injury risk in female athletes. Am J Sports Med 33:492–501

Hewett TE, Torg JS, Boden BP (2009) Video analysis of trunk and knee motion during non-contact anterior cruciate ligament injury in female athletes: lateral trunk and knee abduction motion are combined components of the injury mechanism. Br J Sports Med 43:417–422

Landry SC, McKean KA, Hubley-Kozey CL, Stanish WD, Deluzio KJ (2007a) Neuromuscular and lower limb biomechanical differences exist between male and female elite adolescent soccer players during an unanticipated run and crosscut maneuver. Am J Sports Med 35:1901–1911

Landry SC, McKean KA, Hubley-Kozey CL, Stanish WD, Deluzio KJ (2007b) Neuromuscular and lower limb biomechanical differences exist between male and female elite adolescent soccer players during an unanticipated side-cut maneuver. Am J Sports Med 35:1888–1900

Malinzak RA, Colby SM, Kirkendall DT, Yu B, Garrett WE (2001) A comparison of knee joint motion patterns between men and women in selected athletic tasks. Clin Biomech (Bristol, Avon) 16:438–445

McLean SG, Neal RJ, Myers PT, Walters MR (1999) Knee joint kinematics during the sidestep cutting maneuver: potential for injury in women. Med Sci Sports Exerc 31:959–968

McLean SG, Lipfert SW, van den Bogert AJ (2004) Effect of gender and defensive opponent on the biomechanics of sidestep cutting. Med Sci Sports Exerc 36:1008–1016
McLean SG, Walker KB, van den Bogert AJ (2005a) Effect of gender on lower extremity kinematics during rapid direction changes: an integrated analysis of three sports movements. J Sci Med Sport 8:411–422
McLean SG, Huang X, van den Bogert AJ (2005b) Association between lower extremity posture at contact and peak knee valgus moment during sidestepping: implications for ACL injury. Clin Biomech (Bristol, Avon) 20:863–870
Nagano Y, Ida H, Akai M, Fukubayashi T (2011a) Relationship between three-dimensional kinematics of knee and trunk motion during shuttle run cutting. J Sports Sci 29:1525–1534
Nagano Y, Ishii H, Iwamoto S, Ida H, Akai M et al (2011b) Gender differences of lower limb kinematics and kinetics during shuttle run cutting with the instruction of trunk forward inclination. Jpn J Phys Fitness Sports Med 60:823
Pollard CD, Davis IM, Hamill J (2004) Influence of gender on hip and knee mechanics during a randomly cued cutting maneuver. Clin Biomech (Bristol, Avon) 19:1022–1031
Pollard CD, Sigward SM, Ota S, Langford K, Powers CM (2006) The influence of in-season injury prevention training on lower-extremity kinematics during landing in female soccer players. Clin J Sport Med 16:223–227
Sheehan FT, Sipprell WH 3rd, Boden BP (2012) Dynamic sagittal plane trunk control during anterior cruciate ligament injury. Am J Sports Med 40:1068–1074
Sigward SM, Powers CM (2006) The influence of gender on knee kinematics, kinetics and muscle activation patterns during side-step cutting. Clin Biomech (Bristol, Avon) 21:41–48
Sigward SM, Powers CM (2007) Loading characteristics of females exhibiting excessive valgus moments during cutting. Clin Biomech (Bristol, Avon) 22:827–833

Chapter 12
Biomechanical Risk Factors and Prevention of Anterior Cruciate Ligament Injury

Hirohisa Magoshi and Toru Fukubayashi

Abstract Previous reports have identified increases in the knee abduction angle and knee abduction moment during landing and the cutting maneuver as biomechanical risk factors for anterior cruciate ligament (ACL) injury. Decreasing knee valgus has therefore been an important component of the ACL injury prevention program.

ACL injury prevention training is now conducted worldwide and it has been proven effective. However there is no consensus on the preventative effect of training.

The purpose of this study is to summarize the biomechanical factors for ACL injury as well as efficacy of the ACL injury prevention program. We also offer a suggestion for ACL injury prevention programming in the future.

Keywords Anterior cruciate ligament • Biomechanical risk factor • Prevention

12.1 Introduction

Female athletes who participate in jump-landing and cutting maneuver sports suffer anterior cruciate ligament (ACL) injuries at a four to sixfold greater rate than do male athletes participating in the same sports (Agel et al. 2005; Griffin et al. 2000; Arendt and Dick 1995; Arendt et al. 1999; Myklebust et al. 1998; Deitch et al. 2006). In particular, female soccer players are three times more likely to suffer an ACL injury as compared to their male counterparts (Agel et al. 2005; Griffin et al. 2000). The injury incidence rate for soccer is among the highest in sports, particularly ACL injuries in female soccer players (van Beijsterveldt et al. 2012). ACL injury is a serious injury that forces a long-term withdrawal from competition. Therefore, the establishment of effective preventive measures is required urgently.

H. Magoshi (✉)
Graduate School of Sport Sciences, Waseda University, Saitama, Japan
e-mail: roll14hirogear@gmail.com

T. Fukubayashi
Faculty of Sport Sciences, Waseda University, Saitama, Japan
e-mail: fukubayashi@waseda.jp

K. Kanosue et al. (eds.), *Sports Injuries and Prevention*,
DOI 10.1007/978-4-431-55318-2_12

Understanding the biomechanics of this injury is essential for its prevention. Thus far, jump-landing and cutting maneuver kinematics have been analyzed in laboratory environments. Identifying factors responsible for this injury and the explanation of its pathogenesis have been attempted using such analyses.

ACL injury prevention training is now being conducted worldwide. The configuration is different for each program, but the intended effect of the programs is the same, which is decrease in the incidence of ACL injury. Moreover, in order to present objective information on factors that decrease the ACL injury incidence rate, recent studies have examined the effect of prevention programs on lower extremity kinetics and kinematics. However, Agel et al. (2005) report that, despite widespread education and institution of intervention programs in recent years, the ACL injury incidence rate in female athletes who participate in high-risk sports (soccer and basketball) has not declined. Therefore, currently, there is no consensus on the factors that need to be considered for injury prevention.

Clarification and presentation of conclusive evidence obtained thus far are very important, as these can be applied while creating future preventive programs.

12.2 Biomechanical Risk Factors for ACL Injury

Most ACL injuries in female athletes occur during noncontact situations, typically during deceleration, cutting maneuvers, or jump-landings that are often associated with high external knee joint loads (Besier et al. 2001; Boden et al. 2000). Understanding the biomechanics of this injury is essential for identifying factors responsible for this injury and explaining its pathogenesis. In the first section we aim to summarize biomechanical risk factors during jump-landing and cutting for preventing ACL injuries.

12.2.1 Jump-Landing

Because epidemiological data points to a higher ACL injury incidence in women, motion analysis has been conducted to identify gender differences.

Several studies have investigated the knee flexion angle, knee abduction angle, and knee abduction moment during landing. Lephart et al. (2005) reported that the knee flexion angle when landing on both legs is smaller in women than in men. In addition, Ford et al. (2003) reported that female athletes landed with greater total knee valgus motion and a greater maximum knee valgus angle than male athletes.

From this, we can conclude that knee flexion and knee abduction angles are smaller and that the knee valgus moment is larger in women than in men. However, we cannot conclude that these factors are associated with future occurrences of ACL injuries. Meanwhile, in a series of prospective studies conducted recently, Hewett et al. (Hewett et al. 2005; Myer et al. 2010, 2011a, b) conclude that a large knee abduction angle and knee abduction moment during landing is an indication of the magnitude of ACL injury risk (Fig. 12.1).

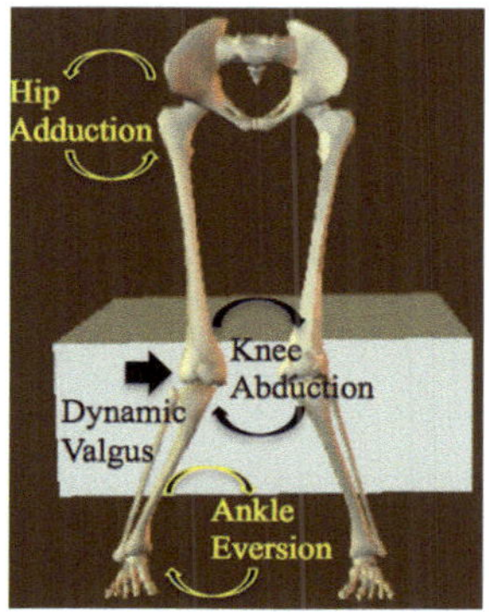

Fig. 12.1 Dynamic valgus during landing

12.2.2 Cutting Maneuver

ACL injury risk was found to be associated with knee abduction angle and knee abduction moment in a series of prospective studies which analyzed landing (Hewett et al. 2005; Myer et al. 2010, 2011a, b). Therefore, knee abduction angle and knee abduction moment during the cutting maneuver may also be risk factors for ACL injury. In fact, knee abduction angle and knee abduction moment during the cutting maneuver are reported to be greater in women than in men (Malinzak et al. 2001; McLean et al. 2004; Sigward and Powers 2006). Malinzak et al. (2001) reported that peak knee flexion angle is smaller in women than in men. Mclean et al. (2004) analyzed the relationship between hip motion and knee joint valgus moment in view of their impact on the knee joint. A positive correlation was found between hip internal rotation and the peak knee abduction moment. On the other hand, no gender difference was observed in the peak knee flexion angle. Also Kristianslund et al. (2014) reported on approach speed and knee abduction angle, and noted that cutting angle affected knee abduction moment during the cutting maneuver.

The above circumstances indicate the possibility that knee abduction angle and knee abduction moment during the cutting maneuver as well as during landing are risk factors for ACL injury. In soccer, ACL injury frequently occurs during the cutting maneuver (Agel et al. 2005; Hewett et al. 2005) (Fig. 12.2) and when stealing the ball (Fig. 12.3). Therefore, a motion analysis of the cutting maneuver is useful in understanding the mechanism of ACL injury in soccer players.

12.3 ACL Injury Prevention Program

In the second section our aim is to summarize the effect of the ACL injury prevention program on ACL incidence rate as well as the effect of the program on body kinematics during the jump-landing and cutting maneuver.

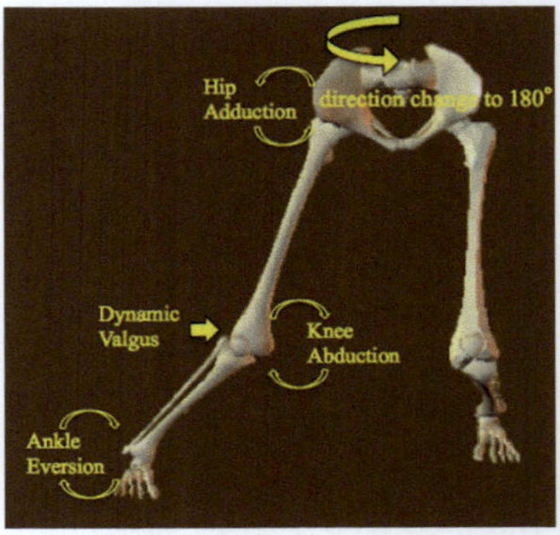

Fig. 12.2 Dynamic valgus during a direction change of 180°

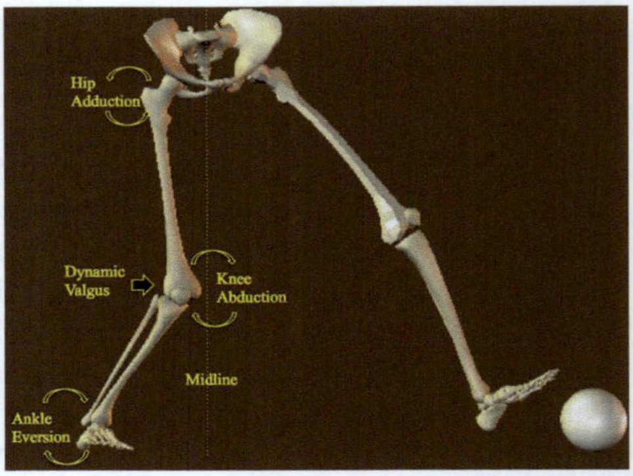

Fig. 12.3 Dynamic valgus when stealing the ball

12.3.1 Effects of the Prevention Program on ACL Injury Incidence Rate

Hewett et al. (1999) report that the incidence rate of knee injury in female athletes decreased after a plyometric training program of 6 weeks. Additionally, Petersen et al. (2005) report that the incidence rate of knee injury decreased in female handball players after a training program that combined balance and jump activities. Mandelbaum et al. (2005) introduced the Prevent Injury, Enhance Performance (PEP) program as an ACL injury prevention program for female football players. An evaluation of the results obtained by PEP program intervention indicated a significantly lower injury incidence rate in the intervention group as compared to the non-intervention group. Another preventive program was introduced by the Federation of International Football Association (FIFA) and was entitled 「The 11+」 program for preventing injuries in soccer. An evaluation of 「The 11+」 program intervention also demonstrated a significantly lower injury incidence rate in the intervention group as compared to the non-intervention group (van Beijsterveldt et al. 2012). The above information as well as other studies (Petersen et al. 2005; Mandelbaum et al. 2005; Myklebust et al. 2003) have documented the effectiveness of ACL injury prevention programs in decreasing the ACL injury incidence rate.

12.3.2 *Effects of ACL Injury Prevention Programs on ACL Injury Risk Factors*

Jump-Landing

Most of the programs for ACL injury prevention in the current literature function to alter lower extremity movement patterns.

Biomechanical analysis of landing characteristics have been used to verify the effect of intervention on ACL injury in cross-sectional studies to date. Pollard et al. (2006) conducted a jump program with a focus jump and agility training for female soccer players and investigated changes in hip and knee joint angles during drop landing. As a result, although no change was observed in knee flexion and abduction angles, female players did demonstrate a lesser hip internal rotation and greater hip abduction. Hewett et al. (1999) reported that knee abduction moment and ground reaction force at the time of landing was reduced by jump training. Furthermore, Myer et al. (2011b) investigated the effect of balance training and plyometric jump training on a non-risk group and risk group for the knee abduction moment. As a result, the knee abduction moment was decreased in the risk group. Additionally, Lephart et al. (2005) reported that jump training produced an increase in hip and knee angles although no change was observed in knee abduction and hip adduction angles.

The overall results of these reports indicate that ACL injury prevention training, at least with respect to biomechanical factors, is effective to some extent. However, such evidence is not conclusive. These prevention programs are similar in that they involve plyometrics, balance, and agility training; however, the details, duration, and the training maneuvers vary depending on the program. Furthermore, the assessment environment (laboratory) is quite different from the actual training environment. Thus, although motion analysis has been carried out in the laboratory environment, it has not been conducted in an actual training environment. To remove this knowledge gap, we conducted a pilot study to clarify body kinematics during the drop vertical jump test. We assessed differences between the laboratory environment and an artificial turf for female collegiate soccer players (Fig. 12.4). On artificial turf displacement of the knee flexion angle was significantly smaller than that of the corresponding laboratory measurement ($p < 0.05$). A statistically significant difference was not observed in knee valgus motion. Thus, laboratory measurements are likely to underestimate values of the actual training environment.

Cutting Maneuver

In addition to the longitudinal studies described earlier, a recent cross-sectional study was performed on the cutting maneuver. Celebrini et al. (2014) conducted prevention programs using videotapes of the jump-landing technique. The groups given feedback showed increased knee angular displacement flexion angles and decreased peak

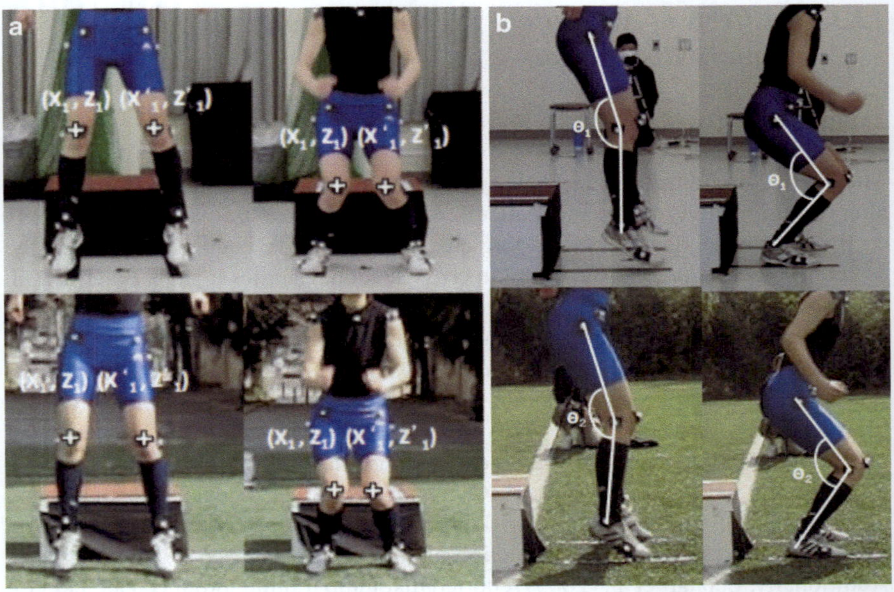

Fig. 12.4 Differences between the laboratory environment and an artificial turf (**a**) frontal plane (**b**) saggital plane

vertical ground reaction forces. This shows that sports techniques with complicated maneuvers can profit from feedback training. However little scientific research on cutting has been conducted to carefully document the effectiveness of prevention programs. Longitudinal studies are needed to evaluate the effect of intervention programs on cutting as well as landing. Additionally, because ACL injuries in soccer frequently occur during the cutting maneuver, it is important to conduct research on this maneuver to elucidate the mechanism of injury occurrence.

12.3.3 Components of ACL Injury Prevention Programs

Plyometric Training

Several studies have investigated the effects of plyometric training on lower extremity motion patterns.

Lephart et al. (2005) reported that hip and knee angles increased following jump training. However, no change was observed in knee abduction and hip adduction angles. Irmischer et al. (2004) reported that after training there was a decrease in ground reaction force during landing. Additionally, Soderman et al. (2000) reported that jump training decreased knee valgus motion.

Balance Training

Caraffa et al. (1996) found that balance training alone was effective, whereas Soderman et al. (2000) did not find it to be effective. Including multiple elements in the balance training program has been shown to have a preventative effect (Myklebust et al. 2003; Olsen et al. 2005). A study conducted by Pfeiffer et al. (2006) did not include balance training failed to show a preventative effect. Therefore, unless a training program is comprehensive and contains a number of effective elements, it may not produce a measurable preventative effect.

Feedback Training

Feedback training is a commonly used method for teaching and learning physical techniques. It can be used as an intervention technique to change motor control strategies such as faulty movement biomechanics that are associated with ACL injury.

Prapavessis and McNair (1999) studied the effects of augmented feedback on lower extremity kinematics and kinetics in drop landing tasks. The augmented feedback group significantly decreased their vertical ground reaction force after training. Onate et al. (2005) investigated the effect of different types of video feedback on landing technique. The 'expert instruction' group viewed a trained expert conducting jump-landing tasks. The 'self-feedback' group viewed their own jump landing tasks. The combined 'instruction and feedback' group viewed both the expert's and their own jump landing tasks. A group which received neither instruction nor self-feedback served as the control group. The self-feedback and combined feedback groups had more knee flexion and less peak ground reaction force than did the 'expert instruction' or control groups.

From the above, it can be seen that sports techniques with complicated maneuvers require feedback training. Simply viewing good technique is not adequate.

Strength Training

Cochrane et al. (2010) reported that strength training alone might not be sufficient to positively alter biomechanical risk factors associated with ACL injury. However, strength training could improve movement patterns if it was combined with feedback training. Cochrane et al. (2010) investigated knee alignment during movement and divided the subjects into a balance training group and a weight training group. The balance training group showed a decrease in peak knee valgus and peak knee internal moments, but the weight training group tended to increase knee load. In contrast, Herman et al. (2009) investigated the effects of feedback training with or without strength training on lower extremity alignment. The group that received a combination of strength and balance training showed an increased hip abduction angle and a decreased peak knee anterior shear force.

While strength training alone may not modify knee alignment, when combined with feedback training and balance training it has the potential to improve knee alignment.

12.4 Discussion

We have presented the relevant literature on landing and the cutting maneuvers as they relate to ACL injury biomechanical risk factors. We have additionally covered the ability of ACL injury prevention programs to alter biomechanical risk factors as well as reduce the incidence rate of ACL injuries.

In previous reports, increased knee abduction angle and knee abduction moment during jump-landing and the cutting maneuver have been identified as biomechanical risk factors for ACL injury (Hewett et al. 2005; Myer et al. 2010; Sigward and Powers 2006; Kristianslund et al. 2014; Landry et al. 2007; Kristianslund and Krosshaug 2013). Therefore, decreasing knee valgus has been an important component of ACL injury prevention programs. ACL injury prevention training is now conducted worldwide. The configuration is different for each program, but the intended effect of the programs is the same. However, a number of studies have reported that it is difficult to achieve preventative effects with training alone (Soderman et al. 2000; Pfeiffer et al. 2006; Heidt et al. 2000). Therefore, development of a program that combines balance, plyometrics, strength, and feedback training is considered important if optimal gains in injury prevention are to be realized (Hewett et al. 1999; Petersen et al. 2005; Mandelbaum et al. 2005; Myklebust et al. 2003). Unfortunately, there is no consensus on the best methods of training. Thus the components of different ACL injury prevention programs are similar, but the details of training content, duration, and object movement are different. Another problem is that the effects of intervention have been studied in a laboratory environment. Due to limited space and differences in surfaces and shoes, it is likely that constrained movement (especially in speed) of the laboratory environment have affected the results of evaluations performed in the laboratory environment. Previous studies have reported that approach speed affected knee abduction moment during the cutting maneuver (Kristianslund et al. 2014). It is possible that biomechanical characteristics in the laboratory environment do not adequately represent what occurs in the field. The results of our study suggested that laboratory results are likely to underestimate values that pertain in the actual training environment. Kinematic measurements are also likely to be affected due to the difficulty of performing at a high running speed in the laboratory.

In summary, we stress that it is necessary to verify the effects of intervention by utilizing movements that are representative of those performed during competition and to simulate, to the greatest degree possible, the actual conditions that occur in the field.

12.5 Conclusion

The aim of this study was to summarize biomechanical factors that are responsible for ACL injuries and on the preventive effects that are produced by ACL injury prevention programs.

Established facts

- A large knee abduction angle and knee abduction moment during landing and cutting maneuvers are valid indicators of the magnitude of ACL injury risk.
- For success in decreasing the ACL injury incidence rate, it is critical that an ACL injury prevention program combine balance, plyometrics, strength, and feedback training.

Important research topics for the future

- It is necessary to verify that the knee abduction angle and knee abduction moment during the cutting maneuver as well as during landing are risk factors for ACL injury.
- It is necessary to verify the effect of intervention for movements that are representative of those performed in an actual competition.
- It is necessary to be aware that results obtained in the laboratory may differ from those obtained under conditions in the field situation.

References

Agel J, Arendt EA, Bershadsky B (2005) Anterior cruciate ligament injury in national collegiate athletic association basketball and soccer: a 13-year review. Am J Sports Med 33(4):524–530

Arendt E, Dick R (1995) Knee injury patterns among men and women in collegiate basketball and soccer. NCAA data and review of literature. Am J Sports Med 23(6):694–701

Arendt EA, Agel J, Dick R (1999) Anterior cruciate ligament injury patterns among collegiate men and women. J Athl Train 34(2):86–92

Besier TF et al (2001) External loading of the knee joint during running and cutting maneuvers. Med Sci Sports Exerc 33(7):1168–1175

Boden BP et al (2000) Mechanisms of anterior cruciate ligament injury. Orthopedics 23(6):573–578

Caraffa A et al (1996) Prevention of anterior cruciate ligament injuries in soccer. A prospective controlled study of proprioceptive training. Knee Surg Sports Traumatol Arthrosc 4(1):19–21

Celebrini RG et al (2014) Effect of a novel movement strategy in decreasing ACL risk factors in female adolescent soccer players: a randomized controlled trial. Clin J Sport Med 24(2):134–141

Cochrane JL et al (2010) Training affects knee kinematics and kinetics in cutting maneuvers in sport. Med Sci Sports Exerc 42(8):1535–1544

Deitch JR et al (2006) Injury risk in professional basketball players: a comparison of Women's National Basketball Association and National Basketball Association athletes. Am J Sports Med 34(7):1077–1083

Ford KR, Myer GD, Hewett TE (2003) Valgus knee motion during landing in high school female and male basketball players. Med Sci Sports Exerc 35(10):1745–1750

Griffin LY et al (2000) Noncontact anterior cruciate ligament injuries: risk factors and prevention strategies. J Am Acad Orthop Surg 8(3):141–150

Heidt RS Jr et al (2000) Avoidance of soccer injuries with preseason conditioning. Am J Sports Med 28(5):659–662
Herman DC et al (2009) The effects of feedback with and without strength training on lower extremity biomechanics. Am J Sports Med 37(7):1301–1308
Hewett TE et al (1999) The effect of neuromuscular training on the incidence of knee injury in female athletes. A prospective study. Am J Sports Med 27(6):699–706
Hewett TE et al (2005) Biomechanical measures of neuromuscular control and valgus loading of the knee predict anterior cruciate ligament injury risk in female athletes: a prospective study. Am J Sports Med 33(4):492–501
Irmischer BS et al (2004) Effects of a knee ligament injury prevention exercise program on impact forces in women. J Strength Cond Res 18(4):703–707
Kristianslund E, Krosshaug T (2013) Comparison of drop jumps and sport-specific sidestep cutting: implications for anterior cruciate ligament injury risk screening. Am J Sports Med 41(3):684–688
Kristianslund E et al (2014) Sidestep cutting technique and knee abduction loading: implications for ACL prevention exercises. Br J Sports Med 48(9):779–783
Landry SC et al (2007) Neuromuscular and lower limb biomechanical differences exist between male and female elite adolescent soccer players during an unanticipated side-cut maneuver. Am J Sports Med 35(11):1888–1900
Lephart SM et al (2005) Neuromuscular and biomechanical characteristic changes in high school athletes: a plyometric versus basic resistance program. Br J Sports Med 39(12):932–938
Malinzak RA et al (2001) A comparison of knee joint motion patterns between men and women in selected athletic tasks. Clin Biomech (Bristol, Avon) 16(5):438–445
Mandelbaum BR et al (2005) Effectiveness of a neuromuscular and proprioceptive training program in preventing anterior cruciate ligament injuries in female athletes: 2-year follow-up. Am J Sports Med 33(7):1003–1010
McLean SG, Lipfert SW, van den Bogert AJ (2004) Effect of gender and defensive opponent on the biomechanics of sidestep cutting. Med Sci Sports Exerc 36(6):1008–1016
Myer GD et al (2010) Clinical correlates to laboratory measures for use in non-contact anterior cruciate ligament injury risk prediction algorithm. Clin Biomech (Bristol, Avon) 25(7):693–699
Myer GD, Ford KR, Hewett TE (2011a) New method to identify athletes at high risk of ACL injury using clinic-based measurements and freeware computer analysis. Br J Sports Med 45(4):238–244
Myer GD et al (2011b) Biomechanics laboratory-based prediction algorithm to identify female athletes with high knee loads that increase risk of ACL injury. Br J Sports Med 45(4):245–252
Myklebust G et al (1998) A prospective cohort study of anterior cruciate ligament injuries in elite Norwegian team handball. Scand J Med Sci Sports 8(3):149–153
Myklebust G et al (2003) Prevention of anterior cruciate ligament injuries in female team handball players: a prospective intervention study over three seasons. Clin J Sport Med 13(2):71–78
Olsen OE et al (2005) Exercises to prevent lower limb injuries in youth sports: cluster randomised controlled trial. BMJ 330(7489):449
Onate JA et al (2005) Instruction of jump-landing technique using videotape feedback: altering lower extremity motion patterns. Am J Sports Med 33(6):831–842
Petersen W et al (2005) A controlled prospective case control study of a prevention training program in female team handball players: the German experience. Arch Orthop Trauma Surg 125(9):614–621
Pfeiffer RP et al (2006) Lack of effect of a knee ligament injury prevention program on the incidence of noncontact anterior cruciate ligament injury. J Bone Joint Surg Am 88(8):1769–1774
Pollard CD et al (2006) The influence of in-season injury prevention training on lower-extremity kinematics during landing in female soccer players. Clin J Sport Med 16(3):223–227

Prapavessis H, McNair PJ (1999) Effects of instruction in jumping technique and experience jumping on ground reaction forces. J Orthop Sports Phys Ther 29(6):352–356
Sigward SM, Powers CM (2006) The influence of gender on knee kinematics, kinetics and muscle activation patterns during side-step cutting. Clin Biomech (Bristol, Avon) 21(1):41–48
Soderman K et al (2000) Balance board training: prevention of traumatic injuries of the lower extremities in female soccer players? A prospective randomized intervention study. Knee Surg Sports Traumatol Arthrosc 8(6):356–363
van Beijsterveldt AM et al (2012) Effectiveness of an injury prevention programme for adult male amateur soccer players: a cluster-randomised controlled trial. Br J Sports Med 46(16):1114–1118

Chapter 13
Risk Factor Analysis of Female Soccer Tournament Players

Shogo Sasaki, Satoshi Kaneko, Takuma Kobayashi, and Toru Fukubayashi

Abstract In this study, a 2-dimensional (2-D) video analysis of anterior cruciate ligament injury events from the sagittal and coronal planes was introduced. Using a sagittal analysis, we analyzed the landing action after heading when no injury occurred and checked variables related to anterior cruciate ligament injury. We found that the trunk angle was similar when injury or no injury occurred, and the hip angle and the distance from the center of mass to the base of support (COM_BOS) were smaller in both injured and control participants compared to those reported in previous studies. Future video studies are planned to analyze not only injury but also non-injury events to develop a screening tool for risk factors.

Keywords Anterior cruciate ligament injury • Video analysis • Screening • Landing • Heading

13.1 Introduction

The Fédération Internationale de Football Association (FIFA) reports that in female soccer players about two-third of all injuries are to the leg. Almost all studies on injury amongst female soccer players stress the high rate of knee injuries, particularly anterior cruciate ligament (ACL) tears. A total of 387 injuries during 174 women's

S. Sasaki (✉)
Faculty of Health Sciences, Tokyo Ariake University of Medical and Health Sciences, Tokyo, Japan
e-mail: sasaki@tau.ac.jp

S. Kaneko
Graduated School of Sport Sciences, Waseda University, Saitama, Japan
e-mail: kaneko.3104@gmail.com

T. Kobayashi
Department of Orthopedic Surgery, Sapporo Medical University, Sapporo, Japan
e-mail: a98m040@yahoo.co.jp

T. Fukubayashi
Faculty of Sport Sciences, Waseda University, Saitama, Japan
e-mail: fukubayashi@waseda.jp

K. Kanosue et al. (eds.), *Sports Injuries and Prevention*,
DOI 10.1007/978-4-431-55318-2_13

soccer matches at 7 international tournaments were reported, which is an average of 2.2 injuries per match (Junge and Dvorak 2007).

ACL injuries in soccer are more frequent in women than in men (Arendt and Dick 1995; Harmon and Dick 1998) and the rate of ACL injuries is three to five times higher for girls than boys (Bjordal et al. 1997; Powell and Barber-Foss 2000). The incidence of match injuries was four to six times higher than the incidence of injuries occurring during training sessions (Junge and Dvorak 2004). One study (Faude et al. 2005) reported that all ACL ruptures occurred during match play in the German national league during the 2003–2004 season. This indicates that identifying risk factors during soccer games will provide the information needed for ACL injury prevention in female players.

Video analysis is ideal for obtaining athlete's actions and behaviors in game situations. Video footage of competition and training provides a lot of information on player performance (Carling et al. 2009). Previous video analyses of ACL injury events have used visual inspection (Olsen et al. 2004; Cochrane et al. 2007; Krosshaug et al. 2007a). However, subjective and qualitative evaluation of human motion is not ideal, because simple visual inspection lacks accuracy and precision (Krosshaug et al. 2007b). To overcome these difficulties, a model-based image-matching (MBIM) technique was developed (Krosshaug and Bahr 2005), and has been used to investigate the mechanisms underlying non-contact ACL injuries (Koga et al. 2010, 2011; Bere et al. 2013). Even though the MBIM technique determines the precision kinematics of human motion, it is quite time consuming (Krosshaug et al. 2007c; Koga et al. 2010; Sasaki et al. 2013). Therefore, it is not a practical method to use for screening risk factors in a large populations of players.

A recent video analysis used a 2-dimensional (2-D) software measurement tool to define the biomechanics of joint angle or position at the time of injury (Boden et al. 2009; Hewett et al. 2009; Sheehan et al. 2012). This 2-D video analysis provides more quantitative information than simple visual inspection and this assessment procedure requires less time and effort as compared with the MBIM method. Therefore, video analysis using 2-D software may offer a screening tool for identifying athletes at risk for sustaining non-contact ACL injuries (Sheehan et al. 2012).

In this article we review the literature on the use of 2-D video to analyze ACL injury events, and we suggest an approach that would facilitate its use as a screening tool.

13.2 2-D Video Analysis of ACL Injury Events

Three studies using 2-D video analysis to investigate ACL injury events have been published (Table 13.1). These studies all determined the kinematics and characteristics at the time of ACL injury, and compared these results with those of uninjured athletes as a control group. The mechanism of non-contact ACL injuries as observed on video had several common components (Olsen et al. 2004; Cochrane et al. 2007; Krosshaug et al. 2007a). The 2-D video analyses using software measurement tools were conducted from both sagittal and coronal planes.

Table 13.1 Video analysis studies using the 2-D software measurement tool during ACL injury events

Study	Video (number)	Subject and sports played (number)	Measurement variables (main results)	Results
Boden et al.	Injury (29)	Injury: basketball (14), handball (7), soccer (3), football (3), cheerleading (1), gymnastics (1)	Sagittal: foot position, ankle angle, knee angle, hip angle	Less planter flexion ankle angle and larger hip flexion angle in injured athletes.
	Control (27)	Control: basketball (27)	Coronal: knee angle, hip angle	
Hewett et al.	Injury (23)	All basketball	Coronal: trunk angle, knee angle	Greater lateral trunk and knee abduction angle in injured female athletes.
	Control (28)			
Sheehan et al.	Injury (20)	Injury: basketball (10), soccer (2), football (5), handball (3)	Sagittal: trunk angle, limb angle, COM_BOS[a]	Greater limb angle and COM_BOS, smaller trunk angle in injured athletes.
	Control (20)	Control: basketball (10), soccer (2), football (5), handball (3)		

[a]COM_BOS indicates the distance from COM (center of mass) to BOS (base of support)

13.2.1 Sagittal View Analysis During ACL Injury

Table 13.1 shows studies (Boden et al. 2009; Sheehan et al. 2012) that used sagittal view analysis to make several important observations. Interestingly, the initial contact with the ground for injured athletes was with the hindfoot or the entire foot, whereas in controls it was with the forefoot or a combination of forefoot and midfoot (Boden et al. 2009). The ankle angle at initial contact was significantly less plantar flexed in the injured group (10.7 ± 9.6 degrees [deg]) than in the control group (22.9 ± 10.1 deg). In addition, the ankle angle of injured athletes changed only a little from the initial contact to the next four frames (120 ms post contact) of the video when compared with a greater change in the uninjured controls. Boden et al. (2009) suggested that the calf musculature may be unable to absorb the ground-reaction forces adequately, following which forces are transmitted directly to the knee in injured athletes because they land flatfooted or with initial hindfoot contact.

Although less knee flexion was observed on recordings involving ACL injury events (Ireland 2002; Krosshaug et al. 2007a), the difference in knee flexion angle between injured and control groups in 2-D video analysis was not statistically significant (Boden et al. 2009). This is a methodological limitation of 2-D video analysis and Boden et al. (2009) stated that the 2-D measurement technique is not sufficiently sensitive to distinguish small differences in joint angle, and knee valgus and rotation may simulate flexion in all subjects. However, the knee was significantly less flexed in the injured athletes at the first frame when the foot was 100 % flat on the ground (average for injured athletes was 17.6 deg and for controls was 39.3 deg), thus causing abnormal force on the knee. This is because the foot becomes completely flat on the ground quicker in injured athletes.

Previous studies investigating hip movement from the sagittal view used two different methods. Boden et al. (2009) defined the angle between a line from the superior tip of the acromioclavicular joint to the superior tip of the greater trochanter and a line from the superior tip of the greater trochanter to the midpoint of the lateral knee at the joint line to define the hip angle. On the other hand, Sheehan et al. (2012) defined the angle between the vertical and the thigh (represented as a line from the center of the knee joint to the center of hip joint) as the limb angle. At initial contact with the ground, Boden et al. (2009) showed that the hip angle was significantly more flexed in the injured athletes (50.1 ± 13.2 deg) than in controls (25.8 ± 14.7 deg), and that the limb angle in the study by Sheehan et al. (2012) was greater in the ACL-injured athletes (48 ± 12 deg) than in controls (31 ± 22 deg). Moreover, Sheehan et al. (2012) reported that the trunk angle, defined as the angle between vertical and the centerline of the trunk, was lower in the injured group (4 ± 14 deg) that in controls (16 ± 13 deg). These hip and trunk positions at initial contact with the ground are important components of the mechanism of non-contact ACL injury.

Sheehan et al. (2012) further evaluated landing by measuring the body's center of mass (COM) to the base of support (BOS). The distance from the COM to the BOS (COM_BOS) was greater in injured athletes (1.5 ± 0.5 pix/pix) than in controls (0.7 ± 0.7 pix/pix) at initial contact. Discriminant analysis of these data showed that no controls landed with the COM_BOS greater than 1.6, and no injuries in female occurred when athletes had a COM_BOS <1.2. In the COM_BOS range of 1.2–1.6, both provocative and safe landing occurred.

13.2.2 Coronal View Analysis During ACL Injury

Table 13.1 shows results from studies involving a coronal view analysis (Boden et al. 2009; Hewett et al. 2009). Boden et al. (2009) reported no significant difference in knee abduction angle at initial contact between injured (29.9 ± 11.0 deg) and control groups (25.7 ± 12.7 deg). Following the time course, however, significant difference in knee abduction angle between injured and control athletes appeared at the third frame after the initial contact (60 ms post contact) (Boden et al. 2009) and 100 ms post contact (Hewett et al. 2009). Meanwhile, no significant differences in hip abduction angle between injured and control athletes were found in any frame (Boden et al. 2009).

Dempsey et al. (2007) investigated how trunk position in the coronal plane changed knee loading. The mean lateral trunk angle relative to vertical was found to be higher in female than male players during ACL injury, and trended toward being greater than female controls at initial contact (Hewett et al. 2009). Hewett et al. (2009) suggested that trunk position and knee load may be mechanically linked, as lateral positioning of the trunk can create abduction load at the knee.

13.3 2-D Video Analysis for Screening ACL Injury Risk; Approach to Soccer Specific Landing

Previous video analysis studies have collected data on ACL injury events from several sports. These studies provide useful information on the multi-risk component for ACL injury and for the development of an ACL injury prevention program. Therefore, some common risk factors will be included along with the soccer specific movements. Faunø and Wulff (2006) reported that landing after jumping for a header was one of the most frequent actions that lead to ACL injury. By observing this specific action using 2-D video analysis, we will be able to identify the aspects of landing behavior that result in the greatest risk for ACL injury. This new approach has the potential to screen large number of players or actions, and help minimize ACL injuries in soccer. Thd following pilot study analyzed the landing action after players jumped for a header. We screened the variables of uninjured landing events using 2-D video analysis.

13.3.1 Data Collection

We recorded a soccer game in the Kanto Women Football League 2012 using four digital video cameras (Sony HDR-CX590V, Japan) with frame rates of 60 Hz. The cameras were located around the soccer field and video recorded in AVCHD format and high definition (1080i).

Three experts in the field of soccer biomechanics and sports medicine formed the analysis team. The analysis was organized into two stages to identify events for analysis. First, one analyst (S.I) identified single-leg landing events for one of the teams from a recorded video sequence in 2× slow motion. A total of 27 landing events were selected for further analysis. In stage 2, the other two analysts (S.S, S.K) reviewed single-leg landing events after the player jumped for a header and there was a skirmish with another player in mid-air. Criteria for inclusion of a particular event were: (1) a good quality video recording, with the camera angle approximating a sagittal view of the athlete; (2) good visibility of the foot contacting the ground; (3) an unobscured view of the athletes. Events were excluded if the athletes landed without a skirmish with another player. Twelve landing events met the study criteria for inclusion in stage 2.

13.3.2 Video Editing and Analysis

The video recordings were edited using Edius Neo (version 3.5, Grass valley llc, USA). For each event, the frame in which the foot initially contacted the ground was captured and stored in TIFF files for analysis. Image-J (National Institutes of Health, USA) was used for all analyses. For consistency, a single author (S.S) performed all measurements.

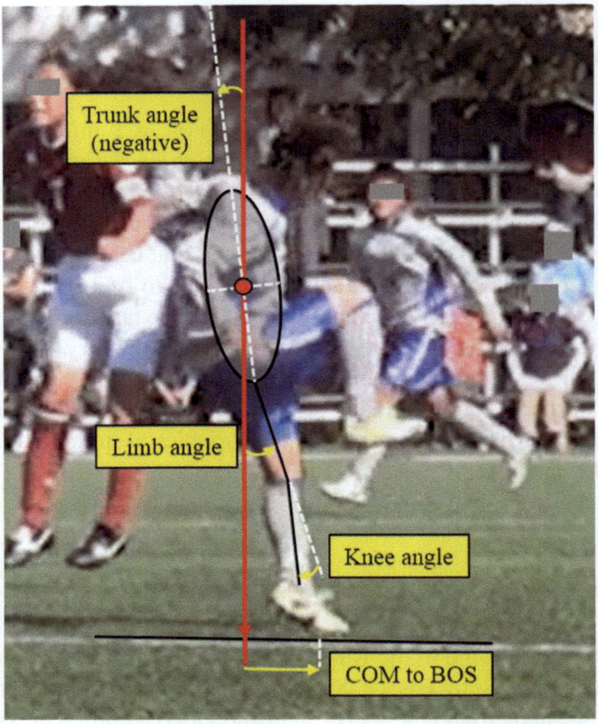

Fig. 13.1 Analysis of landing on the ground after jumping for a header

The trunk angle, limb angle, and the COM_BOS were measured in this study (Fig. 13.1). The trunk angle was defined as the angle from the vertical to the centerline of the trunk, and the limb angle was defined as the angle between the vertical line and the centerline of the thigh. A positive trunk and/or limb angle indicated that the trunk and/or limb were rotated anteriorly relative to the vertical. The COM was defined as the center of an ellipse delineating the athlete's trunk and the BOS was defined as the point bisecting the line of contact between the shoe and the floor at initial contact. The COM_BOS (in pixel) was taken along the anterior/posterior direction and normalized by femur length (in pixel). A more detailed description of the method used has been published (Sheehan et al. 2012).

13.3.3 Statistical Analysis

All the data are expressed as mean ± standard deviation. The relationship among the measurement variables was determined using Pearson's product-moment correlation coefficient. All statistical procedures were performed using PASW Statistics (ver. 18.0 for Windows), and the statistical significance for all tests was set at $p < 0.05$.

Table 13.2 The trunk, limb, and knee angles and the COM_BOS for landing after heading

	Mean ± SD	Range
Trunk angle (deg)	5 ± 10	−15 to −21
Limb angle (deg)	12 ± 8	−1 to −27
COM_BOS/femur (pix/pix)	0.3 ± 0.4	−0.5 to −0.9

Table 13.3 Correlations among each variable for landing after heading

	Trunk angle	Limb angle	COM_BOS
Trunk angle	1.00		
Limb angle	0.07	1.00	
COM_BOS	−0.38	0.81[a]	1.00

[a]$p < 0.05$

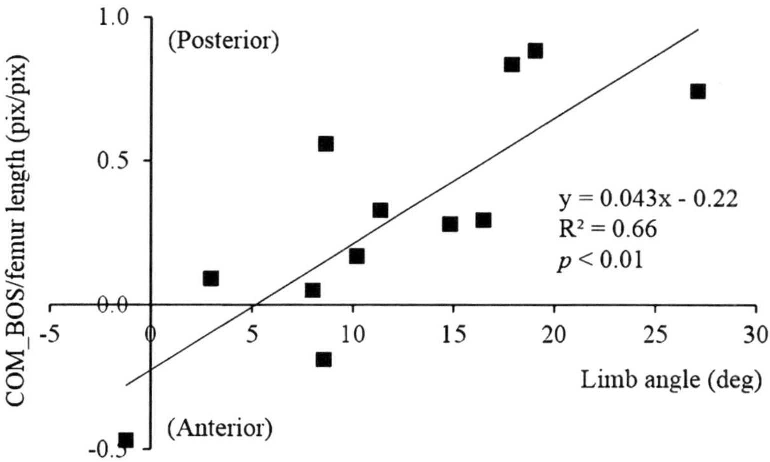

Fig. 13.2 Relationships between the limb angle and the COM_BOS at the time of initial contact

13.3.4 Results

Table 13.2 presents the trunk, limb angle, and the distance from COM to BOS at initial contact. Table 13.3 shows the Pearson's product-moment correlation coefficient for all variables of landing. The limb angle was found to be strongly and positively correlated with COM_BOS at the time of initial contact ($r = 0.81$, $p < 0.01$, Fig. 13.2).

13.3.5 Discussion

In this pilot study, we used a 2-D video analysis of soccer players to screen for landing events related to ACL injury. From a sagittal view analysis, we checked various safe single-leg landing events after a player jumped for a header in a soccer game in order to compare them to the landing events leading to ACL injury. Landing maneuvers differ from one sport to another because certain characteristics of the sport affect the landing movement used by the players (e.g., landing after rebounding in basketball, jumping to shoot in handball, spiking in volleyball, jumping for a smash in badminton). By considering the risk factors for each sport and analyzing the movements related to ACL injury, we hope to develop injury prevention procedures for each sport and possibly develop a broader understanding of all ACL injuries.

Previous 2-D video studies from sagittal and coronal planes have identified risk factors for ACL injury and treatment options. In their 2012 study, Sheehan et al. (2012) suggested that these parameters may help screen athletes at risk. Therefore, in this pilot study we aimed to determine the risk of injury when a player landed after a header by recording and analyzing the landing maneuvers during a woman's soccer game. Future studies will analyze not only injury events but also non-injury events in order to screen for risk factors during a game.

In our study, the average trunk angle (Table 13.2) was similar to that of injured participants but smaller compared to that of control players in a previous study (Sheehan et al. 2012). Soccer players need to maintain body balance when jumping for a header and to land with their trunk perpendicular to the ground. This posture may interrupt the soft landing with hip flexion. Male soccer players created greater hip extension torque during an experimental dynamic jump heading task, which may be a consequence of greater trunk flexion (Butler et al. 2013). The perpendicular trunk position is a feature of the landing posture after jumping for a header by female soccer players.

In this study, the hip angle and the COM_BOS (12 ± 8 deg and 0.3 ± 0.4, respectively; Table 13.2) were less for both injured and control participants compared with the measurements obtained in previous research (Sheehan et al. 2012). The maximum value for the limb angle was 27 deg and 0.9 for the COM_BOS, which indicates that these maneuvers resulted in a safe landing. ACL injury did not occur if COM_BOS <1.2 for female landing, which is close to the distance between the base of support and the trunk (Sheehan et al. 2012). Boden et al. (2009) showed that having a more flexed hip will result in a more provocative landing position. Although most landing events after a header with a skirmish with another player were safe landings, these variables should be checked during screening.

The limb angle positively correlated with COM_BOS ($r = 0.81$, $p < 0.01$) (Fig. 13.2), however several other variables did not correlate with each other (Table 13.3). The positive correlation was in agreement with that of a previous study of discriminant analysis of risk factors for ACL injury (Sheehan et al. 2012). This suggests that the limb angle is greater at landing, thus distancing the COM further from the BOS. Therefore, for injury prevention in woman soccer players a smaller limb angle may be critical when players land on one leg after jumping for a header.

Several limitations must be kept in mind when interpreting the results of this report. Most importantly, this is a pilot study and the sample size was small. We have to record more women's soccer games and screen the landing events. Technical limitations of the analysis system we utilized were discussed in previous studies (Boden et al. 2009; Hewett et al. 2009; Sheehan et al. 2012) and included difficulties with identifying anatomical landmarks in players with clothes on and variations in camera angles that may not have captured a perfect sagittal plane. However, 2-D video analysis is useful and less time consuming than some of the other techniques. We predict that analysis of events during games can help identify an efficient tool for screening players at risk, and thus prevent ACL injury.

13.4 Conclusion

The literature on the use of sagittal view 2-D video analysis of ACL injured athletes shows some abnormalities in lower limb kinematics and trunk location. Athletes who sustain an injury tend to land on their hindfoot or the entire foot, and have increased hip flexion, smaller trunk angle, and COM that is posterior to the BOS. In the analysis utilizing the coronal view, lateral trunk angle relative to the vertical was greater in female athletes who suffered an ACL injury. Thus, these variables may be risk factors for ACL injury during games. As a next step, we attempted to establish a screen to detect injury risk for players from uninjured events during soccer games. Almost all landing of players after jumping for a header during soccer were safe, and the limb angle at initial contact positively correlated with the COM_BOS. These correlations may be critical for injury prevention when female players land on one leg after jumping for a header.

Acknowledgement We thank Dr. Yasuharu Nagano (Niigata University of Health and Welfare) and Mr. Hirohisa Magoshi (Waseda University) for their advice and debate. In addition, we acknowledge the Kanto Gakuen University Female Soccer Club for participating in this research.

References

Arendt E, Dick R (1995) Knee injury patterns among men and women in collegiate basketball and soccer. NCAA data and review of literature. Am J Sports Med 23(6):694–701

Bere T, Mok KM, Koga H, Krosshaug T, Nordsletten L, Bahr R (2013) Kinematics of anterior cruciate ligament ruptures in World Cup alpine skiing: 2 case reports of the slip-catch mechanism. Am J Sports Med 41(5):1067–1073

Bjordal JM, Arnły F, Hannestad B, Strand T (1997) Epidemiology of anterior cruciate ligament injuries in soccer. Am J Sports Med 25(3):341–345

Boden BP, Torg JS, Knowles SB, Hewett TE (2009) Video analysis of anterior cruciate ligament injury: abnormalities in hip and ankle kinematics. Am J Sports Med 37(2):252–259

Butler RJ, Willson JD, Fowler D, Queen RM (2013) Gender differences in landing mechanics vary depending on the type of landing. Clin J Sport Med 23(1):52–57

Carling C, Reilly T, Williams MA (2009) Performance assessment for field sports. Routledge, Oxon

Cochrane JL, Lloyd DG, Buttfield A, Seward H, McGivern J (2007) Characteristics of anterior cruciate ligament injuries in Australian football. J Sci Med Sport 10(2):96–104
Dempsey AR, Lloyd DG, Elliott BC, Steele JR, Munro BJ, Russo KA (2007) The effect of technique change on knee loads during sidestep cutting. Med Sci Sports Exerc 39(10):1765–1773
Faude O, Junge A, Kindermann W, Dvorak J (2005) Injuries in female soccer players: a prospective study in the German national league. Am J Sports Med 33(11):1694–1700
Faunø P, Wulff Jakobsen B (2006) Mechanism of anterior cruciate ligament injuries in soccer. Int J Sports Med 27(1):75–79
Fédération Internationale de Football Association. Injuries to woman-when and why you are risk. Health and fitness for the female football player – a guide for players and coaches. http://www.fifa.com/mm/document/footballdevelopment/medical/59/78/19/ffb_gesamt_e_20035.pdf. Accessed 12 December 2013
Harmon KG, Dick R (1998) The relationship of skill level to anterior cruciate ligament injury. Clin J Sport Med 8(4):260–265
Hewett TE, Torg JS, Boden BP (2009) Video analysis of trunk and knee motion during non-contact anterior cruciate ligament injury in female athletes: lateral trunk and knee abduction motion are combined components of the injury mechanism. Br J Sports Med 43(6):417–422
Ireland ML (2002) The female ACL: why is it more prone to injury? Orthop Clin North Am 33(4):637–651
Junge A, Dvorak J (2004) Soccer injuries: a review on incidence and prevention. Sports Med 34(13):929–938
Junge A, Dvorak J (2007) Injuries in female football players in top-level international tournaments. Br J Sports Med 41(Suppl 1):i3–i7
Koga H, Nakamae A, Shima Y, Iwasa J, Myklebust G, Engebretsen L, Bahr R, Krosshaug T (2010) Mechanisms for noncontact anterior cruciate ligament injuries: knee joint kinematics in 10 injury situations from female team handball and basketball. Am J Sports Med 38(11):2218–2225
Koga H, Bahr R, Myklebust G, Engebretsen L, Grund T, Krosshaug T (2011) Estimating anterior tibial translation from model-based image-matching of a noncontact anterior cruciate ligament injury in professional football: a case report. Clin J Sport Med 21(3):271–274
Krosshaug T, Bahr R (2005) A model-based image-matching technique for three-dimensional reconstruction of human motion from uncalibrated video sequences. J Biomech 38(4):919–929
Krosshaug T, Nakamae A, Boden BP, Engebretsen L, Smith G, Slauterbeck JR, Hewett TE, Bahr R (2007a) Mechanisms of anterior cruciate ligament injury in basketball: video analysis of 39 cases. Am J Sports Med 35(3):359–367
Krosshaug T, Nakamae A, Boden B, Engebretsen L, Smith G, Slauterbeck J, Hewett TE, Bahr R (2007b) Estimating 3D joint kinematics from video sequences of running and cutting maneuvers–assessing the accuracy of simple visual inspection. Gait Posture 26(3):378–385
Krosshaug T, Slauterbeck JR, Engebretsen L, Bahr R (2007c) Biomechanical analysis of anterior cruciate ligament injury mechanisms: three-dimensional motion reconstruction from video sequences. Scand J Med Sci Sports 17(5):508–519
Olsen OE, Myklebust G, Engebretsen L, Bahr R (2004) Injury mechanisms for anterior cruciate ligament injuries in team handball: a systematic video analysis. Am J Sports Med 32(4):1002–1012
Powell JW, Barber-Foss KD (2000) Sex-related injury patterns among selected high school sports. Am J Sports Med 28(3):385–391
Sasaki S, Koga H, Krosshaug T, Sakurai T, Fukubayashi T (2013) A biomechanical approach to evaluating field sports performance using a model-based image-matching technique: a case report of 4 defensive football game situations. Gazz Med Ital 172(10):799–805
Sheehan FT, Sipprell WH 3rd, Boden BP (2012) Dynamic sagittal plane trunk control during anterior cruciate ligament injury. Am J Sports Med 40(5):1068–1074

Chapter 14
Prevention of Anterior Cruciate Ligament (ACL) Injury

Dai Sugimoto and Gregory D. Myer

Abstract Along with the increase in number of female athletic participants, the number of resultant athletic injuries among female athletes has also rapidly increased. Anterior cruciate ligament (ACL) injury is more prevalent in physically active females as compared to their male counterparts in the sports of basketball and soccer. ACL injury was classified into three categorizations based on three common pathomechanics, and the relative distribution of ACL injury was analyzed between sexes. The most common mechanism of ACL injury was noncontact in nature, and such noncontact ACL injury was more prevalent in female athletes as compared to male athletes. Because a majority of ACL injuries occur in noncontact mechanism, especially in female athletes, a few clinical trials were conducted to examine the effectiveness of preventive neuromuscular training on ACL injury reduction in the female population. Recent meta-analytic reports have confirmed significant ACL injury reduction by female athletes who performed preventive neuromuscular training compared to those who did not or only maintained their routine warm-up. Numerous evidence supports the proposition that specific exercises incorporated into preventive neuromuscular training protocols lead to favorable biomechanical alteration in dynamic movements, which likely contribute to ACL injury reduction in female athletes. Future directions of ACL injury research include the

D. Sugimoto, PhD, ATC, CSCS (✉)
The Micheli Center for Sports Injury Prevention, Waltham, MA, USA

Division of Sports Medicine, Boston Children's Hospital, Boston, MA, USA
e-mail: dai.sugimoto@childrens.harvard.edu

G.D. Myer, PhD, FACSM, CSCS*D
Cincinnati Children's Hospital Medical Center, Cincinnati, OH, USA

Sports Medicine Biodynamics Center and Human Performance Laboratory, Cincinnati, OH, USA

Department of Pediatrics, College of Medicine, University of Cincinnati, Cincinnati, OH, USA

Athletic Training Division, School of Allied Medical Professions, The Ohio State University, Columbus, OH, USA

The Micheli Center for Sports Injury Prevention, Waltham, MA, USA
e-mail: greg.myer@cchmc.org

K. Kanosue et al. (eds.), *Sports Injuries and Prevention*,
DOI 10.1007/978-4-431-55318-2_14

development of an effective screening system in order to identify at-risk athletes. Also, a strategy to implement preventive neuromuscular training programs and to enhance compliance and adherence needs to be developed.

Keywords ACL Prevention • ACL pathomechanics • ACL prevention clinical trials • Neuromuscular training • Clinical feasibility

14.1 ACL Injuries – Why We Need to Prevent Them

14.1.1 Historical Background

Since inception of Title IX of The Educational Assistance Act in 1972, the number of female athletes at the high school level has increased more than 11 fold (from 0.3 to 3.2 million); whereas male athlete participation has shown a much smaller increase from 3.7 to 4.5 million in the United States (National Federation of State High School Association 2012). Along with the increase in number of female athletic participants, the number of resultant athletic injuries among female athletes has also rapidly increased, and exposed potential sex specific risk factors.

Among many types of debilitating injuries, female athletes have shown particular susceptibility to knee joint injury, especially anterior cruciate ligament (ACL) injury. In high school basketball, it was reported that one in every 81 female athletes sustained an ACL injury (Messina et al. 1999; Gomez et al. 1996) and the total number of ACL injuries among female athletes was estimated to be approximately 38,000 annually in the United States as of 2001. In recent reports, the estimated number of ACL injuries ranged from 80,000 to 350,000 annually in the US (Cimino et al. 2010; Wojtys and Brower 2010). Epidemiological studies have demonstrated that during landing and cutting/pivoting sports, ACL injury occurs at a 4–6 fold greater incidence in females when compared to males in the sports of basketball and soccer (Arendt and Dick 1995).

14.1.2 Physical Consequences of ACL Injury

The ACL plays an integral role in knee joint stability. When the ACL is ruptured, an individual often experiences various functional difficulties including inability to decelerate, cut and pivot, along with the presence of pain and effusion in the knee joint. (Cimino et al. 2010) Because of these functional difficulties, most individuals with an ACL injury are forced to limit their physical activity level, which may lead to restricted participation in athletics. At the collegiate level, ACL injury results in the greatest loss of time from athletic participation when compared to ankle and traumatic head injuries (Hootman et al. 2007). In addition, once an athlete sustains a primary ACL injury, they may have up to a 24–29 % increased risk for a subsequent ACL injury if they return back to the sport. (Webster et al. 2014)

In order to restore knee joint stability and functionality, most patients choose to have ACL reconstruction surgery. However, ACL reconstruction surgery is not a solution that resolves all deficits or prevents long term morbidity. After ACL reconstruction surgery, return to play rate is approximately 65 % (McCullough et al. 2012), and female athletes are even less likely to return to previous athletic activities (Ardern et al. 2011). One study found 12 % of ACL reconstructed individuals experienced a subsequent ACL injury on their ipsilateral (reconstructed/involved) or contralateral (uninvolved) ACL within 5 years (Salmon et al. 2005). Another study also reported 24 % of ACL reconstructed high school age athletes who had an ACL reconstruction experienced a subsequent ACL injury within 1 year from return to play (Paterno et al. 2010). In short, the low rate of return to play and high risk for the second ACL injury are major concerns.

Over the long term, a high prevalence of pre-mature osteoarthritis (OA) has been reported 10–15 years after ACL reconstruction surgery (Oiestad et al. 2010, 2011). Several studies concluded that 50 % of those who had an ACL reconstruction surgery were likely to develop OA within 10 years (Maffulli et al. 2010; Lohmander et al. 2007). In addition, a longitudinal study that followed female soccer players with a reconstructed ACL reported a negative knee-related quality of life along with the OA development (Lohmander et al. 2004). The mean age of this cohort was 31 (range between 26 and 40) years old (Lohmander et al. 2004). At this point, no effective OA treatment has been introduced.

14.2 Common ACL Injury Pathomechanics and the Relative Distribution

14.2.1 Introduction

Because of the sudden impact and continuous negative health consequences of ACL injury, especially for the female population, a number of individuals and research teams have made a commitment to understanding the mechanism underlying the occurrence of ACL injuries. From their reports, the basis for several common ACL injury mechanisms have emerged. Identifying the mechanism of ACL injury is important because it provides a pathway in the development of an effective preventive intervention strategy. Hence, this chapter focuses on identifying common pathomechanics of ACL injury as well as providing distribution of the ACL injury based on the pathomechanics.

14.2.2 Pathomechanics Classifications

At a recent ACL research retreat (Research Retreat 2008) agreement was reached on a pathomechanic definition introduced by Hewett and colleagues (2007a). According to this definition, ACL injuries can be classified into three different schemes based on the type of contact mechanisms: Type I: Direct contact (an external force was directly applied to the injured knee and which is probably the proximate cause of injury), Type II: Indirect contact (an external force was applied to the athlete but not directly to the injured knee. The force is involved in the injury process but is probably not the proximate cause), and Type III: Noncontact (forces applied to the knee at the time of injury result from the athlete's own movements and do not involve contact with another athlete or object). Previously published studies that investigated the pathomechanics of ACL injuries have typically utilized the above classification schemes (Arendt and Dick 1995; Agel et al. 2005; Boden et al. 2000; Myklebust et al. 1997; Walden et al. 2011). However, they have not made a distinction between direct and indirect contact ACL injuries (Arendt and Dick 1995; Agel et al. 2005; Boden et al. 2000; Myklebust et al. 1997; Walden et al. 2011). Rather, the two mechanisms (direct and indirect mechanisms) were combined and expressed as a contact ACL injury (Arendt and Dick 1995; Agel et al. 2005; Boden et al. 2000; Myklebust et al. 1997; Walden et al. 2011). In research focused on injury prevention, it is critical to differentiate between contact and noncontact mechanisms in order to adopt the most effective design strategy.

14.2.3 Direct Contact Mechanism

A contact ACL injury is defined as an external force application to the injured knee, which is likely to be the proximate cause of injury (Hewett et al. 2007a). Several studies reported that external force application is likely to come from other players rather than balls, goals, or floors (Agel et al. 2007; Dick et al. 2007). A recent descriptive laboratory study reported that an external force that induces knee abduction and anterior tibial translation increases ACL strain by 4.6 fold (Quatman et al. 2011). Also, anterior tibial translation with internal and external rotations increases ACL strain by 3.9 fold and 3.7 fold, respectively (Quatman et al. 2011). Therefore, direct external force applications to the knee joint, which cause knee abduction, and anterior tibial translation with internal and external rotations, likely elevate the risk of tearing the ACL.

14.2.4 Indirect Contact Mechanism

In an indirect injury mechanism, an external force is applied to body parts besides the knee joint and results in an ACL tear. For instance, an external force applied to the trunk can create a loss of control over the trunk, which may generate an excessive ground reaction force (GRF) because of the asymmetrical dynamic motions. Since the produced GRF runs laterally approximately to the center of the femoral head, an external hip abductor torque is created. In order to compensate for the torque, the hip adductor needs to be activated, which is likely needed to bring the knee joint into a more abducted or valgus position (Hewett and Myer 2011).

Another potential indirect contact ACL injury mechanism involves a contact force on the thigh. During cutting, pivoting, and landing activities, if external force is applied to the lateral side of the thigh with the foot fixed, the knee is forced toward an abducted direction, which can result in an ACL tear. Similarly, if the external force comes from the anterior side of the thigh, the knee is forced into hyperextention. A hyperextended knee with a combination of anterior tibial translation and rotation (Quatman et al. 2011) is likely to cause a strain on the ACL, which may result in an ACL tear.

14.2.5 Noncontact Mechanism

Numerous studies have sought the underlying mechanism and risk factors for noncontact ACL injuries. Using a prospective cohort study design, Hewett and colleagues discovered several biomechanical differences between high school female athletes who suffered noncontact ACL injuries and those who did not (Hewett et al. 2005). The observed biomechanical risk factors were: elevated knee abduction moment, limited knee flexion, excessive GRF, and an asymmetrical landing pattern in the drop phase of a vertical jump. In addition to the knee kinematic differences, recent studies reported an influence of trunk neuromuscular control on knee injury (Hewett et al. 2009; Zazulak et al. 2007). A 3 year prospective study that initially aimed to find a trunk neuromuscular deficit involving low back pain found a link between trunk control and ACL injury (Zazulak et al. 2007). The study reported that the inability to control lateral trunk displacement accurately predicted future ACL injury. With a logistic regression model, trunk displacement, proprioception, and history of low back pain indicate future ACL injury risk with 91 % accuracy in female athletes, but not male athletes. Additionally, using National Basketball Association (NBA) and Women's National Basketball Association (WNBA) players, a video analysis study reported that lateral trunk flexion and knee abduction angles were greater in cutting and landing tasks in female players who suffered ACL injuries as compared to male and control female players (Hewett et al. 2009).

Available evidence indicates that ACL rupture is a resultant product of kinetic (neuromuscular) and kinematic (biomechanical) events that occur during dynamic human movements. The following provides a potential explanation of the underlying anatomical structures and how they sequentially interact in the production of a noncontact ACL injury: upper body strength differences exist between males and females (Miyaguchi and Demura 2009; Evans et al. 2007). Because of limited upper body strength, females may present more lateral displacement of the trunk in cutting, pivoting, and landing activities. When the trunk is flexed laterally without sufficient control during dynamic movements, weight can be shifted to one leg. If one leg contacts the ground before the other leg, the GRF tends to be higher than when both legs make contact at the same time, since only one leg is available to absorb the GRF. The GRF, as the result of asymmetrical movements, runs approximately lateral to the center of the femoral head, which produces external hip abductor torque.

The Newton's third law states: For every action, there is an equal and opposite reaction. Thus, in order to balance the external hip abduction torque created by the lateral GRF, the hip adductor is activated. The activated hip adductor musculature brings the thigh an inward direction and potentially contributes to hip adduction and internal rotation of the femur. The increased hip adduction angles create increased knee abduction angles in weight-bearing activities (Ford et al. 2006). As a result, the ACL is strained and occasionally torn if the external knee abduction moment is beyond the ACL's yielding point.

14.2.6 Distribution of ACL Injuries Based on a Pathomechanical Classifications

Arendt and Dick reviewed 5 years of epidemiological data from the NCAA men's and women's soccer and basketball athletes and found that the most frequently observed ACL injury mechanism was noncontact (Arendt and Dick 1995). They concluded that 68.7 % of all ACL injuries were noncontact in nature based, on an analysis of the injury mechanisms. However, definite sex differences exist within the mechanisms. Although 55.4 % of all ACL injuries were noncontact in nature in male athletes, 74.5 % of such ACL injuries occurred in female athletes. More specifically, the data can be organized by sex and sports. In the sport of soccer, the number of contact and noncontact ACL injuries were about the same in males (34 contact vs. 32 noncontact ACL injuries), but females demonstrated a 1.7 times higher noncontact mechanisms as compared to contact mechanisms (31 contact vs. 53 noncontact ACL injuries). Similarly, in basketball, males had 1.9 times higher noncontact mechanisms relative to contact mechanism (16 contact vs. 30 noncontact ACL injuries), whereas females suffered ACL injuries involving noncontact mechanism at 4 times the rate of ACL injuries involving contact (34 contact vs. 137 noncontact ACL injuries).

Another study that analyzed the NCAA men's and women's soccer and basketball epidemiological data for 13 years (1990–2002) generally supported the above findings (Agel et al. 2005). This study reported 65.3 % of all the ACL injuries involved the noncontact mechanism. Comparing the mechanisms based on the sexes, 56.9 % of male soccer and basketball athletes suffered noncontact ACL injuries, and 68.4 % of the female athletes who competed in the same sports experienced noncontact ACL injuries. When the data was analyzed by sex and sports, the results were very similar to the above study done by Arendt and Dick. Male soccer players suffered slightly more contact than noncontact ACL injuries (72 contact vs. 66 noncontact ACL injuries). On the other hand, female soccer players had a 1.4 times higher noncontact ACL injury rate as compared to contact ACL injuries (115 contact vs. 161 noncontact ACL injuries). In the sport of basketball, male basketball players suffered noncontact ACL injuries 2.1 times higher than contact ACL injuries (37 contact vs. 78 noncontact ACL injuries). However, female basketball players suffered noncontact ACL injuries at a rate 3.1 times higher than contact mechanism (100 contact vs. 305 noncontact ACL injuries). Comparing noncontact ACL injury rates solely based on the sex, a combination of college female soccer and basketball athletes suffered ACL injury incidents 2.7 times higher than their male counterparts (253 total ACL injuries from male soccer and basketball athletes vs. 681 total ACL injuries from female soccer and basketball athletes).

Other studies have reported comparable findings. A prospective study which analyzed elite European male and female soccer players documented that 62 % of all the observed ACL injuries were noncontact in nature (Walden et al. 2011). A study examining the type of ACL injury mechanisms in women's college basketball games reported that 64 % of all ACL injuries mechanisms occurred in noncontact situations (Agel et al. 2007). Another report in which retrospective interviews were conducted with ACL injured individuals noted that 72 % of their injuries involved no contact (Boden et al. 2000). For male and female European handball players, a study analyzed the number of ACL injuries based on the player's position (Agel et al. 2007). According to the study, 84 % of all ACL injuries come from back and wing position players and 95 % of those ACL injuries were non-contact in nature.

14.2.7 Summary of Common ACL Injury Pathomechanics and the Relative Distribution

ACL pathomechanics are classified into three schemes: direct, indirect, and noncontact mechanisms. Based on NCAA epidemiological data and other sources, 60–70 % of ACL injuries involve a noncontact mechanism. When there is no contact, ACL is likely to be strained or torn by a resultant force created by human movements, which can be triggered by a loss of trunk control. The loss of trunk control, especially lateral flexion, can lead to asymmetrical movement patterns, such as sudden single leg landing or cutting, which cause an excessive GRF. Lower extremity muscles, in particular, hip musculature, attempt compensating the excessive GRF. However, if the force is not well attenuated, ACL bundles can be strained or torn. For the indirect mechanism, an external force application to body parts other than the knee joint may trigger the whole cascade of events similar to those described above and result in a torn ACL. External force application to a knee joint directly, which promotes knee abduction and anterior tibial translation with internal rotations, is also likely to cause a direct contact ACL injury.

Females are 10–20 % more susceptible to noncontact ACL injuries than their male counterparts. Because the studies we reviewed lumped direct and indirect mechanisms together as contact ACL injuries, it was difficult to estimate the rate of direct and indirect ACL injuries separately. Nevertheless, noncontact situations appeared to account for a majority of the ACL injuries. Since the noncontact situation does not involve a contact and the external forces were created within the individuals own movements, altering the causative forces by an intervention program may be possible. Future studies need to focus on what type of interventions would alter the dynamic movements in a beneficial manner.

14.3 ACL Injury Prevention Clinical Trials

14.3.1 Introduction

The available data confirmed that approximately 60–70 % of all the ACL injury are noncontact in nature (Arendt and Dick 1995; Agel et al. 2005; Boden et al. 2000; Myklebust et al. 1997; Walden et al. 2011). NCAA female soccer and basketball athletes demonstrated a 2.7 higher overall ACL injury incidence rate as compared to male athletes participating in the same sports (Agel et al. 2005). Additionally, epidemiological data indicated that the prevalence of noncontact ACL injury mechanism in female athletes is approximately 10–20 % higher than their male counterparts (Arendt and Dick 1995; Agel et al. 2005). From the evidence, it is abundantly clear that young female athletes are the most susceptible to noncontact ACL injuries.

In order to explain the higher ACL injury incidence in adolescent female athletes, numerous research studies have been conducted from anatomical, hormonal, neuromuscular/biomechanical, and recently genetic (Naraoka et al. 2012; Posthumus et al. 2009; Lo et al. 1998) viewpoints. However, to establish prevention protocol, the focus needs to be on identifying ***modifiable*** risk factor(s) instead of non-modifiable risk factor(s) (Hewett et al. 2007b). While some studies did report an association between ACL injury and anatomical, hormonal, and DNA risk factors (Posthumus et al. 2009), those risk factors are not modifiable. Currently, the only risk factor considered to be modifiable is neuromuscular/biomechanical component. Therefore, in this chapter, we review the literature which covers the utilization of neuromuscular training (NMT) as an intervention method to reduce ACL injury in young female athletes.

14.3.2 Literature Search Method

A literature search was performed using PubMed and EBSCO host (CINAHL, MEDLINE, and SPORTDiscus) with a date range from 1995 to 2011 on May 31, 2012. The key words search was performed with the following words: “knee”, “anterior cruciate ligament”, “ACL”, “prospective”, “neuromuscular”, “training”, “female”, and “prevention.” The language was limited to English, and subjects were all human. The following inclusion criteria were applied: (1) the number of ACL injury incidents were reported, (2) a NMT intervention aimed to reduce ACL incidence was applied, (3) a control group was used, (4) a prospective controlled trial study design was employed, and (5) females were included as subjects.

14.3.3 Documented Clinical Trials from 1995 to 2012

Hewett et al. (1999)

This research team used a prospective cluster study design and provided 6 weeks of neuromuscular training, consisting of weight training, plyometrics, and flexibility, to a total of 43 teams (volleyball, soccer and basketball) from area high schools. Each neuromuscular training session lasted from 60 to 90 min and took place three times per week for 6 weeks. Certified athletic trainers and physical therapists gave technique instructions and the training sessions progressed through three phases: (I) Technique phase, (II) Fundamental phase, and (III) Performance phase. The 15 girls' teams that received the intervention (6 weeks of neuromuscular training) consisted of 366 athletes: 185 volleyball (50.5 %), 97 soccer (26.5 %), and 84 basketball (23.0 %) players. An ACL incidence rate of the intervention group was 0.06 per 1,000 h of Athletic-Exposure (1,000 h AE) in the intervention group and 0.11 per 1,000 h AE in the control group.

Soderman et al. (2000)

This prospective randomized controlled trial provided 10–15 min of balance training utilizing dynadiscs and balance boards to a total of 221 soccer players for 6 months. After randomization, 121 athletes (seven teams) were assigned to the intervention group and 100 athletes (six teams) were assigned to the control group. The athletes in the intervention group were asked to perform the balance training with balance boards every day for the 1st month. After the 1st month, training was decreased to 3 days per week. This study reported an ACL incidence rate of 0.68 per 1,000 h AE in the intervention group and 0.12 per 1,000 h AE in the control group.

Heidt et al. (2000)

This research group employed a 75 min long custom-made speed and agility program which was given to 42 randomly selected high school age soccer players for a total of 21 sessions (1st session was an orientation) over 7 weeks. The randomly selected subjects in the intervention group commuted to a local fitness gym to perform the intervention program in the pre-season. Over the course of 4 months, the ACL injury rate in the intervention group was 2.38, and 3.10 % in the control group. This study did not record or report exposure data.

Myklebust et al. (2003)

This study involved a 3 year prospective cross-over protocol (the 1st year was an observational year, with the two subsequent years involving intervention) The study subjects consisted of 1,705 female handball athletes playing for the top three

Norwegian handball leagues. A 15 min session of balance exercises with mats and wobble boards was implemented 3 days per week in the initial 5–7 weeks, which was subsequently reduced to once a week for the remainder of the handball season (approximately 5 months). During the study period, the ACL incidence rates were: control (Year 1) 0.14 per 1,000 h AE, intervention (Year 2) 0.13 per 1,000 h AE, and intervention (Year 3) 0.09 per 1,000 h AE.

Mandelbaum et al. (2005)

Using a prospective cluster cohort study design, the research team applied a neuromuscular and proprioceptive program to a total of 1,885 female soccer players (1,041 subjects in the 1st year and 844 in the 2nd year) and compared the number of ACL injuries to age- and skill- matched controls. The program was 20 min in duration and consisted of education, basic warm-up, stretching for trunk and lower extremities, trunk and lower extremity strengthening and plyometrics. The exercises were performed at a pace of two to three sessions per week. This investigation reported an ACL incidence rate of 0.04 per 1,000 h AE in the intervention group and 0.24 per 1,000 h AE in the control group over two competitive soccer seasons.

Olsen et al. (2005)

Utilizing a cluster randomized controlled trial design, a 15–20 min long structured warm-up program was implemented to improve neuromuscular control, balance, knee and ankle strength, and running, cutting, and landing technique for Norwegian handball players (808 subjects in the intervention group). The structured warm-up program had four different exercises (warm-up, technique, balance, and strength and power), and each exercise progressed through increasing the levels of difficulty. The structured warm-up program was performed over 15 consecutive sessions and then once a week during the competitive Norwegian handball season. During the study period, the ACL incidence rate in the intervention group was 0.03 per 1,000 AE, whereas in the control group the incidence was 0.10 per 1,000 h AE.

Petersen et al. (2005)

This prospective cohort study incorporated 10 min of injury prevention training into a team warm-up. The injury prevention training program consisted of improving awareness of injury mechanisms and prevention strategies, balance-board exercises and jump training and was executed three times per week in the preseason and once a week during the competition period. Lower extremity injuries were tracked in the 134 female handball players who performed the training program and compared with age- and skill- matched controls. The study reported an ACL incidence rate 0.04 per 1,000 h AE in the intervention group and 0.21 per 1,000 AE in the control group.

Pfeiffer et al. (2006)

This research team implemented a 20 min plyometric-based exercise program twice a week for high school female soccer, volleyball, and basketball athletes over a 2 year period. The plyometric-based exercise program, "Knee Ligament Injury Prevention" (KLIP), was developed by various healthcare practitioners and experts. A total of 577 athletes (43 teams) were placed in the intervention group, and 862 athletes (69 teams) in the control group. During the investigation, the ACL incidence rate was 0.08 per 1,000 h AE in the intervention group and 0.04 per 1,000 AE in the control group.

Pasanen et al. (2008)

This cluster-randomized controlled study provided 20–30 min of NMT to top-level floorball athletes several times during the 6-month season. The NMT was designed to enhance the athletes' motor skills and body-control awareness with sport-specific maneuvers that consisted of balance, body control, plyometrics, and strengthening. Players worked in pairs and were instructed to give each other feedback during the training. In total, 256 athletes (14 teams) maintained their training routine. The study reported an ACL incidence rate of 0.19 per 1,000 h AE in the intervention group and 0.16 per 1,000 h AE in the control group.

Steffen et al. (2008a)

Using a cluster-randomized controlled trial (113 teams, 2,100 players), this research team prescribed a 15 min structured warm-up program called "11". The 11, which consisted of core stability, balance, plyometrics, and hamstring strengthening exercises, was applied to 1,073 young female soccer players (51 teams) for the first 15 consecutive sessions and once a week for the remaining seven and half months. The study documented an ACL incidence rate of 0.06 per 1,000 h AE in the intervention and 0.08 per 1,000 h AE in the control group.

Gilchrist et al. (2008)

In a randomized cluster controlled study, investigators applied the 20 min long program, previously reported by Mandelbaum et al. (Mandelbaum et al. 2005) to high level college female soccer teams. The intervention and control groups were paired and formed a cluster. The clustered pairs were purposefully allocated to different geographic regions throughout the US. Then, one cluster of each region was randomly selected for the study. Soccer players (583 players, 26 teams) classified in the intervention group performed the program three times per week for the entire fall soccer season (12 weeks). An ACL incidence rate in the intervention group was 0.20 per 1,000 h AE, whereas the ACL incidence rate in the control group was 0.34 per 1,000 h AE.

Kiani et al. (2010)

This prospective cluster control trial experiment (97 teams, 1,506 players) included a 20–25 min neuromuscular regimen consisting of a running warm-up, isometric contraction of lower extremity muscle groups, balance exercises with jump components, strengthening of lower extremities and core stability utilized by 777 young soccer players (48 teams) 2 days per week for the 2 month pre-season and once a week during 6 months of in-season sessions. This study reported an ACL incidence rate of 0 per 1,000 h AE in the intervention group and 0.08 per 1,000 h AE in the control group.

LaBella et al. (2011)

Using a randomized cluster controlled design, investigators applied a program called the "Knee Injury Prevention Program" (KIPP). A total of 737 athletes: 321 soccer (43.6 %), and 416 basketball (56.4 %) players practiced the KIPP, which was comprised 20 min of progressive strengthening, plyometric, balance, and agility exercises three times per week for one competitive season. Over the course of the study, an ACL incidence rate of 0.10 per 1,000 h AE was documented in the intervention group and 0.48 per 1,000 h AE was documented in the control group.

Walden et al. (2012)

This research team applied a cluster-randomized control design to a total of 4,564 young soccer players (230 teams) and intervened with a 15-min neuromuscular warm-up focused on the development of core stability, balance, and jump landing with proper knee alignment. Study therapists provided specific instruction for implementing the exercises to 1 coach and a player from each team before the initiation of intervention. Coaches were responsible primarily for instruction of exercise executions. A therapist made 2 unannounced visits to assist in the correction of training errors. A total of 2,479 young soccer players (121 teams) trained two times per week for the entire competitive season (7 months). During the course of the study, an ACL incidence rate of 0.05 per 1,000 h AE was documented in the intervention group and 0.11 per 1,000 h AE in the control group.

14.3.4 Summary of Prior ACL Injury Prevention Clinical Trials

A total of 14 clinical trials were reviewed. Recent meta-analytic studies utilizing these studies have demonstrated the effectiveness of neuromuscular training intervention for ACL injury reduction (Sadoghi et al. 2012; Gagnier et al. 2013). One study that incorporated all 14 clinical trials reported a significantly greater ACL injury reduction (56 %) in female athletes who performed preventive NMT compared with those who did not or continued traditional warm-up procedures (Myer et al. 2013). Although the prophylactic effectiveness of preventive NMT was identified, a large diversity existed in the preventive NMT programs including age, dosage, exercises, and sports. The aforementioned meta-analytic study pointed out influence of age on the prophylactic effectiveness (Myer et al. 2013). Future projects are warranted to find the most effective aspects within preventive NMT programs.

14.4 Biomechanical Alterations Achieved through Specific Neuromuscular Training Exercises

14.4.1 Introduction

As previously mentioned, preventive NMT exercises vary widely. In this chapter, we discuss a framework of common exercises in preventive NMT programs and evaluate their effects on ACL injury in the physically active female population. Although many different exercise modes are utilized in preventive NMT, the exercises listed below are those most commonly observed in preventive NMT programs. Each exercise is categorized and accompanied by research evidence and rationale for the exercises implementation. Also, listed are the expected biomechanical alterations and the effects on ACL injury reduction.

14.4.2 Strengthening Exercise

It has been reported that females have a lower hamstring peak torque and also a slower activation as compared to their male counterparts (Urabe et al. 2005; Chappell et al. 2007; Ahmad et al. 2006; Hewett et al. 1996). The primary movement function of the hamstring is knee flexion. However, the hamstrings also have a critical role in active knee stabilization. A reduction of hamstring strength relative to quadriceps strength is related to an increased risk of ACL injury in female athletes (Myer et al. 2009). Hamstrings with a lower peak torque and slower activation may not flex the knee optimally in dynamic movements. In fact, one of several biomechanical differences identified between noncontact ACL injured and non-ACL injured athletes was limited knee flexion in a drop box jump task (Hewett et al. 2005; Padua et al. 2009). Additionally, an anatomical study indicated that ACL bundles produces maximum tension in the final 30° of knee extension (Amis and Dawkins 1991). Therefore, a greater degree of knee flexion in dynamic movements is considered as a protective function against traumatic knee injuries such as an ACL rupture.

Another muscle that needs to be mentioned relative to its relationship to the hamstrings is the quadriceps. The quadriceps, which collectively insert to the tibial tuberosity is a major contributor for an anterior shear force on the tibia. The anterior shear force on the tibia is largely determined by the quadriceps muscle force via the patella tendon-tibia shaft angle (Yu and Garrett 2007). A cadaver study reported that ACL fibers can be torn with a 4,500 N anterior shear force produced by the quadriceps muscle at 20° of knee flexion (DeMorat et al. 2004). The hamstrings, which is an antagonist muscle in relation to the quadriceps, provides a posterior shear force on the tibia. The posterior shear force produced by hamstrings counteracts the anterior shear force created by the quadriceps. Therefore, it is theorized that strengthening the hamstrings to counterbalance the anterior shear force produced by the quadriceps is important. In fact, the entire hamstring/posterior chain musculature is considered to work as an agonist to protect against the anterior shear loads to the ACL produced by the quadriceps.

One of the most commonly incorporated strength exercises of the NMT programs was the Russian/Nordic hamstring curl. A recent study reported physiological changes in the hamstrings following Russian/Nordic hamstring curl exercises (Mendiguchia et al. 2013). This physiological changes led by the Russian/Nordic hamstring curl is important because ACL bundles produce maximum tension in the final 30° of knee extension (Amis and Dawkins 1991). The anterior shear force generated during knee extension is mainly determined by the quadriceps muscle action via the patella tendon-tibia shaft angle (Yu and Garrett 2007). Therefore, it is theorized that strengthening the hamstrings can counterbalance the anterior shear force produced by the quadriceps and is thus considered a contributor to the ACL's protective mechanism. Consistent evidence was documented for enhancing hamstring peak torque and increasing the knee flexion angle (de Marche Baldon et al. 2012; Lim et al. 2009). One study (Lim et al. 2009) compared hamstring strength and knee flexion angles before and after implementing an NMT program which was developed based on a prevent injury and enhance performance (PEP) program used in the clinical trials performed by Mandelbaum et al. (2005) and Gilchrist et al. (2008). After 8 weeks of the modified PEP program training, female participants demonstrated a 9.8 % hamstring isokinetic strength increase as well as a 1.6° knee flexion angle increase during the drop vertical jump maneuver (Lim et al. 2009). Two clinical trials (Mandelbaum et al. 2005; Gilchrist et al. 2008) that implemented the PEP program demonstrated a 74–88 % reduction of ACL risk compared to the control group; however, another study (Steffen et al. 2008a) which did not document a strength improvement in the hamstring musculature only demonstrated a 29 % risk reduction for ACL injury risk reduction in relation to the control group (Steffen et al. 2008b).

14.4.3 Plyometric Exercise

A prospective cohort study that collected biomechanical data from a total of 205 high school female athletes compared the biomechanical data between nine ACL injured and 196 non-ACL injured athletes (Hewett et al. 2005). The study determined that the knee abduction moment predicts future ACL injury with a sensitivity of 78 % and a specificity of 73 % (Hewett et al. 2005). Along with the knee abduction moment, the study reported that limited knee flexion angles, an elevated GRF, and asymmetrical landing patterns were significantly different between the two groups (Hewett et al. 2005).

A number of NMT protocols instituted plyometric exercises, such as jumping forward and backward, jumping side to side, and tuck and scissor jumps (Hewett et al. 1999; Heidt et al. 2000; Mandelbaum et al. 2005; Petersen et al. 2005; Pfeiffer et al. 2006; Pasanen et al. 2008; Steffen et al. 2008a; Gilchrist et al. 2008; LaBella et al. 2011) The underlying reason why those clinical trials included plyometric exercises was that such exercises were aimed to reduce the GRF. Laboratory controlled studies that incorporated those exercises demonstrated a reduction in GRF upon landing (Hewett et al. 1999; Irmischer et al. 2004; Vescovi et al. 2008). A group of physically active females performed plyometric exercises two times per week for 9 weeks, and GRF values were compared pre- and post- plyometric exercise intervention (Irmischer et al. 2004). The participants of this study demonstrated a 26.4 % reduction in GRF in post-testing landing as compared to pre-test values (Irmischer et al. 2004). Another study that incorporated a series of plyometric exercises into female high school volleyball teams' workouts three times per week for 6 weeks demonstrated a 22.0 % reduction in GRF for a drop vertical jump task as compared to pre-intervention values (Hewett et al. 1999). It is crucial to note that both studies increased the difficulty of the plyometric exercises periodically and provided supervision during each training session (Hewett et al. 1999; Irmischer et al. 2004). A majority of the clinical trials that incorporated plyometrics stressed the importance of educating knee alignment upon landing (Hewett et al. 1999; Petersen et al. 2005; Pfeiffer et al. 2006; Pasanen et al. 2008; Steffen et al. 2008a; Gilchrist et al. 2008; LaBella et al. 2011). A high GRF was identified as one of the risk factors for future non-contact ACL injuries in female athletes (Hewett et al. 2005). Therefore, implementing a set of plyometric exercises into the NMT protocol to reduce the GRF in landing maneuvers may contribute to providing a protective mechanism against ACL injury. Additionally, another study demonstrated a reduction in side by side asymmetry landing force following plyometric exercises (Myer et al. 2006a). Yet, the reviewed clinical trials, which employed plyometric exercises incorporated verbal cues, such as, "land like feather," to emphasize attenuating GRF. Recent studies report greater biomechanical alterations for athletes who receive verbal feedback during exercises as compared to athletes who do not (Myer et al. 2013a; Stroube et al. 2013).

14.4.4 Proximal Control Exercise

Using college male and female athletes, a 3 year prospective study found that athletes who sustained severe knee ligamentous injuries, including those of the ACL, demonstrated greater deficits in trunk neuromuscular control compared to athletes who did not (Zazulak et al. 2007). Among the trunk neuromuscular control deficits, a lack of lateral trunk control was the most accurate predictor of future ACL injury (Zazulak et al. 2007). A logistic regression model that used trunk displacements, proprioception, and a history of low back pain was able to predict ACL injury risk in female athletes with a 91 % accuracy. Additionally, Hewett et al. conducted a video analysis of cutting and landing patterns of players from the NBA and WNBA and found a connection between lateral trunk control deficits and ACL injuries in female athletes (Hewett et al. 2009). The video analysis study revealed that lateral trunk flexion and knee abduction angles were greater in females who tore their ACL during cutting and landing tasks. During dynamic movements, when the trunk flexes laterally, vector of GRF shifts laterally as well (Hewett and Myer 2011). If the GRF runs laterally in relation to the center of the head of the femur, an external hip abductor torque is likely to be generated. In order to counteract the external hip abductor torque, hip adductor torque needs to be produced (Hewett and Myer 2011). This increased hip adductor torque is likely to create greater knee abduction angles.

Two intervention studies measured the effects of trunk focused training on the knee joint. A prospective cohort study that implemented a progressive functional stabilization training protocol (80 min per session three times per week for 8 weeks, incorporated many trunk/hip exercises) for healthy young females reported significantly decreased knee abduction angles (−6.86° ± 5.20° at baseline, 1.49° ± 3.56° at 8 weeks) in a single leg task compared to pre-intervention values. This study also reported a 10.9 % eccentric hip abductor torque increase (1.31 ± 0.16 $N.m.kg^{-1}$ at baseline, 1.45 ± 0.19 $N.m.kg^{-1}$ at 8 weeks). Since several studies found an association between decreased hip abductor strength and increased knee abduction, especially in the female population (Heinert et al. 2008; Jacobs et al. 2007), increased hip abductor strength may be protective against excessive knee abduction (de Marche Baldon et al. 2012). Using 20 high school female soccer players, another prospective cohort study measured a star excursion balance test after the implementation of a core focused NMT program (60 min per session two times per week for 8 weeks) that was designed to enhance core stability (Filipa et al. 2010). In this study, core stability was defined as dynamic trunk control that allowed the production, transfer, and control of force and motion to distal segments of the kinetic chain. This study reported an improvement of the star excursion balance test score (right limb: 96.4 ± 11.7 % at baseline, 104.6 ± 6.1 % at 8 weeks; left limb 96.9 ± 10.1 % at baseline,103.4 ± 8.0 % at 8 weeks) for the subjects who performed the core focused NMT.

14.4.5 Balance and Postural Stability Exercise

Several studies have identified mechanoreceptors in ACL fibers (Schultz et al. 1984; Schutte et al. 1987). Therefore, once an ACL is torn and even if the ACL is reconstructed, afferent information from this ligament is disrupted and ACL reconstructed individuals experience neuromuscular control deficits, especially in the detection of passive joint motion or joint repositioning (Adachi et al. 2002; Reider et al. 2003). From this standpoint, training (actually retraining) the sensorimotor system, which includes both afferent and efferent neural pathways, is imperative. Balance training that uses exercise tools such as dynadiscs, airrex, and other balance devices that create perturbation are effective. Maintaining postural stability in balance training requires feeding peripheral sensory input via the central nervous system (CNS) to the brain and/or spinal cord, and in turn the brain and/or spinal cord directs motor output to various local motor systems such as those involved with the hip, knee, and ankle joints. In short, balance training focuses on improving the efficacy of the sensorimotor system.

Several large scale controlled studies have been conducted to determine the effectiveness of balance training for the prevention of ACL injury (Soderman et al. 2000; Myklebust et al. 2003; Caraffa et al. 1996). Applying balance training to semi-professional and amateur soccer players, a 3 years of prospective cohort study documented a significant ACL injury reduction in athletes who performed balance training as compared to athletes who maintained normal training procedure (10 ACL injuries in balance training vs. 70 ACL injuries in control group) (Caraffa et al. 1996). Another prospective intervention study measured the effects of balance training in female handball players for 3 years. Compared to the control year (1st year: 29 ACL injuries (18 noncontact mechanism)), the study demonstrated decreases in ACL injuries in the experimental years (2nd year: 23 ACL injuries (10 noncontact mechanism) and 3rd year: 17 ACL injuries (7 noncontact mechanism)) (Myklebust et al. 2003). However, a study on female players that utilized a randomized clinical trial design involving 6 months of balance training actually had more ACL injuries in athletes who had the balance training (4 ACL injuries) as compared to subjects who did not have the balance training (1 ACL injury) (Soderman et al. 2000). Although the study utilized a randomized controlled design, it should be noted that the study instituted balance training as a home program; how rigorously the players actually followed the program guidelines is questionable.

In addition to the above large scale intervention clinical trials, several control studies in the laboratory controlled studies have investigated the effect of balance training. A prospective randomized intervention trial that provided balance training (90 min per session several times per week for 7 weeks) to high school female athletes showed several biomechanical alterations (Myer et al. 2006b). Compared to pre-testing values, the athletes who had balance training demonstrated a decrease in knee abduction angles and also an increase in maximum knee flexion (53.9° at baseline, 62.5° at 7 weeks) in the medial drop landing test. Another prospective randomized study in which the same balance training protocol was implemented (90 min per session several times per week for 7 weeks) to the same population documented a 7 % decrease in GRF in the drop vertical jump and also a significant increase in hamstring peak torque (dominant limb: 0.23 Torque/Body Weight (BW) at baseline, 0.27 Torque/Body Weight (BW) at 7 weeks; non-dominant limb 0.24 Torque/Body Weight (BW) at baseline, 0.26 Torque/Body Weight (BW) at 7 weeks) (Myer et al. 2006a). Another study that measured the effects of balance training (15 min per session for 3 days per week for 5–7 weeks; then, once a week for 5 months) in female handball players documented a significant improvement in their balance index scores (924.1 ± 225.5 at baseline, 778.9 ± 174.4 at 8 weeks, 730.3 ± 156.2 at 1 year) in their post-test values as compared to pre-test values (Holm et al. 2004).

A meta-analysis performed by Hewett and his colleagues stated that balance training alone may not be sufficient to produce significant ACL injury prevention effects (Hewett et al. 2006). Therefore, it is suggested that the balance training is employed with other exercise modalities. In fact, in the same manuscript, the following recommendations were written: plyometrics, balance, and strengthening exercises are incorporated into a comprehensive training protocol. (Hewett et al. 2006)

14.4.6 Summary of Biomechanical Alterations Achieved through Specific Neuromuscular Training Exercises

Although this chapter addressed how each exercise component alters the biomechanics of dynamic movements in the female population, it is imperative to consider how those exercises function synergistically. For instance, performing plyometric exercises not only assists in reducing GRF and provides an opportunity for providing correct knee alignment through feedback, but probably also enhances muscular strength. In fact, several studies demonstrated lower extremity muscular strength increases (Hewett et al. 1996; Lephart et al. 2005). Furthermore, plyometric exercises entail balance sustaining skills, which may promote improved balance ability. An explanation of how each exercise component contributes to the overall preventive effect of NMT programs need further investigation. However, it is important to recognize that the exercises discussed above work synergistically and generate the prophylactic effectiveness of preventive NMT programs.

14.5 Future Directions of ACL Injury Research

14.5.1 Identification of At-Risk Athletes

Based on recent comprehensive relative risk reduction (RRR) studies, RRR of 73.4 % (95 % confidence interval (CI): 62.5–81.1 %) and 43.8 % (95 % CI: 28.9–55.5 %) for noncontact and overall ACL injuries has been documented (Sugimoto et al. 2012a). From at numbers-needed-to treat (NNT) analysis, it was determined that 108 (95 % CI: 86–150) and 120 (95 % CI: 74–316) individuals would need to be trained to prevent one noncontact or one overall ACL injury during the course of one competitive athletic season. Although the RRR analysis indicated the prophylactic effectiveness and benefits of neuromuscular training, the relatively large NNT indicated that many athletes are needed to be trained in order to prevent one ACL injury. A future direction to reduce NNT and improve the efficiency of the ACL injury prevention strategy would be to develop a screening system for identifying at risk athletes. Several screening tools were recently introduced by using two-dimensional cameras. A clinical screening tool that utilized a landing error scoring system (LESS) was designed by Padua et al. (2009) In this system, two standard video cameras are placed at the front and on the side to capture the athlete's landing kinematics from sagittal and frontal plane views. The landing patterns captured by the video are utilized to provide a total error score. Another tool developed by Myer and colleagues used a nomogram scale, which consists of a point total generated from a combination of static (body mass and tibia length) and dynamic (knee valgus motion, knee flexion range of motion, quadriceps/hamstring strength ratio) measures (Myer et al. 2011). Two standard video cameras are needed to measure knee valgus motion and knee flexion range of motion. A third study was conducted by Stensrud et al. in which they utilized a standard video camera to capture images of knee joint alignment in the frontal view during dynamic movements (Stensrud et al. 2011). Validation of the efficacy of these video screening tools awaits further study.

14.5.2 Clinical Feasibility of Neuromuscular Training Intervention

Compliance with the preventive NMT intervention protocol needs to be improved. Among the previously published studies, several studies pointed out that a low compliance with the preventive NMT protocol was a major limitation (Myklebust et al. 2003; Steffen et al. 2008a). An inverse dose-response relationship between participation in a preventive NMT program and ACL reduction in female athletes was reported (Sugimoto et al. 2012b). In this study, participants with a low compliance showed a 4.9 times greater ACL injury risk as compared to participants with a high compliance. In another study, the incidence of acute and overall soccer injuries were evaluated in relation to NMT compliance rates (Soligard et al. 2010). Compliance was divided into high, intermediate, and low rates. High compliance athletes showed a 35 % lower risk rate as compared to those with an intermediate compliance rate in overall soccer related injuries. Similarly, in terms of acute soccer injuries, high compliance athletes experienced a 39 % risk reduction in relation to athletes with an intermediate compliance rate (Soligard et al. 2010).

In order to improve compliance with the preventive NMT protocol, healthcare providers need to educate coaching personnel. One study reported that the difference between the 20 % of high school soccer coaches who implement the preventive

NMT and those who do not is the unavailability of healthcare providers and the lack of coaching experience (Joy et al. 2013). Additionally, the same study identified the three biggest barriers for NMT intervention, lack of knowledge, time restriction, and lack of understanding and support. (Joy et al. 2013) Thus, educating coaching personnel will be a critical step for enhancing implementation of preventive NMT. After a couple years of a preventive NMT intervention study, a study reported decreases of ACL injury in women's handball teams in Norway (Myklebust et al. 2013). However, several years after the completion of the preventive NMT intervention studies, number of ACL injuries has gradually increased and surpassed the baseline ACL injury incidence rate, which is before the implementation of the preventive NMT intervention (Myklebust et al. 2013). Thus, this research group provided an educational DVD that contains a series of preventive NMT programs to coaching staff and observed a sudden decline of ACL injury (Myklebust et al. 2013). Therefore, healthcare providers who work in clinical settings are an extremely valuable facet in the promotion of a preventive NMT and thus a reduction of ACL injuries.

References

Adachi N, Ochi M, Uchio Y, Iwasa J, Ryoke K, Kuriwaka M (2002) Mechanoreceptors in the anterior cruciate ligament contribute to the joint position sense. Acta Orthop Scand 73(3):330–334

Agel J, Arendt EA, Bershadsky B (2005) Anterior cruciate ligament injury in national collegiate athletic association basketball and soccer: a 13-year review. Am J Sports Med 33(4):524–530

Agel J, Olson DE, Dick R, Arendt EA, Marshall SW, Sikka RS (2007) Descriptive epidemiology of collegiate women's basketball injuries: National Collegiate Athletic Association Injury Surveillance System, 1988–1989 through 2003–2004. J Athl Train 42(2):202–210

Ahmad CS, Clark AM, Heilmann N, Schoeb JS, Gardner TR, Levine WN (2006) Effect of gender and maturity on quadriceps-to-hamstring strength ratio and anterior cruciate ligament laxity. Am J Sports Med 34(3):370–374

Amis AA, Dawkins GP (1991) Functional anatomy of the anterior cruciate ligament. Fibre bundle actions related to ligament replacements and injuries. J Bone Joint Surg Br 73(2):260–267

Ardern CL, Webster KE, Taylor NF, Feller JA (2011) Return to sport following anterior cruciate ligament reconstruction surgery: a systematic review and meta-analysis of the state of play. Br J Sports Med 45(7):596–606

Arendt E, Dick R (1995) Knee injury patterns among men and women in collegiate basketball and soccer. NCAA data and review of literature. Am J Sports Med 23(6):694–701

Boden BP, Dean GS, Feagin JA Jr, Garrett WE Jr (2000) Mechanisms of anterior cruciate ligament injury. Orthopedics 23(6):573–578

Caraffa A, Cerulli G, Projetti M, Aisa G, Rizzo A (1996) Prevention of anterior cruciate ligament injuries in soccer. A prospective controlled study of proprioceptive training. Knee Surg Sports Traumatol Arthrosc 4(1):19–21

Chappell JD, Creighton RA, Giuliani C, Yu B, Garrett WE (2007) Kinematics and electromyography of landing preparation in vertical stop-jump: risks for noncontact anterior cruciate ligament injury. Am J Sports Med 35(2):235–241

Cimino F, Volk BS, Setter D (2010) Anterior cruciate ligament injury: diagnosis, management, and prevention. Am Fam Physician 82(8):917–922

de Marche Baldon R, Lobato DF, Carvalho LP, Lam Wun PY, Santiago PR, Serrao FV (2012) Effect of functional stabilization training on lower limb biomechanics in women. Med Sci Sports Exerc 44:135–145
DeMorat G, Weinhold P, Blackburn T, Chudik S, Garrett W (2004) Aggressive quadriceps loading can induce noncontact anterior cruciate ligament injury. Am J Sports Med 32(2):477–483
Dick R, Putukian M, Agel J, Evans TA, Marshall SW (2007) Descriptive epidemiology of collegiate women's soccer injuries: National Collegiate Athletic Association Injury Surveillance System, 1988–1989 through 2002–2003. J Athl Train 42(2):278–285
Evans K, Refshauge KM, Adams R (2007) Trunk muscle endurance tests: reliability, and gender differences in athletes. J Sci Med Sport 10(6):447–455
Filipa A, Byrnes R, Paterno MV, Myer GD, Hewett TE (2010) Neuromuscular training improves performance on the star excursion balance test in young female athletes. J Orthop Sports Phys Ther 40(9):551–558
Ford KR, Myer GD, Smith RL, Vianello RM, Seiwert SL, Hewett TE (2006) A comparison of dynamic coronal plane excursion between matched male and female athletes when performing single leg landings. Clin Biomech (Bristol, Avon) 21(1):33–40
Gagnier JJ, Morgenstern H, Chess L (2013) Interventions designed to prevent anterior cruciate ligament injuries in adolescents and adults: a systematic review and meta-analysis. Am J Sports Med 41(8):1952–1962
Gilchrist J, Mandelbaum BR, Melancon H et al (2008) A randomized controlled trial to prevent noncontact anterior cruciate ligament injury in female collegiate soccer players. Am J Sports Med 36(8):1476–1483
Gomez E, DeLee JC, Farney WC (1996) Incidence of injury in texas girls' high school basketball. Am J Sports Med 24(5):684–687
Heidt RS Jr, Sweeterman LM, Carlonas RL, Traub JA, Tekulve FX (2000) Avoidance of soccer injuries with preseason conditioning. Am J Sports Med 28(5):659–662
Heinert BL, Kernozek TW, Greany JF, Fater DC (2008) Hip abductor weakness and lower extremity kinematics during running. J Sport Rehabil 17(3):243–256
Hewett TE, Myer GD (2011) The mechanistic connection between the trunk, knee, and ACL injury. Exerc Sport Sci Rev 39(4):161–166
Hewett TE, Stroupe AL, Nance TA, Noyes FR (1996) Plyometric training in female athletes. Decreased impact forces and increased hamstring torques. Am J Sports Med 24(6):765–773
Hewett TE, Lindenfeld TN, Riccobene JV, Noyes FR (1999) The effect of neuromuscular training on the incidence of knee injury in female athletes. A prospective study. Am J Sports Med 27(6):699–706
Hewett TE, Myer GD, Ford KR et al (2005) Biomechanical measures of neuromuscular control and valgus loading of the knee predict anterior cruciate ligament injury risk in female athletes: a prospective study. Am J Sports Med 33(4):492–501
Hewett TE, Ford KR, Myer GD (2006) Anterior cruciate ligament injuries in female athletes: part 2, a meta-analysis of neuromuscular interventions aimed at injury prevention. Am J Sports Med 34(3):490–498
Hewett TE, Shultz SJ, Girffin LY (2007a) Understanding and preventing noncontact ACL injuries. In: Kinetics H (ed) Understanding and preventing noncontact ACL injuries. 1st edn. Human Kinetics, Cincinnati
Hewett TE, Myer GD, Ford KR, Slauterbeck JR (2007b) Dynamic neuromuscular analysis training for preventing anterior cruciate ligament injury in female athletes. Instr Course Lect 56:397–406
Hewett TE, Torg JS, Boden BP (2009) Video analysis of trunk and knee motion during non-contact anterior cruciate ligament injury in female athletes: lateral trunk and knee abduction motion are combined components of the injury mechanism. Br J Sports Med 43(6):417–422
Holm I, Fosdahl MA, Friis A, Risberg MA, Myklebust G, Steen H (2004) Effect of neuromuscular training on proprioception, balance, muscle strength, and lower limb function in female team handball players. Clin J Sport Med 14(2):88–94

Hootman JM, Dick R, Agel J (2007) Epidemiology of collegiate injuries for 15 sports: summary and recommendations for injury prevention initiatives. J Athl Train 42(2):311–319

Irmischer BS, Harris C, Pfeiffer RP, DeBeliso MA, Adams KJ, Shea KG (2004) Effects of a knee ligament injury prevention exercise program on impact forces in women. J Strength Cond Res 18(4):703–707

Jacobs CA, Uhl TL, Mattacola CG, Shapiro R, Rayens WS (2007) Hip abductor function and lower extremity landing kinematics: sex differences. J Athl Train 42(1):76–83

Joy E, Taylor JR, Novak M, Chen M, Fink B, Porucznik C (2013) Factors influencing the implementation of ACL injury prevention strategies by girls soccer oaches. J Strength Cond Res 27:2263–2269

Kiani A, Hellquist E, Ahlqvist K, Gedeborg R, Michaelsson K, Byberg L (2010) Prevention of soccer-related knee injuries in teenaged girls. Arch Intern Med 170(1):43–49

LaBella CR, Huxford MR, Grissom J, Kim KY, Peng J, Christoffel KK (2011) Effect of neuromuscular warm-up on injuries in female soccer and basketball athletes in urban public high schools: cluster randomized controlled trial. Arch Pediatr Adolesc Med 165(11):1033–1040

Lephart SM, Abt JP, Ferris CM et al (2005) Neuromuscular and biomechanical characteristic changes in high school athletes: a plyometric versus basic resistance program. Br J Sports Med 39(12):932–938

Lim BO, Lee YS, Kim JG, An KO, Yoo J, Kwon YH (2009) Effects of sports injury prevention training on the biomechanical risk factors of anterior cruciate ligament injury in high school female basketball players. Am J Sports Med 37(9):1728–1734

Lo IK, Marchuk LL, Hart DA, Frank CB (1998) Comparison of mRNA levels for matrix molecules in normal and disrupted human anterior cruciate ligaments using reverse transcription-polymerase chain reaction. J Orthop Res 16(4):421–428

Lohmander LS, Ostenberg A, Englund M, Roos H (2004) High prevalence of knee osteoarthritis, pain, and functional limitations in female soccer players twelve years after anterior cruciate ligament injury. Arthritis Rheum 50(10):3145–3152

Lohmander LS, Englund PM, Dahl LL, Roos EM (2007) The long-term consequence of anterior cruciate ligament and meniscus injuries: osteoarthritis. Am J Sports Med 35(10):1756–1769

Maffulli N, Longo UG, Gougoulias N, Loppini M, Denaro V (2010) Long-term health outcomes of youth sports injuries. Br J Sports Med 44(1):21–25

Mandelbaum BR, Silvers HJ, Watanabe DS et al (2005) Effectiveness of a neuromuscular and proprioceptive training program in preventing anterior cruciate ligament injuries in female athletes: 2-year follow-up. Am J Sports Med 33(7):1003–1010

McCullough KA, Phelps KD, Spindler KP et al (2012) Return to high school- and college-level football after anterior cruciate ligament reconstruction: a Multicenter Orthopaedic Outcomes Network (MOON) cohort study. Am J Sports Med 40(11):2523–2529

Mendiguchia J, Arcos AL, Garrues MA, Myer GD, Yanci J, Idoate F (2013) The use of MRI to evaluate posterior thigh muscle activity and damage during nordic hamstring exercise. J Strength Cond Res 27(12):3426–3435

Messina DF, Farney WC, DeLee JC (1999) The incidence of injury in Texas high school basketball. A prospective study among male and female athletes. Am J Sports Med 27(3):294–299

Miyaguchi K, Demura S (2009) Gender difference in ability using the stretch-shortening cycle in the upper extremities. J Strength Cond Res 23(1):231–236

Myer GD, Ford KR, Brent JL, Hewett TE (2006a) The effects of plyometric vs. dynamic stabilization and balance training on power, balance, and landing force in female athletes. J Strength Cond Res 20(2):345–353

Myer GD, Ford KR, McLean SG, Hewett TE (2006b) The effects of plyometric versus dynamic stabilization and balance training on lower extremity biomechanics. Am J Sports Med 34(3):445–455

Myer GD, Ford KR, Barber Foss KD, Liu C, Nick TG, Hewett TE (2009) The relationship of hamstrings and quadriceps strength to anterior cruciate ligament injury in female athletes. Clin J Sport Med 19(1):3–8

Myer GD, Ford KR, Khoury J, Hewett TE (2011) Three-dimensional motion analysis validation of a clinic-based nomogram designed to identify high ACL injury risk in female athletes. Phys Sportsmed 39(1):19–28
Myer GD, Sugimoto D, Thomas S, Hewett TE (2013) The influence of age on the effectiveness of neuromuscular training to reduce anterior cruciate ligament injury in female athletesa A meta-analysis. Am J Sports Med 41(1):203–215
Myer GD, Stroube BW, DiCesare CA et al (2013a) Augmented feedback supports skill transfer and reduces high-risk injury landing mechanics: a double-blind, randomized controlled laboratory study. Am J Sports Med 41(3):669–677
Myklebust G, Maehlum S, Engebretsen L, Strand T, Solheim E (1997) Registration of cruciate ligament injuries in Norwegian top level team handball. A prospective study covering two seasons. Scand J Med Sci Sports 7(5):289–292
Myklebust G, Engebretsen L, Braekken IH, Skjolberg A, Olsen OE, Bahr R (2003) Prevention of anterior cruciate ligament injuries in female team handball players: a prospective intervention study over three seasons. Clin J Sport Med 13(2):71–78
Myklebust G, Skjolberg A, Bahr R (2013) ACL injury incidence in female handball 10 years after the Norwegian ACL prevention study: important lessons learned. Br J Sports Med 47(8):476–479
Naraoka T, Ishibashi Y, Tsuda E et al (2012) Time-dependent gene expression and immunohistochemical analysis of the injured anterior cruciate ligament. Bone Joint Res 1(10):238–244
National Federation of State High School Association (2012) 2002 high school participation survey. National Federation of State High School Association, Indianapolis
Oiestad BE, Holm I, Aune AK et al (2010) Knee function and prevalence of knee osteoarthritis after anterior cruciate ligament reconstruction: a prospective study with 10 to 15 years of follow-up. Am J Sports Med 38(11):2201–2210
Oiestad BE, Holm I, Engebretsen L, Risberg MA (2011) The association between radiographic knee osteoarthritis and knee symptoms, function and quality of life 10-15 years after anterior cruciate ligament reconstruction. Br J Sports Med 45(7):583–588
Olsen OE, Myklebust G, Engebretsen L, Holme I, Bahr R (2005) Exercises to prevent lower limb injuries in youth sports: cluster randomised controlled trial. BMJ 330(7489):449
Padua DA, Marshall SW, Boling MC, Thigpen CA, Garrett WE Jr, Beutler AI (2009) The Landing Error Scoring System (LESS) is a valid and reliable clinical assessment tool of jump-landing biomechanics: the JUMP-ACL study. Am J Sports Med 37(10):1996–2002
Pasanen K, Parkkari J, Pasanen M et al (2008) Neuromuscular training and the risk of leg injuries in female floorball players: cluster randomised controlled study. BMJ 337:a295
Paterno MV, Schmitt LC, Ford KR et al (2010) Biomechanical measures during landing and postural stability predict second anterior cruciate ligament injury after anterior cruciate ligament reconstruction and return to sport. Am J Sports Med 38(10):1968–1978
Petersen W, Braun C, Bock W et al (2005) A controlled prospective case control study of a prevention training program in female team handball players: the German experience. Arch Orthop Trauma Surg 125(9):614–621
Pfeiffer RP, Shea KG, Roberts D, Grandstrand S, Bond L (2006) Lack of effect of a knee ligament injury prevention program on the incidence of noncontact anterior cruciate ligament injury. J Bone Joint Surg Am 88(8):1769–1774
Posthumus M, September AV, Keegan M et al (2009) Genetic risk factors for anterior cruciate ligament ruptures: COL1A1 gene variant. Br J Sports Med 43(5):352–356
Quatman CE, Kiapour A, Myer GD et al (2011) Cartilage pressure distributions provide a footprint to define female anterior cruciate ligament injury mechanisms. Am J Sports Med 39(8):1706–1713
Reider B, Arcand MA, Diehl LH et al (2003) Proprioception of the knee before and after anterior cruciate ligament reconstruction. Arthroscopy 19(1):2–12

Research Retreat IV (2008) ACL Injuries-The Gender Bias, April 3–5, 2008, Greensboro, NC, USA. Proceedings and abstracts. J Athl Train 43(5):530–560

Sadoghi P, von Keudell A, Vavken P (2012) Effectiveness of anterior cruciate ligament injury prevention training programs. J Bone Joint Surg Am 94(9):769–776

Salmon L, Russell V, Musgrove T, Pinczewski L, Refshauge K (2005) Incidence and risk factors for graft rupture and contralateral rupture after anterior cruciate ligament reconstruction. Arthroscopy 21(8):948–957

Schultz RA, Miller DC, Kerr CS, Micheli L (1984) Mechanoreceptors in human cruciate ligaments. A histological study. J Bone Joint Surg Am 66(7):1072–1076

Schutte MJ, Dabezies EJ, Zimny ML, Happel LT (1987) Neural anatomy of the human anterior cruciate ligament. J Bone Joint Surg Am 69(2):243–247

Soderman K, Werner S, Pietila T, Engstrom B, Alfredson H (2000) Balance board training: prevention of traumatic injuries of the lower extremities in female soccer players? A prospective randomized intervention study. Knee Surg Sports Traumatol Arthrosc 8(6):356–363

Soligard T, Nilstad A, Steffen K et al (2010) Compliance with a comprehensive warm-up programme to prevent injuries in youth football. Br J Sports Med 44(11):787–793

Steffen K, Myklebust G, Olsen OE, Holme I, Bahr R (2008a) Preventing injuries in female youth football–a cluster-randomized controlled trial. Scand J Med Sci Sports 18(5):605–614

Steffen K, Bakka HM, Myklebust G, Bahr R (2008b) Performance aspects of an injury prevention program: a ten-week intervention in adolescent female football players. Scand J Med Sci Sports 18(5):596–604

Stensrud S, Myklebust G, Kristianslund E, Bahr R, Krosshaug T (2011) Correlation between two-dimensional video analysis and subjective assessment in evaluating knee control among elite female team handball players. Br J Sports Med 45(7):589–595

Stroube BW, Myer GD, Brent JL, Ford KR, Heidt RS Jr, Hewett TE (2013) Effects of task-specific augmented feedback on deficit modification during performance of the tuck-jump exercise. J Sport Rehabil 22(1):7–18

Sugimoto D, Myer GD, McKeon JM, Hewett TE (2012a) Evaluation of the effectiveness of neuromuscular training to reduce anterior cruciate ligament injury in female athletes: a critical review of relative risk reduction and numbers-needed-to-treat analyses. Br J Sports Med 46(14):979–988. doi:10.1136/bjsports-2011-090895, Epub 092012 Jun 090828

Sugimoto D, Myer GD, Bush HM, Klugman MF, Medina McKeon JM, Hewett TE (2012b) Compliance with neuromuscular training and anterior cruciate ligament injury risk reduction in female athletes: a meta-analysis. J Athl Train 47(6):714–723

Urabe Y, Kobayashi R, Sumida S et al (2005) Electromyographic analysis of the knee during jump landing in male and female athletes. Knee 12(2):129–134

Vescovi JD, Canavan PK, Hasson S (2008) Effects of a plyometric program on vertical landing force and jumping performance in college women. Phys Ther Sport 9(4):185–192

Walden M, Hagglund M, Werner J, Ekstrand J (2011) The epidemiology of anterior cruciate ligament injury in football (soccer): a review of the literature from a gender-related perspective. Knee Surg Sports Traumatol Arthrosc 19(1):3–10

Walden M, Atroshi I, Magnusson H, Wagner P, Hagglund M (2012) Prevention of acute knee injuries in adolescent female football players: cluster randomised controlled trial. BMJ 344:e3042

Webster KE, Feller JA, Leigh WB, Richmond KE (2014) Younger patients are at increased risk for graft rupture and contralateral injury after anterior cruciate ligament reconstruction. Am J Sports Med 42(3):641–647

Wojtys EM, Brower AM (2010) Anterior cruciate ligament injuries in the prepubescent and adolescent athlete: clinical and research considerations. J Athl Train 45(5):509–512

Yu B, Garrett WE (2007) Mechanisms of non-contact ACL injuries. Br J Sports Med 41(Suppl 1):i47–i51

Zazulak BT, Hewett TE, Reeves NP, Goldberg B, Cholewicki J (2007) Deficits in neuromuscular control of the trunk predict knee injury risk: a prospective biomechanical-epidemiologic study. Am J Sports Med 35(7):1123–1130

Chapter 15
Anterior Cruciate Ligament Injury Prevention in Female Adolescents

Reiko Otsuki and Toru Fukubayashi

Abstract Anterior cruciate ligament (ACL) injuries are frequently seen in female adolescents. Lower extremity mechanics change as female adolescents mature. The biomechanical changes associated with female pubertal growth could be contributing factors to the increased occurrence of ACL injury after puberty. Several studies have demonstrated that injury prevention training was effective in improving lower extremity kinematics. In addition, some intervention studies have reported that implementing injury prevention training can reduce the incidence of ACL injury in female adolescents. Intervening at a younger age and limiting the changes in mechanics that occur during puberty might be important in the prevention of ACL injuries.

Keywords Anterior cruciate ligament • Injury prevention • Female adolescents

15.1 Introduction

Engaging in sports activities at a young age has numerous health benefits. However, youth sports have become more competitive and young athletes are required to participate in more prolonged and intense training programs (Caine et al. 2006). As a result, increased sports-related injuries in young athletes have become a problematic issue in the health care system. Health benefits from participating in sports activities may be diminished by an increase in injuries related to athletic participation (Leddy et al. 1994; Marshall and Guskiewicz 2003); thus, the risk management of injuries in young athletes is critical.

Anterior cruciate ligament (ACL) injuries are frequently seen in female adolescents (Shea et al. 2011). The rate of ACL injuries is three to five times higher in girls than in boys (Powell and Barber-Foss 2000). Injuries to the ACL are problematic

R. Otsuki (✉)
Graduate School of Sport Sciences, Waseda University, Saitama, Japan
e-mail: rotsuki@akane.waseda.jp

T. Fukubayashi
Faculty of Sport Sciences, Waseda University, Saitama, Japan

K. Kanosue et al. (eds.), *Sports Injuries and Prevention*,
DOI 10.1007/978-4-431-55318-2_15

because they often require surgery and extensive rehabilitation before athletes can return to sports activities. In addition, the risk of developing osteoarthritis can be tripled after an ACL injury, regardless of management (Drawer and Fuller 2001; Roos 2005). The premature development of osteoarthritis decreases the overall quality of life.

One of the risk factors for developing an ACL injury is poor kinematics of the lower extremity, such as increased knee valgus and decreased knee flexion. A cadaveric study evaluated ACL strain under physiological loading conditions and found that a combination of anterior tibial shear force, knee abduction, and internal tibial rotation led to ACL failure (Levine et al. 2013). In addition, video analysis of actual injury situations revealed that the main mechanisms of ACL injury were increased knee valgus motion with internal tibial rotation (Krosshaug et al. 2007; Koga et al. 2010). A reduced knee flexion angle was also commonly observed in injury situations (Krosshaug et al. 2007; Koga et al. 2010). A low knee flexion angle would increase quadriceps drawer and anterior tibial shear force, which strains the ACL. Moreover, it has been reported that female athletes with an increased dynamic knee valgus motion and reduced knee flexion angle were at a high risk of developing ACL injury (Hewett et al. 2005). Accordingly, an increased knee valgus motion and a decreased knee flexion angle seem to be important factors associated with the occurrence of ACL injury. Since the incidence of ACL injuries increase significantly after pubertal onset, changes in lower extremity kinematics might occur during puberty as well. This chapter reviews the changes in these biomechanical risk factors for ACL injury during maturation and also ACL injury prevention strategies in female adolescents.

15.2 Changes in ACL Injury Risk Factors in Female Adolescents

15.2.1 Knee Valgus

Female adolescents increase knee valgus motion following the onset of the pubertal growth spurt. Hewett et al. (2004) investigated, in a cross-sectional study, whether musculoskeletal changes that accompany maturation were associated with reduced control of the knee joint. Eighty-one male and one hundred female youth soccer and basketball players were categorized based on their maturational stage: prepubertal, early pubertal, or late/postpubertal. Three dimensional knee valgus angle during a drop vertical jump, and the isokinetic strength of quadriceps and hamstrings were compared among maturational stages. In the prepubertal and early pubertal stages, females and males had similar landing mechanics. However, in the late and post pubertal stages, females displayed a greater knee valgus motion than did males. This study also found that males exhibited an increase in leg strength following the growth spurt, but females did not increase their strength. Therefore, biomechanical changes associated with maturation might be due to the changes in muscular development.

Ford et al. (2010) conducted a longitudinal study investigating changes in biomechanical risk factors during maturation. Three hundred and fifteen subjects were categorized based on their maturational stage: pubertal or postpobertal. Knee valgus angle during a drop vertical jump was measured during two testing sessions approximately 1 year apart. Changes in knee abduction angle were compared among maturational stages. This study found that pubertal females increased knee abduction angle from the first year to the second year, while pubertal males did not display a similar change.

15.2.2 Knee Flexion

Knee flexion angle also is reported to change during maturation. Yu et al. (2005) investigated age and gender effects on lower extremity kinematics of youth soccer players in the stop-jump task. Thirty male and female youth soccer players between 11 and 16 years of age performed a stop-jump task. The age effects on knee flexion angle at initial foot contact and maximum knee flexion were compared between genders using multiple regression analyses. The results of this study found that female soccer players had reduced knee flexion angles at initial foot contact and during the landing of the stop-jump task as compared to male counterparts. This gender difference was observed after 12 years of age and increased with age before the age of 16. Although no other studies have reported changes in knee flexion angle associated with growth, it seems likely that knee flexion angle during landing decreases in females during maturation.

15.2.3 ACL Injury Risk

Based on above studies which investigated biomechanical risk factors, we attempted to evaluate overall ACL injury risk among different maturational stages in females (Otsuki et al. 2014). Ninety-two female youth basketball players were categorized into early, middle, late, or post pubertal stages. ACL injury risk was evaluated with an ACL injury prediction algorithm (Myer et al. 2011) which utilized tibia length, body mass, quadriceps: hamstrings strength ratio, knee valgus motion, and knee flexion range of motion. The knee valgus motion, knee flexion range of motion, and ACL injury risk were compared among the four maturational stages. The results of this study showed that the knee valgus motion increased significantly after early pubertal stage (Fig. 15.1). The knee flexion range of motion was gradually decreased

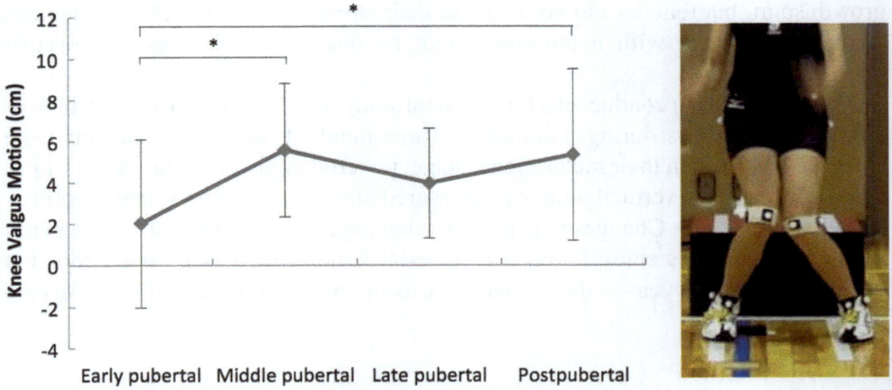

Fig. 15.1 Knee valgus motion in the early, middle, late, and post pubertal stages (*:p < 0.05, **:p < 0.01, †:p < 0.001)

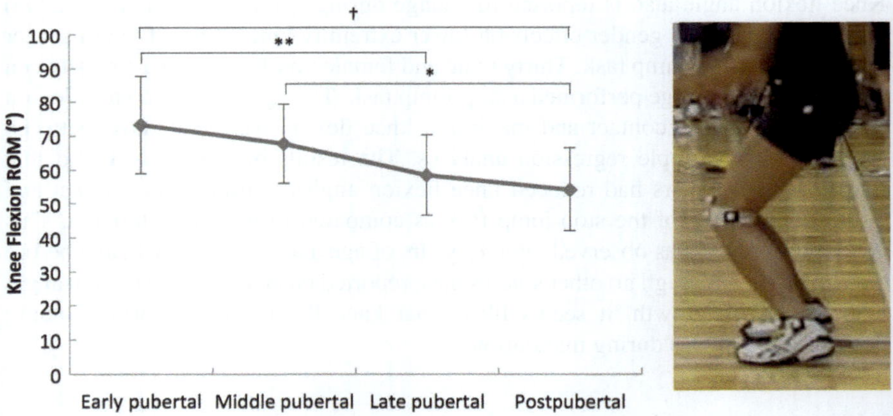

Fig. 15.2 Knee flexion range of motion in the early, middle, late, and post pubertal stages (*:p < 0.05, **:p < 0.01, †:p < 0.001)

during maturation (Fig. 15.2). The ACL injury risk was significantly increased after early pubertal stage (Fig. 15.3). Between early and middle pubertal stages, height and weight increased significantly. These physical growth factors might have played a role in an increase in ACL injury risk, since an increase in height and weight affects the moment at the knee joint (Quatman et al. 2006). This study indicated that early and middle pubertal stages might be a critical time period related to ACL injury; thus, implementation of an injury prevention program during this period might be of significant value.

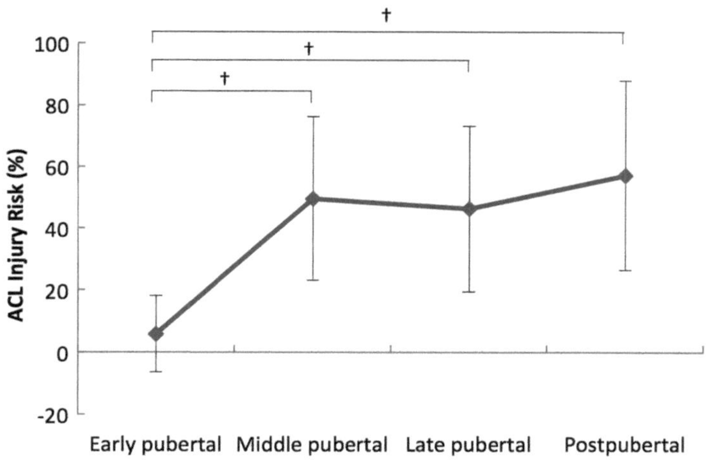

Fig. 15.3 ACL injury risk in the early, middle, late, and post pubertal stages (*:p < 0.05, **:p < 0.01, †:p < 0.001)

15.3 Effect of Injury Prevention Training on Lower Extremity Kinematics

Several studies have evaluated whether implementing injury prevention program in female adolescents can alter the high-risk movement patterns. Hewett et al. (1996) first investigated the effect of a jump-training program on landing mechanics and lower extremity strength in female adolescents. Eleven female high school volleyball players and nine male control subjects participated in the study. The female athletes underwent a 2-h jump-training program for 6 weeks. Lower extremity mechanics, muscle strength, and vertical jump height were measured before and after the training period. After the training period, female athletes significantly decreased peak landing force and knee abduction and adduction moments. Hamstring-to-quadriceps muscle peak torque ratio increased on both dominant and nondominant sides. Mean vertical jump height increased after the training period. This jump-training program was thus effective in improving knee mechanics and muscle strength.

Myer et al. (2005) investigated the effect of neuromuscular training on lower extremity biomechanics in female adolescents. Forty-one female basketball, soccer, and volleyball players from high school participated in a 6-week neuromuscular training program. Age, weight, and height-matched control also participated in the study. Three-dimensional lower extremity biomechanical testing was conducted before and after the training period in both groups. The results of this study showed that neuromuscular training decreased knee valgus and varus torques compared to the control group. Therefore, this study suggested that a 6-week neuromuscular training program was effective in improving lower extremity biomechanics.

Myer et al. (2006) further evaluated the effect of plyometric and balance training on lower extremity kinematics in high school female volleyball players. Eighteen high school female athletes participated in 7-week long training sessions. Subjects were divided into a plyometric training group and a balance training group. Lower extremity kinematics were measured during a drop vertical jump and medial drop landing before and after the training sessions. On frontal plane measurements, both groups decreased the initial contact and the maximum hip adduction angle during the drop vertical jump after the training period. During the medial drop landing, both groups decreased the initial contact and the maximum knee abduction angle. In the sagittal plane, the plyometric training group demonstarated increased knee flexion at initial contact and maximum angle during the drop vertical jump. The balance training group, on the other hand, demonstrated increased maximum knee flexion angle during the medial drop landing. This study suggested that both training methods were effective in improving lower extremity kinematics in the frontal plane, whereas training effects on the sagittal plane knee kinematics were task specific.

Pollard et al. (2006) investigated the influence of in-season injury prevention training on hip and knee kinematics during a landing task. Eighteen female soccer players between the age of 14 and 17 participated in the study. Subjects underwent the Prevent Injury and Enhance Performance (PEP) injury prevention training during the soccer season. Three-dimensional kinematics were measured during a drop-landing task before and after the season. After the season, subjects demonstrated decreased hip internal rotation and increased hip abduction, although there were no differences in knee valgus or knee flexion angles. This study suggested that in-season injury prevention training was effective in improving lower extremity kinematics.

Lim et al. (2009) evaluated the effects of injury prevention training on muscle strength, flexibility, and lower extremity biomechanics in female high school basketball players. Twenty-two high school female basketball players participated in the study. Subjects were divided into training and control groups. The training group underwent an 8-week injury prevention training regime, while the controp group maintained their regular routine. Lower extremity kinematics during a rebound-jump task were measured before and after the training period. The training group demonstrated increased strength of hip abduction, hip extension, and knee flexion, as well as increased flexibility of knee flexion. The training group also showed an increased knee flexion angle and inter-knee distances, and decreased maximum knee extension torques during the rebound-jump task after the training period. This injury prevention program was effective in modifying the strength, flexibility, and biomechanical properties that are associated with ACL injuries.

Although the methods of the various injury prevention training regimes were slightly different, above studies have demonstrated that injury prevention programs are effective in improving lower extremity mechanics in both frontal and sagittal planes. The improved kinematic functions involve those which are associated with ACL injury in female adolescents.

15.4 Effect of Injury Prevention Training on Incidence of ACL Injury

A number of studies have investigated the effect of various injury prevention programs on the incidence of ACL injury in female adolescents (Hewett et al. 1999, 2000; Mandelbaum et al. 2005; Olsen et al. 2005; Pfeiffer et al. 2006; Steffen et al. 2008; Kiani et al. 2010; Lebella et al. 2011; Walden et al. 2012). Table 15.1 summarizes the results of these studies. Although some studies involve problematical methodologies, the overall conclusion is clearly that injury prevention training is effective in reducing the number of ACL injuries in female adolescents. A meta-analysis by Myer et al. (2012) investigated the influence of age on the effectiveness of injury prevention training for reducing ACL injuries. Myer et al. found that implementing injury prevention training in the mid-teen years was more effective in reducing ACL injury than such training in the late teen or early adult period. This finding suggests that injury prevention training should be encouraged in female adolescents – specifically in early adolescents.

15.5 Future Directions

Recent evidence has greatly increased knowledge about ACL injury prevention in female adolescents. Future studies should investigate how to utilize this knowledge to prevent ACL injury even more effectively. One approach might be to target younger female athletes. As mentioned earlier, female adolescents demonstrate high-risk movement patterns following the onset of the pubertal growth. Since these changes in female adolescents coincide with an increase in the number of ACL injuries, changes in the lower leg and hip mechanics might be a factor contributing to the increased incidence of ACL injuries. Targeting the period when female adolescents develop a change in their lower body mechanics could potentially reduce the occurrence of ACL injuries. Currently, it is known that injury prevention training is effective in improving knee mechanics in high school female athletes These athletes are most likely in postpuberty. However, it is still unknown whether implementing injury prevention training during puberty could limit the changes that develop in movement patterns; thus, reducing the risk of ACL injury. Also, no studies have evaluated whether implementing injury prevention training in pubertal females is effective in reducing the incidence of future ACL injury. Intervening at an earlier stage of development and might well greatly lower the risk of ACL injury and thus prove to be the most effective strategy for the prevention of ACL injury.

Table 15.1 Summary of ACL injury prevention intervention studies

Author (year)	Study design	Sports	Age	No. of subjects (intervention vs. control)	Strength	Flexibility	Balance	Plyometrics	Agility	Skill	No. of ACL injury (intervention vs. control)
Hewett et al. (1999)	Non-randomized cohort	Soccer, volleyball, basketball	14–18	366 vs. 463	○	○		○			2 vs. 5 0.12 vs. 0.43/1,000 AE
Heidt et al. (2000)	Randomized controlled	Soccer	14–18	42 vs. 258	○	○		○	○	○	1 vs. 8 2.4% vs. 3.0%
Mandelbaum et al. (2005)	Non-randomized cohort	Soccer	14–18	1,885 vs. 3,818	○	○		○	○		6 vs. 67 0.49 vs. 0.99/1,000 AE; RR = 0.18
Olsen et al. (2005)	Cluster randomized controlled	Handball	15–17	958 vs. 879	○		○			○	3 vs. 9 RR = 0.2
Pfeifferet et al. (2006)	Non-randomized cohort	Soccer, volleyball, basketball	14–18	577 vs. 862				○			3 vs. 3 0.167 vs. 0.078/1,000 AE

Steffen et al. (2008)	Cluster randomized controlled	Soccer	13–17	1,073 vs. 947	○		○	○			4 vs. 5
Kiani et al. (2010)	Cluster non-randomized cohort	Soccer	13–19	777 vs. 729	○		○	○			0 vs. 5
LeBella et al. (2011)	Cluster randomized controlled	Soccer, basketball	16.2 (mean)	737 vs. 755	○		○	○	○		2 vs. 6 0.07 vs. 0.26/1,000 AE; RR = 0.20
Walden et al. (2012)	Cluster randomized controlled	Soccer	12–17	2,479 vs. 2,085	○		○	○			7 vs. 14 RR = 0.36

References

Caine D, DiFiori J, Maffuli N (2006) Physical injuries in children's and youth sports: reason for concern? Br J Sports Med 40:749–760

Drawer S, Fuller CW (2001) Propensity for osteoarthritis and lower limb joint pain in retired professional soccer players. Br J Sports Med 35(6):402–408

Ford KR, Shapiro R, Myer GD et al (2010) Longitudinal sex differences during landing in knee abduction in young athletes. Med Sci Sports Exerc 42:1923–1931

Heidt RS, Sweeterman LM, Carlonas RL et al (2000) Avoidance of soccer injuries with preseason conditioning. Am J Sports Med 28(5):659–662

Hewett TE, Stroupe AL, Nance AT et al (1996) Plyometric training in female athletes. Decreased impact forces and increased hamstring torques. Am J Sports Med 24(6):765–773

Hewett TE, Lindenfeld TN, Riccobene JV et al (1999) The effect of neuromuscular training on the incidence of knee injury in female athletes. A prospective study. Am J Sports Med 27(6):699–706

Hewett TE, Myer GD, Ford KR (2004) Decrease in neuromuscular control about the knee with maturation in female athletes. J Bone Joint Surg Am 86-A:1601–1608

Hewett TE, Myer GD, Ford KR et al (2005) Biomechanical measures of neuromuscular control and valgus loading of the knee predict anterior cruciate ligament injury risk in female athletes: a prospective study. Am J Sports Med 33:492–501

Kiani A, Hellquist E, Ahlqvist K et al (2010) Prevention of soccer-related knee injuries in teenaged girls. Arch Intern Med 170(1):43–49

Koga H, Nakamae A, Shima Y et al (2010) Mechanisms for noncontact anterior cruciate ligament injuries: Knee joint kinematics in 10 injury situations from female team handball and basketball. Am J Sports Med 38:2218–2225

Krosshaug T, Nakamae A, Boden BP et al (2007) Mechanisms of anterior cruciate ligament injury in basketball: video analysis of 39 cases. Am J Sports Med 35:359–367

LeBella CR, Huxford MR, Grissom J et al (2011) Effect of neuromuscular warm-up on injuries in female soccer and basketball athletes in urban public high schools: cluster randomized controlled trial. Arch Pediatr Adolesc Med 165(11):1033–1040

Leddy M, Lambert M, Ogles B (1994) Psychological consequences of athletic injury among high-level competitors. Res Q Exerc Sport 65:347–354

Levine JW, Kiapour AM, Quatman CE et al (2013) Clinically relevant injury patterns after an anterior cruciate ligament injury provide insight into injury mechanisms. Am J Sports Med 41:385–395

Lim B, Lee YS, Kim JG et al (2009) Effects of sports injury prevention training on the biomechanical risk factors of anterior cruciate ligament injury in high school female basketball players. Am J Sports Med 37:1728–1734

Mandelbaum BR, Silvers HJ, Watanabe DS et al (2005) Effectiveness of a neuromuscular and proprioceptive training program in preventing anterior cruciate ligament injuries in female athletes. 2-year follow-up. Am J Sports Med 33:1003–1010

Marshall S, Guskiewicz K (2003) Sports and recreational injury: the hidden cost of a healthy life-style. Inj Prev 9:100–102

Myer GD, Ford KR, Palumbo JP et al (2005) Neuromuscular training improves performance and lower-extremity biomechanics in female athletes. J Strength Cond Res 19(1):51–60

Myer GD, Ford KR, McLean SG et al (2006) The effects of plyometric versus dynamic stabilization and balance training on lower extremity biomechanics. Am J Sports Med 34(3):445–455

Myer GD, Ford KR, Hewett TE et al (2011) New method to identify athletes at high risk of ACL injury using clinic-based measurements and freeware computer analysis. Br J Sports Med 45:238–244

Myer GD, Sugimoto D, Thomas S et al (2012) The influence of age on the effectiveness of neuromuscular training to reduce anterior cruciate ligament injury in female athletes: a meta-analysis. Am J Sports Med 41:203–215

Olsen OE, Myklebust G, Engebretsen L et al (2005) Exercises to prevent lower limb injuries in youth sports: cluster randomised controlled trial. BMJ 330:449
Otsuki R, Magoshi H, Fukubayashi T (2014) ACL injury risk in female basketball players during maturation. Jpn J Clin Sports Med 22(1):51–58
Pfeiffer RP, Shea KG, Roberts D et al (2006) Lack of effect of a knee ligament injury prevention program on the incidence of noncontact anterior cruciate ligament injury. J Bone Joint Surg Am 88(8):1769–1774
Pollard CD, Sigward SM, Ota S et al (2006) The influence of in-season injury prevention training on lower-extremity kinematics during landing in female soccer players. Clin J Sports Med 16:223–227
Powell JW, Barber-Foss KD (2000) Sex-related injury patterns among selected high school sports. Am J Sports Med 28(3):385–391
Quatman CE, Ford KR, Myer GD et al (2006) Maturation leads to gender differences in landing force and vertical jump performance. A longitudinal study. Am J Sports Med 34:806–813
Roos EM (2005) Joint injury causes knee osteoarthritis in young adults. Curr Opin Rheumatol 17(2):195–200
Shea KG, Grimm NL, Ewing CK et al (2011) Youth sports anterior cruciate ligament and knee injury epidemiology: who is getting injured? In what sports? When? Clin Sports Med 30(4):691–706
Steffen K, Myklebust G, Olsen E et al (2008) Preventing injuries in female youth football-a cluster-randomized controlled trial. Scand J Med Sci Sports 18(5):605–614
Walden M, Atroshi I, Magnusson H et al (2012) Prevention of acute knee injuries in adolescent female football players: cluster randomised controlled trial. BMJ 344:e3042
Yu B, McClure SB, Onate JA et al (2005) Age and gender effects on lower extremity kinematics of youth soccer players in a stop-jump task. Am J Sports Med 33:1356–1364

Olsen OE, Myklebust G, Engebretsen L et al (2005) Exercises to prevent lower limb injuries in youth sports: cluster randomised controlled trial. BMJ 330:449
Otsuki R, Matsuo H, Fukubayashi T (2014) ACL injury risk of female basketball players during maturation. Jpn J Clin Sports Med 22(1):[illegible]–58
Pfeiffer RP, Shea KG, Roberts D et al (2006) Lack of effect of a knee ligament injury prevention program on the incidence of noncontact anterior cruciate ligament injury. J Bone Joint Surg Am 88(8):1769–1774
Pollard CD, Sigward SM, Ota S et al (2006) The influence of in-season injury prevention training on lower-extremity kinematics during landing in female soccer players. Clin J Sport Med 16:223–227
Powell JW, Barber-Foss KD (2000) Sex-related injury patterns among selected high school sports. Am J Sports Med 28(3):385–391
Quatman CE, Ford KR, Myer GD et al (2006) Maturation leads to gender differences in landing force and vertical jump performance: a longitudinal study. Am J Sports Med 34(5):806–813
Roos EM (2005) Joint injury causes knee osteoarthritis in young adults. Curr Opin [illegible] 17(2):195–200
Shea KG, Grimm NL, Ewing CK et al (2011) Youth sports anterior cruciate ligament and knee injury epidemiology: who is getting injured? In what sports? When? Clin Sports Med 30(4):691–706
Steffen K, Myklebust G, Olsen OE et al (2008) Preventing injuries in female youth football—a cluster-randomized controlled trial. Scand J Med Sci Sports 18(5):605–614
Waldén M, Atroshi I, Magnusson H et al (2012) Prevention of acute knee injuries in adolescent female football players: cluster randomised controlled study. BMJ 344:e3042
Yu B, McClure SB, Onate JA et al (2005) Age and gender effects on lower extremity kinematics of youth soccer players in a stop-jump task. Am J Sports Med 33(9):1356–[illegible]

Chapter 16
FIFA 11+ Injury Prevention in Amateur Football from Development to Worldwide Dissemination

Mario Bizzini, Astrid Junge, and Jiri Dvorak

Abstract In the last decade injury prevention has received a lot of attention in sports medicine. Recently, international sports governing bodies such as the International Olympic Committee (IOC) have declared the protection of the athletes' health as one of their major objectives.

In 1994 the Fédération Internationale de Football Association (FIFA) established its Medical Assessment and Research Centre (F-MARC) with the aim "to prevent football injuries and to promote football as a health-enhancing leisure activity, improving social behaviour". Since then, FIFA has developed and evaluated its injury prevention programmes "The 11" and "FIFA 11+" in several scientific studies, demonstrating how simple exercise-based programmes can decrease the incidence of injuries in amateur football players.

This paper summarises 18 years of scientific and on-field work in injury prevention by an international sports federation (FIFA), from formulating the aim to make its sport safer, to the worldwide dissemination of its injury prevention programme in amateur football.

Keywords Injury prevention • Soccer • Prevention programme • Dissemination

---

This paper is an adaptation of the original publication "Implementation of the FIFA 11+ football warm up program: how to approach and convince the Football Associations to invest in prevention", "open access" by the *British Journal of Sports Medicine* (Br J Sports Med. 2013; 47(12):803–6. Epub 2013/06/27).

Electronic supplementary material The online version of this chapter (doi: 10.1007/978-4-431-55318-2_16) contains supplementary material, which is available to authorized users.

M. Bizzini (✉) • A. Junge • J. Dvorak
FIFA – Medical Assessment and Research Centre (F-MARC), Schulthess Clinic, Zurich, Switzerland
e-mail: mario.bizzini@f-marc.com

K. Kanosue et al. (eds.), *Sports Injuries and Prevention*,
DOI 10.1007/978-4-431-55318-2_16

16.1 Introduction

Football is played on an amateur or recreational level by almost 300 million people all over the world (FIFA 2006). Besides being a healthy leisure activity, football, as a contact team sport, has a certain risk of injury. The medical treatment of football related injuries can have a significant socio-economic impact: for example, in Switzerland (with 7.9 million inhabitants) the healthcare costs for football injuries were nearly 170 million US Dollars in 2010 (SUVA).

There is extensive literature on the frequency and characteristics of football injuries (Agel et al. 2007; Dick et al. 2007; Ekstrand et al. 2011; Giza and Micheli 2005; Junge and Dvorak 2004), and several scientific studies on injury prevention programmes in amateur football players have been published (Emery and Meeuwisse 2010; Engebretsen et al. 2008; Gilchrist et al. 2008; Hagglund et al. 2007; Heidt et al. 2000; Junge et al. 2002; Kiani et al. 2010; LaBella et al. 2011; Mandelbaum et al. 2005; Mohammadi 2007; Sharpe et al. 1997; Soderman et al. 2000; Soligard et al. 2008; Surve et al. 1994; Tropp et al. 1985; van Beijsterveldt et al. 2012; Walden et al. 2012; Gatterer et al. 2012). However, the implementation of injury prevention programmes in the real-world of sports represents a major challenge (Finch 2006; 2011). Nevertheless, the effectiveness of country-wide campaigns to reduce the incidence of football injuries has been proven (Dick et al. 2009; Junge et al. 2011).

In 1994 Fédération Internationale de Football Association (FIFA) founded its Medical Assessment and Research Centre (F-MARC) in order to create and disseminate scientific knowledge on various medical topics in football, to reduce football injuries and to promote football as a health enhancing leisure activity.

Eighteen years later, this paper summarises the historical background, development, scientific evaluation, and dissemination strategies of FIFA's injury prevention programmes ("FIFA 11+") in order to provide a model of how an international sports federation can make its sport safer.

16.2 Development of Injury Prevention Programmes

The first scientific study on injury prevention in football was Jan Ekstrand's thesis in the 1980s (Ekstrand et al. 1983). For about 20 years, no other author published a research paper on the prevention of football injuries in general, and only few studies investigated the prevention of recurrent ankle sprains (Sharpe et al. 1997; Surve et al. 1994; Tropp et al. 1985) and/or severe knee injuries (Soderman et al. 2000; Caraffa et al. 1996). In 2000 Heidt et al. showed that the 42 female players who participated in the Frappier Acceleration Training Programme had less time-loss injuries than the control group (Heidt et al. 2000). At the same time, F-MARC conducted its first study on the prevention of football injuries, showing 21 % fewer injuries in the intervention as compared to the control group (Junge et al. 2002). The interventions were focused on improving the structure and content of the training by educating and supervising the coaches and players. The programme included preventive interventions such as improvement of warm-up, regular cool-down, taping of unstable ankles, adequate rehabilitation, promotion of the spirit of fair play and ten sets of exercises designed to improve coordination, stability of the ankle and knee, flexibility and strength of the trunk, hip and leg muscles. Based on the experiences with this pilot study and in cooperation with international experts, F-MARC developed a basic injury prevention programme for amateur football players called "The 11".

"**The 11**" comprises 10 evidence-based or best-practice exercises (core stability, balance, dynamic stabilisation, and eccentric hamstring strength) and the promotion of fair play. The programme was designed to reduce the most common football injuries (ankle and knee sprains, hamstring and groin strains). It can be completed in 10–15 min and requires no equipment other than a ball. "The 11" was implemented in two countrywide campaigns (Switzerland and New Zealand) in cooperation with the national accident insurance company and the national football association (Dick et al. 2009; Junge et al. 2011).

In Switzerland, implementation of "The 11" and its effect on the injury rate were carefully evaluated by an independent research company (Junge et al. 2011). Four years after the launch of the programme, teams that included "The 11" as a part of their warm-up had 11.5 % fewer match injuries and 25.3 % fewer training injuries than teams that warmed-up as usual. In New Zealand, implementation of "The 11" resulted in a 2.4 dollar return of investment for the national accident insurance company after 2 years (Dick et al. 2009). In a controlled randomised study (RCT) compliance of the intervention group was too poor to demonstrate a statistically significant effect of the programme (Steffen et al. 2008).

Based on experiences with "The 11", "PEP" (Prevent Injury and Enhance Performance programme) (Gilchrist et al. 2008; Mandelbaum et al. 2005) and other exercise-based programmes (Hewett et al. 1999; 2000; Soderman et al. 2000; Caraffa et al. 1996) to prevent football injuries, an advanced version ("**FIFA 11+**") was developed in 2006 together with the OSTRC and the Santa Monica Orthopaedic and Sports Medicine Research Foundation. "FIFA 11+" is a complete warm-up programme with running exercises at the beginning and end to activate the cardiovascular system, and specific preventive exercises focussing on core and leg strength, balance and agility, and contains three levels of increasing difficulty to provide variation and progression. It takes about 20 min to be completed, and requires a minimum of equipment (a set of cones and balls). "FIFA 11+" is time-efficient because it replaces the usual warm-up.

A RCT, conducted by the Oslo Sports Trauma and Research Center, showed that young female teams performing the "FIFA 11+" at least twice a week (as a standard warm up before their training) had 37 % fewer training injuries and 29 % fewer match injuries, and severe injuries were reduced by almost 50 % (Soligard et al. 2008). The compliance with "FIFA 11+" was high, and players with a higher compliance had a significantly lower injury risk than others (Soligard et al. 2010). Recently, a large RCT in male soccer players found significantly lower training and match injuries in teams utilising "FIFA 11+" as a warm-up routine (yet to be published; Silvers, Mandelbaum et al. Santa Monica Sports Medicine Foundation, CA, USA).

In another RCT study on Canadian youth female football players, Steffen et al. found that players with higher adherence to "FIFA 11+" showed significant improvements in functional balance and reduced injury risk (Steffen et al. 2013a). Two recent studies on Italian amateur football players showed that the physiological warm up effects of "FIFA 11+" are similar to or even better than a standard warm up routine, and that it enhances neuromuscular control (core/lower extremity) and knee flexor strength (Bizzini et al. 2013a; Impellizzeri et al. 2013).

Other authors have found improvements in static/dynamic balance, and thigh muscle strength in male football and futsal players after performing "FIFA 11+" (Daneshjoo et al. 2012a; b; Reis et al. 2013; Brito et al. 2010).

16.3 Development of a Dissemination Strategy

For F-MARC, the **coach** is the key person to promote injury prevention to his/her players. While the coach, especially at a low level, has to regard various aspects in the training (e.g. physical preparation, tactics, fair play, team success), it is important to raise his/her motivation to implement an injury prevention programme with his team. It is important to stress that regular and correct performance of the exercises is crucial for the preventive effect.

Information material on "FIFA 11+" was developed, produced and made available for coaches and players. The material includes a detailed manual, an instructional DVD, a poster, a website and a promotional booklet which includes a DVD. All material is available in the four FIFA languages (English, Spanish, German and French) and can be accessed on www.F-MARC.com/11plus.

See ESM 1.

"FIFA 11+" is best taught to coaches in a **workshop** that includes theoretical background knowledge and practical demonstration of the exercises as recently shown by Steffen et al. (2013b) in a cluster randomised trial on the utilisation of different implementation strategies. The authors found that a pre-season coach education workshop was more effective in terms of better compliance and decreased injury risk in players than other delivery methods (unsupervised website, additional supervision by a physical therapist) of the programme. After increasing the motivation of the coach and raising awareness of injury prevention, the exercises should be briefly explained and demonstrated. It is helpful to select a participant to perform the exercise, while the instructor highlights the correct execution of the exercises. The participants should then perform the exercises and be corrected by the instructor(s). The participants should get "a feeling" for the exercises and appreciate the challenges behind each exercise. In the second half of the workshop each of the participants should teach at least one of the exercises to the group, and get feedback on this performance from the instructor.

For the **countrywide campaign** in Switzerland, "The 11" was integrated into the coach education of the Swiss Football Association (Schweizerischer Fussballverband (SFV)) using a "teach the teacher" strategy. All instructor coaches of the SFV were educated by sports physical therapists on how to deliver the programme to the coaches in their licensing or refresher courses. During a period of 3 years, 5,000 licensed amateur coaches were subsequently instructed on how to perform "The 11" with their teams and received the information material (Junge et al. 2011). The same strategy was used in New Zealand, where "The 11" was implemented as part of the "SoccerSmart Program" (Dick et al. 2009; Gianotti et al. 2010).

16.4 Worldwide Dissemination of "FIFA 11+"

In 2009 FIFA started the dissemination of "FIFA 11+" in its 209 Member Associations (MAs). Based on the experience with the countrywide implementation in Switzerland and New Zealand, and on the evaluations of other sports injury prevention programme implementations (i.e. rugby) (Chalmers et al. 2004; Gianotti et al. 2009), a guideline on how to implement the "FIFA 11+" injury prevention programme at a larger scale in amateur football was developed (Bizzini et al. 2013b) (Fig. 16.1). The implementation is conducted either in close cooperation with MAs or via FIFA Coaching Instructor courses.

The national Football Associations of Spain, Japan, Italy, Brazil and Germany integrated "FIFA 11+" in their coaching curriculum or in their physical training/education curriculum. Thus, the world football champions took the lead and acted as role models, and other MAs (Costa Rica, Hong Kong, Netherlands, Denmark, Poland, Hungary, Australia, England, Thailand, and Singapore) followed. Up to now about 6,000 coaches from more than 70 countries have been instructed on how to implement the "FIFA 11+" (Bizzini et al. 2013b) (Fig. 16.2). Even if this represents just about 25 % of FIFA's MAs and a much lower percentage of all football coaches, it is an important step for the worldwide dissemination of the programme. F-MARC supports the MAs in the preparation of the educational material in the local language, and the workshops for the first group of instructors to initiate the cascade training.

See ESM 2.

16.5 Lessons Learned

The F-MARC team has gained experience during the years of dissemination of the injury prevention programmes. It has been found that understanding the coach's character and highlighting the importance of the programme to the coach is especially important. By preventing injuries and therefore reducing the number of injured players, the coach will have more players available for his/her ideal team.

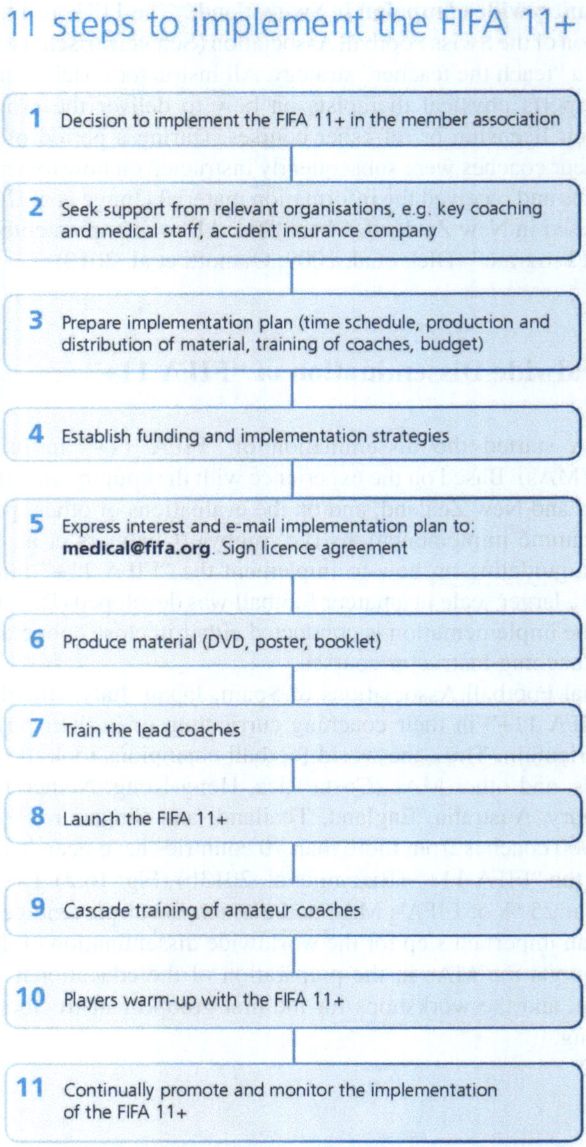

Fig. 16.1 FIFA 11+ implementation guideline. "Eleven steps to implement FIFA 11+"

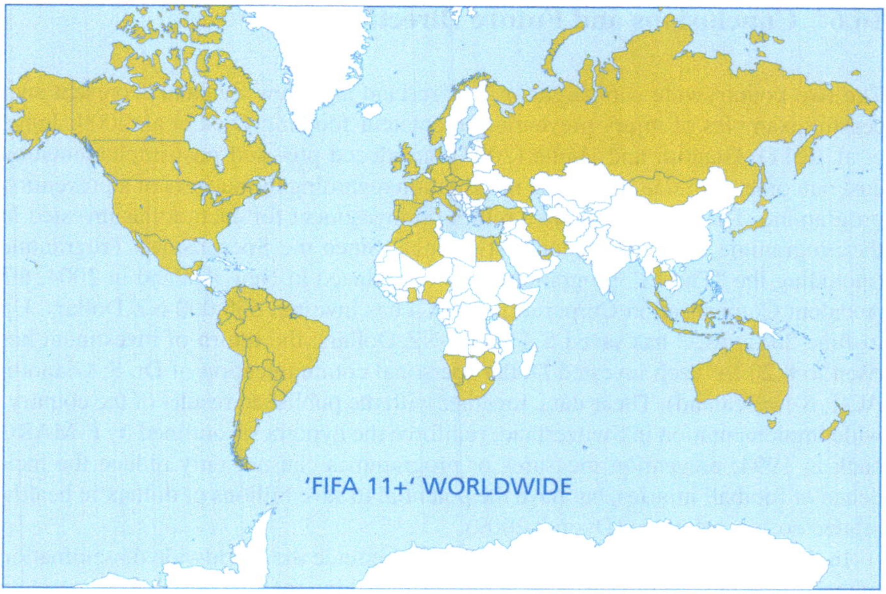

Fig. 16.2 FIFA 11+ worldwide. The countries in the yellow are the ones whose coaches attended a FIFA 11+ instructor course (2008–2013)

Therefore, it is not only information and education about the role of injury prevention which is important, but moreover to speak the same language as the coach. One of keys while conducting a course is “proposing” rather than “imposing” “FIFA 11+”. The dialogue on the pitch with coaches is often more important than the distributed materials, thus allowing for friendly discussion and practical work with the preventive programme. Therefore, the choice of the instructors is crucial and F-MARC’s best experiences have been with sports physiotherapists or athletic trainers who have an active involvement in football, because they already “live and speak the football language”. Additionally, cooperation with famous players and coaches acting as “FIFA 11+” ambassadors (see teaser on https://vimeo.com/45562029 and www.f-marc.com/11plus) has helped significantly in the communication with coaches. “FIFA 11+” has also been presented to the delegates of all MAs at the last two FIFA Medical Conferences (Zürich 2009; Budapest 2012). After initial enthusiasm from the interested MAs, F-MARC has experienced a wide range of dedication and compliance from those MAs to the proposed “FIFA 11+” implementation guidelines. At the MA level, it has to be acknowledged that highly motivated people are needed, in order to successfully plan, realise, and constantly monitor a countrywide implementation.

16.6 Conclusions and Future Directions

The two countrywide campaigns in Switzerland and New Zealand represent successful examples of injury prevention in amateur football (Dick et al. 2009; Junge et al. 2011). Gianotti and Hume (2007) introduced pre- and post-implementation cost outcome formulae to provide information regarding the success of a prevention programme. These data provide a return on investment for each dollar invested in the programme and subsequent cost-savings. Since the SoccerSmart Programme (including the "The 11" programme) was introduced in New Zealand in 2004, the Accident Compensation Corporation (ACC) has invested 650,000 NZ Dollars. Up to June 2011, ACC has saved 5,331,000 NZ Dollars: the return of investment has risen to 8.20 for each invested Dollar (personal communication of Dr. S. Gianotti, ACC, New Zealand). These data, together with the published results of the countrywide implementation in Switzerland, reinforce the hypothesis outlined by F-MARC back in 1994: prevention measures or programmes can not only reduce the incidence of football injuries, but have the potential to save billions of dollars in health-related costs worldwide (Dvorak 2009).

In the next years FIFA and F-MARC will continue the worldwide dissemination of the "FIFA 11+", with particular attention to seeking the best possible cooperation with the MAs adopting the "FIFA 11+" injury prevention programme.

References

Agel J, Evans TA, Dick R et al (2007) Descriptive epidemiology of collegiate men's soccer injuries: National Collegiate Athletic Association Injury Surveillance System, 1988–1989 through 2002–2003. J Athl Train 42(2):270–277. Epub 2007/08/22

Bizzini M, Impellizzeri F, Dvorak J et al (2013a) Physiological and performance responses to the FIFA 11+ (part 2): is it a good warm up? J Sports Sci 31(13):1491–502

Bizzini M, Junge A, Dvorak J (2013b) Implementation of the FIFA 11+ football warm up program: how to approach and convince the Football associations to invest in prevention. Br J Sports Med 47(12):803–806. Epub 2013/06/27

Brito J, Figueiredo P, Fernandes L et al (2010) Isokinetic strength effects of FIFA's "The 11+" injury prevention training programme. Isokinet Exerc Sci 18:211–215

Caraffa A, Cerulli G, Projetti M et al (1996) Prevention of anterior cruciate ligament injuries in soccer. A prospective controlled study of proprioceptive training. Knee Surg Sports Traumatol Arthrosc 4(1):19–21. Epub 1996/01/01

Chalmers DJ, Simpson JC, Depree R (2004) Tackling Rugby injury: lessons learned from the implementation of a five-year sports injury prevention program. J Sci Med Sport 7(1):74–84. Epub 2004/05/14

Daneshjoo A, Mokhtar AH, Rahnama N et al (2012a) The effects of comprehensive warm-up programs on proprioception, static and dynamic balance on male soccer players. PLoS One 7(12):e51568. Epub 2012/12/20

Daneshjoo A, Mokhtar AH, Rahnama N et al (2012b) The effects of injury preventive warm-up programs on knee strength ratio in young male professional soccer players. PLoS One 7(12):e50979. Epub 2012/12/12

Dick R, Putukian M, Agel J et al (2007) Descriptive epidemiology of collegiate women's soccer injuries: National Collegiate Athletic Association Injury Surveillance System, 1988–1989 through 2002–2003. J Athl Train 42(2):278–285. Epub 2007/08/22

Dick R, Gaulet C, Gianotti S (2009) Implementing large-scale injury prevention programs. In: Bahr R, Engebretsen L (eds) Injury prevention. Wiley-Blackwell, Chichester

Dvorak J (2009) Give Hippocrates a jersey: promoting health through football/sport. Br J Sports Med 43(5):317–322. Epub 2009/03/25

Ekstrand J, Gillquist J, Liljedahl SO (1983) Prevention of soccer injuries. Supervision by doctor and physiotherapist. Am J Sports Med 11(3):116–120. Epub 1983/05/01

Ekstrand J, Hagglund M, Walden M (2011) Injury incidence and injury patterns in professional football: the UEFA injury study. Br J Sports Med 45(7):553–558. Epub 2009/06/26

Emery CA, Meeuwisse WH (2010) The effectiveness of a neuromuscular prevention strategy to reduce injuries in youth soccer: a cluster-randomised controlled trial. Br J Sports Med 44(8):555–562. Epub 2010/06/16

Engebretsen AH, Myklebust G, Holme I et al (2008) Prevention of injuries among male soccer players: a prospective, randomized intervention study targeting players with previous injuries or reduced function. Am J Sports Med 36(6):1052–1060. Epub 2008/04/09

FIFA (2006) Big count: www.fifa.com/worldfootball/bigcount/. [Cited 2013 21.02.2013]

Finch C (2006) A new framework for research leading to sports injury prevention. J Sci Med Sport 9(1–2):3–9; discussion 10. Epub 2006/04/18

Finch CF (2011) No longer lost in translation: the art and science of sports injury prevention implementation research. Br J Sports Med 45(16):1253–1257. Epub 2011/06/24

Gatterer H, Ruedl G, Faulhaber M et al (2012) Effects of the performance level and the FIFA "11" injury prevention program on the injury rate in Italian male amateur soccer players. J Sports Med Phys Fitness 52(1):80–84. Epub 2012/02/14

Gianotti S, Hume PA (2007) A cost-outcome approach to pre and post-implementation of national sports injury prevention programmes. J Sci Med Sport 10(6):436–446. Epub 2007/03/14

Gianotti SM, Quarrie KL, Hume PA (2009) Evaluation of RugbySmart: a rugby union community injury prevention programme. J Sci Med Sport 12(3):371–375. Epub 2008/03/22

Gianotti S, Hume PA, Tunstall H (2010) Efficacy of injury prevention related coach education within netball and soccer. J Sci Med Sport 13(1):32–35. Epub 2008/12/09

Gilchrist J, Mandelbaum BR, Melancon H et al (2008) A randomized controlled trial to prevent noncontact anterior cruciate ligament injury in female collegiate soccer players. Am J Sports Med 36(8):1476–1483. Epub 2008/07/29

Giza E, Micheli LJ (2005) Soccer injuries. Med Sport Sci 49:140–169. Epub 2005/10/26

Hagglund M, Walden M, Ekstrand J (2007) Lower reinjury rate with a coach-controlled rehabilitation program in amateur male soccer: a randomized controlled trial. Am J Sports Med 35(9):1433–1442. Epub 2007/03/21

Heidt RS Jr, Sweeterman LM, Carlonas RL et al (2000) Avoidance of soccer injuries with preseason conditioning. Am J Sports Med 28(5):659–662. Epub 2000/10/14

Hewett TE, Lindenfeld TN, Riccobene JV et al (1999) The effect of neuromuscular training on the incidence of knee injury in female athletes. A prospective study. Am J Sports Med 27(6):699–706. Epub 1999/11/24

Impellizzeri F, Bizzini M, Dvorak J et al (2013) Physiological and performance responses to the FIFA 11+ (part 1): a randomized controlled trial on the training effects. J Sports Sci 31(13):1491–502

Junge A, Dvorak J (2004) Soccer injuries: a review on incidence and prevention. Sports Med 34(13):929–938. Epub 2004/10/19

Junge A, Rosch D, Peterson L et al (2002) Prevention of soccer injuries: a prospective intervention study in youth amateur players. Am J Sports Med 30(5):652–659. Epub 2002/09/20

Junge A, Lamprecht M, Stamm H et al (2011) Countrywide campaign to prevent soccer injuries in Swiss amateur players. Am J Sports Med 39(1):57–63. Epub 2010/10/20

Kiani A, Hellquist E, Ahlqvist K et al (2010) Prevention of soccer-related knee injuries in teenaged girls. Arch Intern Med 170(1):43–49. Epub 2010/01/13

LaBella CR, Huxford MR, Grissom J et al (2011) Effect of neuromuscular warm-up on injuries in female soccer and basketball athletes in urban public high schools: cluster randomized controlled trial. Arch Pediatr Adolesc Med 165(11):1033–1040. Epub 2011/11/09

Mandelbaum BR, Silvers HJ, Watanabe DS et al (2005) Effectiveness of a neuromuscular and proprioceptive training program in preventing anterior cruciate ligament injuries in female athletes: 2-year follow-up. Am J Sports Med 33(7):1003–1010. Epub 2005/05/13

Mohammadi F (2007) Comparison of 3 preventive methods to reduce the recurrence of ankle inversion sprains in male soccer players. Am J Sports Med 35(6):922–926. Epub 2007/03/24

Reis I, Rebelo A, Krustrup P et al (2013) Performance enhancement effects of Federation Internationale de Football Association's "The 11+" injury prevention training program in youth futsal players. Clin J Sport Med Off J Can Acad Sport Med 23(4):318–20

Sharpe SR, Knapik J, Jones B (1997) Ankle braces effectively reduce recurrence of ankle sprains in female soccer players. J Athl Train 32(1):21–24. Epub 1997/01/01

Soderman K, Werner S, Pietila T et al (2000) Balance board training: prevention of traumatic injuries of the lower extremities in female soccer players? A prospective randomized intervention study. Knee Surg Sports Traumatol Arthrosc 8(6):356–363. Epub 2001/01/09

Soligard T, Myklebust G, Steffen K et al (2008) Comprehensive warm-up programme to prevent injuries in young female footballers: cluster randomised controlled trial. BMJ 337:a2469. Epub 2008/12/11

Soligard T, Nilstad A, Steffen K et al (2010) Compliance with a comprehensive warm-up programme to prevent injuries in youth football. Br J Sports Med 44(11):787–793. Epub 2010/06/17

Steffen K, Myklebust G, Olsen OE et al (2008) Preventing injuries in female youth football – a cluster-randomized controlled trial. Scand J Med Sci Sports 18(5):605–614. Epub 2008/01/23

Steffen K, Emery CA, Romiti M et al (2013a) High adherence to a neuromuscular injury prevention programme (FIFA 11+) improves functional balance and reduces injury risk in Canadian youth female football players: a cluster randomised trial. Br J Sports Med 47:794–802

Steffen K, Meeuwisse WH, Romiti M et al (2013b) Evaluation of how different implementation strategies of an injury prevention program impact team adherence and injury risk in Canadian female youth football players. Br J Sports Med 47(8):480–7

Surve I, Schwellnus MP, Noakes T et al (1994) A fivefold reduction in the incidence of recurrent ankle sprains in soccer players using the Sport-Stirrup orthosis. Am J Sports Med 22(5):601–606. Epub 1994/09/01

SUVA. Ball sport statistics. http://www.suva.ch/startseite-suva/praevention-suva/sichere-freizeit-suva/kampagne-ballsport-fitness-suva/statistik-ballsport-suva.htm. Available from: http://www.suva.ch/startseite-suva/praevention-suva/sichere-freizeit-suva/kampagne-ballsport-fitness-suva/statistik-ballsport-suva.htm

Tropp H, Askling C, Gillquist J (1985) Prevention of ankle sprains. Am J Sports Med 13(4):259–262. Epub 1985/07/01

van Beijsterveldt AM, van de Port IG, Krist MR et al (2012) Effectiveness of an injury prevention programme for adult male amateur soccer players: a cluster-randomised controlled trial. Br J Sports Med 46(16):1114–1118. Epub 2012/08/11

Walden M, Atroshi I, Magnusson H et al (2012) Prevention of acute knee injuries in adolescent female football players: cluster randomised controlled trial. BMJ 344:e3042. Epub 2012/05/05

Chapter 17
Influence of Changing Direction on the Center of Gravity and Knee Joint Angle in Rugby Players

Norihiko Sunagawa and Toru Fukubayashi

Abstract While rugby is a collision sport, many anterior cruciate ligament (ACL) injuries of the knee involve noncontact changes of direction. This study investigates the risk factors of ACL injury in rugby players by establishing the relationship between noncontact directional change by the players and the consequent alteration of the player's center of gravity (CG) and knee joint motion. Methods: Photographs were taken when players at a university rugby football club changed their direction of movement by 45° or 90° (n = 10). The player's motions were assessed to determine 3- and 2-dimensional coordinate values of markers on the players by using the direct linear transformation method. The values were used to calculate knee joint flexion 40 ms after foot grounding, the CG position (height and rearward position), and the abduction angle of the knee joint. A paired *t* test was used for statistical analysis; a probability level of less than 5 % was considered significant. Results and Discussion: No significant differences were observed in the knee flexion angle between a 45° and a 90° change in direction. However, the CG position (height and rearward position) and the abduction angle of the knee joint increased as the magnitude of the change in direction increased ($p < 0.01$). Because increases in the rearward CG and abduction angle of the knee joint are considered risk factors for ACL injury, the risk of ACL injury is predicted to go up as the magnitude of the change in direction increases.

Keywords Knee joint • Angle of direction change • Rugby union • ACL

N. Sunagawa (✉)
Graduate School of Sport Sciences, Waseda University, Saitama, Japan
e-mail: n.sunagawa@akane.waseda.j

T. Fukubayashi
Faculty of Sport Sciences, Waseda University, Saitama, Japan

K. Kanosue et al. (eds.), *Sports Injuries and Prevention*,
DOI 10.1007/978-4-431-55318-2_17

17.1 Introduction

Rugby is characterized by fierce collisions, and many injuries occur during the scrum, maul, ruck, and while tackling. Based on their survey of injuries in rugby games, Fuller et al. (2008) reported that knee joints are frequently injured. With knee ligament injuries being the most severe. Likewise, Brooks et al. (2005) found that among rugby injuries, those involving the anterior cruciate ligament (ACL) were the most severe. While there is no doubt that a large number of ACL injuries occur during contact situations, a study in Australia (Cochrane et al. 2007), which has many rugby and Australian football teams, noted that 56 % of the ACL injuries in that country occurred in noncontact sports.

The ACL, a major ligament of the knee joint, is significantly involved in knee stability and plays an important role in sports that involve rapid motion. According to a study on noncontact ACL injuries sustained in players of sports such as soccer, handball, and basketball, light flexion of the knee joint, the position of the knee joint during abduction, a rearward center of gravity (CG), single-leg grounding, and abrupt direction changing can be risky movements or positions. It is of note that at approximately 40 ms after grounding, the highest tension is loaded on the ACL (Alentorn-Geli et al. 2009; Boden et al. 2000; Griffin et al. 2000; Hewett et al. 2005, 2009; Koga et al. 2010; McLean et al. 2003; McNair et al. 1990; Olsen et al. 2004).

An abrupt change in direction, which is regarded as one of the major causes of ACL injury, frequently occurs in rugby. This maneuver is often used to shake off the opposing player; subsequently, another quick motion is often required in order to respond to a readjustment of the opposing player. In addition to the acute effects of repeated and frequent changes of direction, many rugby players experience after effects which include knee pain and discomfort during knee motion. Few studies of rugby players have evaluated how rapid changes in motion affect such ACL injury risk factors as knee joint position and altered CG.

17.2 Purpose

This study is an investigation of how rapid changes in direction affect the CG and knee joint motion in rugby players.

17.3 Method

17.3.1 Subjects

The subjects were ten male rugby players belonging to a university rugby club in group A of the Kanto Varsity Matches championship. The subjects were healthy and had no history of ACL injury, lower limb fracture, joint instability, or injuries interfering with exercise. We explained the objectives of the experiment to the

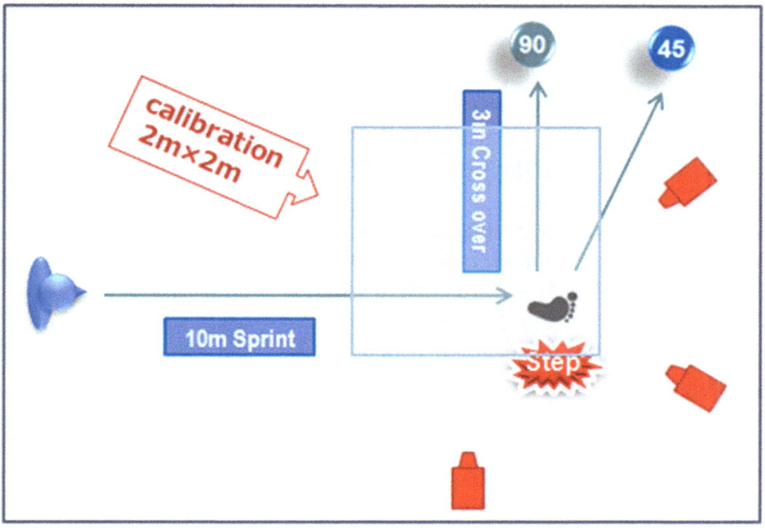

Fig. 17.1 Diagram of experiment field

subjects and obtained their consent for participation. This study was conducted after approval was obtained from the Ethical Review Board for Studies Using Humans as Subjects of Waseda University.

17.3.2 Motion Task

The subjects were requested to perform a 10-m linear sprint followed by one of two changes in direction, and then a final 5-m sprint. The changes in direction were angles of (1) 45° and (2) 90°; both turns were performed without a ball (Fig. 17.1).

The subjects used their right leg as a pivot for both the 45° and 90° turn, both of which were to the left Each trial was performed three times, each with a maximum effort. Before the performance the players were given a sufficient warm up and confirmation of the motion they were to do. All data were recorded. Sufficient rest time was given between trials to eliminate any effect of fatigue.

17.3.3 Measurement of Motions

Three-dimensional position coordinates of markers on the players were measured from three directions using three high-speed cameras (High Speed Exilim Ex-F1; Casio Computer Co., Ltd. Japan). Sampling frequency was set at 300 Hz.

All camera tripods were equipped with a leveling gauge so that the motions could be photographed horizontally. Sensors that started the cameras at a light signal were used to synchronize the three high-speed cameras. To closely simulate the conditions under which injuries typically occur, the experiment was conducted on a rugby training field of artificial turf that was used to prepare for regular games. The subjects wore the same "used" spike shoes that they wore for regular rugby play.

A bar equipped with a leveling gauge was used for calibration. The calibration matrix involved a 2×2 m area, in which a total of 45 points were set, including 9 sites in 4 quarters at 1-m intervals, each of which had 5 positions in height at 50-cm intervals (0, 50, 100, 150, and 200 cm).

Before conducting measurements, markers were attached to bony landmarks on the subject to calculate angles and CG position. There were 23 attachment sites in total, which were the acromions (right and left), the manubrium of the sternum, the first thoracic spine, the lower end of the rib (right and left), the superior anterior iliac spines (right and left), the greater trochanters (right and left), the tibial tuberosities (right and left), the knee joint fissure gap (internal and external), the centers of the medial and lateral malleoli (right and left), the medial and lateral malleoli (right and left), and the fifth lumbar vertebra (Fig. 17.2). The subjects wore a body-contacting suit to which the markers were attached using double-sided tape.

Fig. 17.2 The image of the locations of markers

17.3.4 Analysis

The photographs were taken using high-speed cameras equipped with image-processing software (Edius Neo3, Grass Valley USA). The images were further processed utilizing fetching by motion analysis software (Frame-DIASIV, DKH Co., Ltd. Japan) to calculate the 3- and 2-dimensional coordinate values of the markers using the direct linear transformation method.

Analysis of the knee motion and the CG was done at the time period of 40 ms immediately after foot grounding (which was set as 0 m), in accordance with a study (McLean et al. 2003) in which ground reaction force by side-step peaks was reported to peak at approximately 40 ms after grounding and another study (Koga et al. 2010) in which ACL injuries were reported to occur at 40 ms. The mean value calculated from the three trials was used in the data analysis as a representative value for each individual.

17.3.5 Analysis Items

The calculated dependent variables and the landmarks as well as the method used in their determination follow:

- Flexion angle of the knee joint
  The knee flexion angle was calculated from the 3-dimensional coordinate value of the greater trochanter of the grounding foot (right foot), the external fissure of the knee joint, and the lateral malleolus (Fig. 17.3).

Fig. 17.3 The image of knee flexion

- CG (rearward position: *Y*-coordinate; height: *Z*-coordinate)
  The CG position was calculated from 23 points, which were the tip of the hand (right and left); the center of the hand joint (right and left); elbow joints (right and left); acromions (right and left); the tip of the toe (right and left); hallux balls (right and left); heels (right and left); center of the foot joint (right and left); knee joints (right and left); greater trochanters (right and left); and the upper end of the sternum, vertex, and tragion.

  The rearward position was calculated from the *Y*-axis distance (centimeters) between the heel of the grounding foot and the CG (Fig. 17.4), whereas the height was calculated from the *Z*-axis distance (cm) between the ground and the CG (Fig. 17.5).
- Abduction angle of the knee joint
  The angle formed by the superior anterior iliac spine, center of kneecap, and center of foot joint (center of the line connecting both malleoli) was calculated on the frontal plane using the location's 2-dimensional coordinate values, as reported in a previous study (Nagano et al. 2008) (Fig. 17.6).

Fig. 17.4 The distance of backward center of gravity

Fig. 17.5 The height of backward center of gravity

Fig. 17.6 The angle of knee abduction

17.3.6 *Statistical Processing*

The results are expressed as mean ± SD. Statistical calculations used the IBM SPSS statistics package (ver. 19.0 for Windows). A paired *t* test was used for significance testing; a probability of less than 5 % was considered sufficient to reject the null hypothesis.

17.4 Results

17.4.1 *Knee Flexion Angle*

No significant difference was observed in the knee flexion angle between at the 90° and 45° direction changes 40 ms after grounding (45° cutting: 40.5° ± 4.7°; 90° cutting: 42.4° ± 4.0°) (Fig. 17.7).

17.4.2 *CG (Backward Position)*

The rearward position of the CG 40 ms after grounding showed a significantly higher value for the 90° direction change than for the 45° direction change ($p < 0.01$; 45° cutting: −8.4 ± 5.6 cm; 90° cutting: −27.4 ± 4.0 cm; Fig. 17.8).

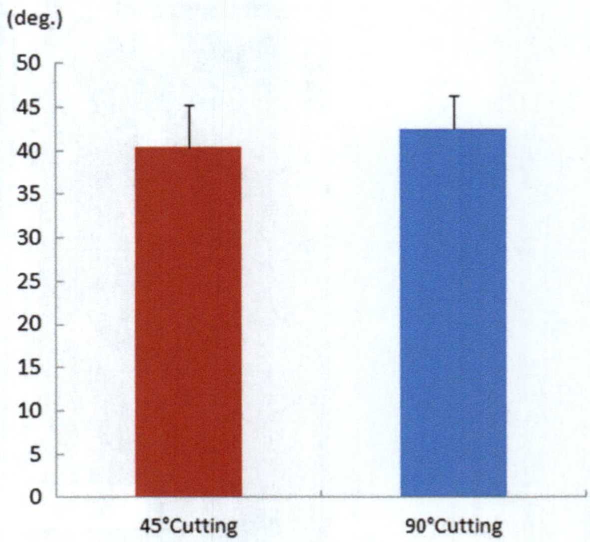

Fig. 17.7 Knee flexion angle at 40 ms in the different angle cutting

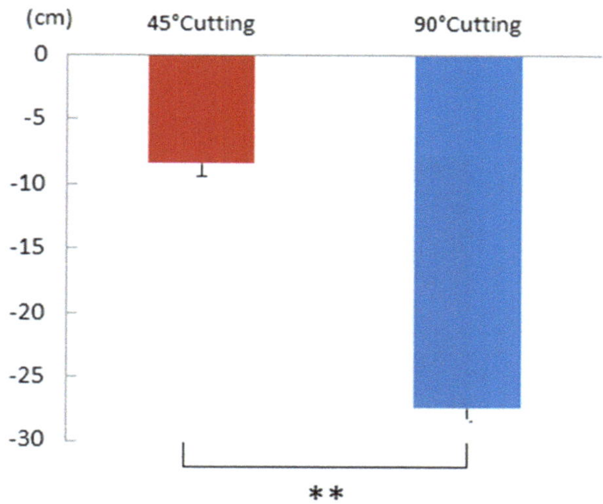

Fig. 17.8 Distance of backward center of gravity at 40 ms in the different angle cutting. **$p < 0.01$

17.4.3 CG (Height)

The CG height 40 ms after grounding showed a significantly lower value while changing direction for the 90° direction change than for the 45° direction change ($p<0.01$; 45° cutting: 73.2±4.5 cm; 90° cutting: 67.7±3.7 cm; Fig. 17.9).

17.4.4 Abduction Angle of Knee Joint

The abduction angle of knee joint 40 ms after grounding was significantly higher for the 90° direction change than for the 45° direction change (45° cutting: 6.7°±4.9°; 90° cutting: 15.7°±4.8°; Fig. 17.10).

17.5 Discussion

This study investigated how changes in the direction of movement influenced rugby players CG and knee joint motion.

This study revealed that the CG shifts rearward, the CG height lowers, and the abduction angle of the knee joint increases as the change in direction becomes greater. Boden et al. (2000) reported that a rearward CG is a risk factor for ACL injury. Hewett et al. (2005), in a study comparing the motions of ACL injury and

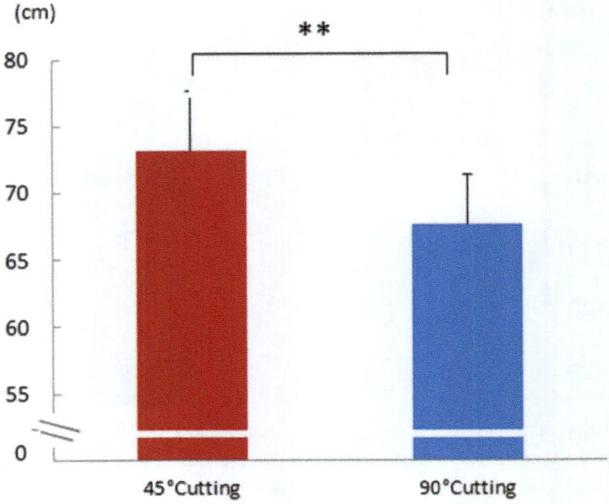

Fig. 17.9 Height of center of gravity at 40 ms in the different angle cutting. $^{**}p < 0.01$

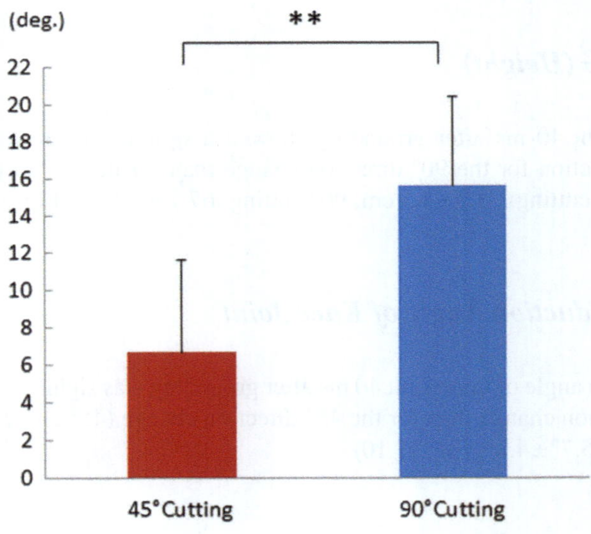

Fig. 17.10 Knee abduction angle at 40 ms in the different angle cutting. $^{**}p < 0.01$

non-injury groups, found a significantly higher abduction angle in the knee joint of the injury group. This indicates that increases in the abduction angle of the knee joint and a rearward CG are likely important risk factors for ACL injury. The results of the present study, in which the subjects showed a rearward CG and an increase in the abduction angle of the knee joint as the change in direction became greater, indicate that an increase in the change of direction of enhances the risk of ACL injury At the time of analysis, knee joint flexion showed no greater change when the change of direction was increased from 45° to 90°, whereas the CG height reduced. Although many studies have mentioned the influence of a rearward CG on ACL injury, the nature of the combined influence of rearward CG, CG height, and flexion angle of knee joint on ACL injury have not been clarified. For this, clarification, further studies will be required. Regarding the fact that the CG shifts rearward as the angle of direction change increases, there remains the possibility that deceleration immediately before changing direction may be important. This is plausable, since the deceleration required for accomplishing a direction change increases as the direction of change increases. Thus, we need to focus on the motion at deceleration in order to reduce the risk of ACL injury during foot grounding while changing direction.

In soccer and basketball training programs, positive effects from ACL injury prevention programs have been reported. These programs typically involve improving malpostures in foot grounding, including light knee joint flexion, knee joint abduction, and rearward CG. It is likely that these programs will be useful in rugby as well.

In rugby play, Besier et al. (2001) report that the abduction moment and extorsion moment of the knee joint during grounding increase significantly greater in "the presence" of a reaction than in "the absence" thereof. However, in the present study, we did not clarify the influences on the change of motion using reaction tasks. These latter points are likely to be very important for preventing injuries of the knee joint in rugby. In future investigations, we will investigate how these points influence the sequelae of changes in direction..

17.6 Conclusion

In rugby players, abrupt changes of direction influence the players' CG position and the abduction angle of their knee joint.

References

Alentorn-Geli E, Myer GD, Silvers HJ, Samitier G, Romero D, Lázaro-Haro C, Cugat R (2009) Prevention of non-contact anterior cruciate ligament injuries in soccer players. Part 1: Mechanisms of injury and underlying risk factors. Knee Surg Sports Traumatol Arthrosc 17(7):705–729

Besier TF, Lloyd DG, Ackland TR, Cochrane JL (2001) Anticipatory effects on knee joint loading during running and cutting maneuvers. Med Sci Sports Exerc 33(7):1176–1181

Boden BP, Dean GS, Feagin JA Jr, Garrett WE Jr (2000) Mechanisms of anterior cruciate ligament injury. Orthopedics 23(6):573–578

Brooks JH, Fuller CW, Kemp SP, Reddin DB (2005) Epidemiology of injuries in English professional rugby union: part 1 match injuries. Br J Sports Med 39(10):757–766

Cochrane JL, Lloyd DG, Buttfield A, Seward H, McGivern J (2007) Characteristics of anterior cruciate ligament injuries in Australian Football. J Sci Med Sport 10:69–104

Fuller CW, Laborde F, Leather RJ et al (2008) International Rugby Board Rugby World Cup 2007 injury surveillance study. Br J Sports Med 42:452–459

Griffin LY, Agel J, Albohm MJ, Arendt EA, Dick RW, Garrett WE, Garrick JG, Hewett TE, Huston L, Ireland ML, Johnson RJ, Kibler WB, Lephart S, Lewis JL, Lindenfeld TN, Mandelbaum BR, Marchak P, Teitz CC, Wojtys EM (2000) Noncontact anterior cruciate ligament injuries: risk factors and prevention strategies. J Am Acad Orthop Surg 8(3):141–150

Hewett TE, Myer GD, Ford KR, Heidt RS Jr, Colosimo AJ, McLean SG, van den Bogert AJ, Paterno MV, Succop P (2005) Biomechanical measures of neuromuscular control and valgus loading of the knee predict anterior cruciate ligament injury risk in female athletes: a prospective study. Am J Sports Med 33(4):492–501

Hewett TE, Torg JS, Boden BP (2009) Video analysis of trunk and knee motion during non-contact anterior cruciate ligament injury in female athletes: lateral trunk and knee abduction motion are combined components of the injury mechanism. Br J Sports Med 43(6):417–422

Koga H, Nakamae A, Shima Y, Iwasa J, Myklebust G, Engebretsen L, Bahr R, Krosshaug T (2010) Mechanisms for noncontact anterior cruciate ligament injuries: knee joint kinematics in 10 injury situations from female team handball and basketball. Am J Sports Med 38(11):2218–2225

McLean SG, Su A, van den Bogert AJ (2003) Development and validation of a 3-D model to predict knee joint loading during dynamic movement. J Biomech Eng 125(6):864–874

McNair PJ, Marshall RN, Matheson JA (1990) Important features associated with acute anterior cruciate ligament injury. N Z Med J 103(901):537–539

Nagano Y, Sakagami M, Ida H, Akai M, Fukubayashi T (2008) Statistical modelling of knee valgus during a continuous jump test. Sports Biomech 7(3):342–350

Olsen OE, Myklebust G, Engebretsen L, Bahr R (2004) Injury mechanisms for anterior cruciate ligament injuries in team handball: a systematic video analysis. Am J Sports Med 32(4):1002–1012

Chapter 18
Anterior Cruciate Ligament Injury Prevention in Female Lacrosse Players Based on an Examination of Knee and Hip Joint Mechanics During Drop Vertical Jumps Performed While Holding a Lacrosse Stick

Manabu Sanomura and Toru Fukubayashi

Abstract Anterior cruciate ligament (ACL) injuries occur during jumping and cutting maneuvers. Female lacrosse is associated with a higher incidence rate of ACL injuries than other female sports. Video analysis of ACL injuries has shown that the valgus knee angle is larger in injured athletes compared with controls, and the knee flexion angle tends to be smaller. In recent years, there has been focus on dynamic lower limb alignment, and the drop vertical jump (DVJ) task has been recommended as a method for assessing the risk of injuries to knee ligaments such as the ACL. The purpose of this study was to examine knee and hip joint mechanics during performance of the DVJ task while holding a lacrosse stick. Nine female collegiate lacrosse players participated in this study. Thirty-six reflective markers were placed on the upper limbs, trunk, pelvis, and right lower limb. An eight camera motion analysis system in conjunction with a force plate were used to record the three-dimensional marker positions and ground reaction forces. Three drop jump tasks were performed while not holding a lacrosse stick (NH), holding a lacrosse stick in the right hand (RH), and holding a lacrosse stick in the right hand with a target ball (RHT). The peak values of knee flexion and abduction angle, hip flexion and adduction angle, knee abduction and hip adduction moment from initial contact to 30 % phase in the landing phase were analyzed. Peak knee flexion angle in the RHT condition was significantly smaller than that in the RH condition (109.7° ± 10.4° vs. 113.5° ± 9.8°, $p < 0.05$). The results of this study suggest that when screening to prevent ACL injury in female lacrosse players, the DVJ task while holding a stick and a target may be a useful means of examining the risk of lower limb injury such as an ACL tear. Additionally, when teaching movements to prevent ACL injury, it is important to emphasize landing from a jump with the knees deeply bent.

M. Sanomura (✉)
Faculty of Medical Technology, Teikyo University, Tokyo, Japan
e-mail: sanatcc@mx52.tiki.ne.jp

T. Fukubayashi
Faculty of Sport Sciences, Waseda University, Saitama, Japan

K. Kanosue et al. (eds.), *Sports Injuries and Prevention*,
DOI 10.1007/978-4-431-55318-2_18

Keywords Anterior cruciate ligament • Drop vertical jump • Lower limb alignment • Knee machanics • Lacrosse

18.1 Introduction

In recent years, the number of lacrosse players has been increasing overseas (US Lacrosse 2011) and has also increased considerably in Japan. In Japan, it is estimated that the number of college lacrosse players is particularly high (Sanomura et al. 2013). Lacrosse is a sport where players use a crosse (stick) to pass a ball, catch it, and score points by shooting the ball at the opposing team's goal. The sport also involves movements to some degree unique to lacrosse such as particular types of running, cutting, and jumping (Dick et al. 2007).

According to previous studies on injuries suffered by female lacrosse players, lower limb injuries are common with injuries to the knee or ankle joints being particularly common (Dick et al. 2007; Hinton et al. 2005; Matz and Nibbelink 2004). While many of these injuries are sprains (Dick et al. 2007; Hinton et al. 2005), there are also many reports of anterior cruciate ligament (ACL) injuries (Hootman et al. 2007; Mihata et al. 2006). Due to unique movements in lacrosse such as throwing and catching the ball with the stick, many studies also report collision-induced contusions to a player's body (Diamond and Gale 2001; Hinton et al. 2005; Matz and Nibbelink 2004; Lincoln et al. 2007), as well as contusions to the head and face (Caswell et al. 2012; Goldenberg and Hossler 1995; Otago et al. 2007; Waicus and Smith 2002). There are also reports of concussions (Marar et al. 2012). In a previous 2-year-long study we conducted on injuries suffered by female lacrosse players at Japanese universities (Sanomura et al. 2013), we found that the most common site injured was the ankle, followed by the hip. Lower limb injuries accounted for approximately 70 % of all injuries. Sprains were the most common type of injury, followed by muscle cramps/spasms and tendinosis/tendinopathy. Approximately 70 % of contusions were caused by contact with an object such as the stick or ball. The three main causes of injury were "overuse" "contact" and "non-contact". Recurring injuries accounted for approximately 25 % of all injuries, with the most common recurring injury being an ankle sprain. Non-contact ankle sprain was the most characteristic injury suffered by female university lacrosse players (Sanomura et al. 2013).

Recent epidemiological studies have shown the incidence of ACL injury to be two to eight times higher in female athletes than in male athletes, particularly for sports such as basketball, soccer, and handball (Agel et al. 2005; Myklebust et al. 1998). The ACL injury incidence is also high in female lacrosse players (Hootman et al. 2007). Landing after a jump and cutting maneuver is most commonly associated with ACL injury (Boden et al. 2000), and lacrosse is unique in that players hold a stick (crosse) with both hands while jumping and cutting (Dick et al. 2007). Video analysis of ACL injuries has shown that the valgus knee angle and the angle of hip adduction are larger in injured athletes as compared with controls, and that in sagittal views, the angle of knee flexion tends to be smaller (Boden et al. 2009). Evidence from biomechanical studies conducted in laboratories also indicates that the same unique lower limb kinematics are highly apparent during motions such as the drop vertical jump (DVJ) and cuts, and many studies have shown them to be the main cause of ACL injury in female athletes (Hewett et al. 2006). In recent years, with focus on these issues by examining dynamic lower limb alignment, the DVJ task has been recommended as a method to assess injury risk to knee ligaments such as the ACL (Noyes et al. 2005; Renstrom et al. 2008).

The purpose of this study was to examine female lacrosse players to determine how holding a lacrosse stick and performing certain motions during a DVJ affects lower limb kinematics and kinetics. The ultimate purpose was to determine whether DVJ tasks, which are used in sports as a preventive screening for ACL injuries and other knee joint injuries, could provide a more specific evaluation of the unique changes in lower limb alignment that are a risk factor for injury if a method that incorporates the unique aspects of lacrosse were used in addition to the traditional method. In this paper the authors examine the angles and moments of the knee and hip joints seen during drop vertical jumps performed while holding a stick and discuss strategies for preventing ACL injuries in female lacrosse players.

18.2 Materials and Methods

18.2.1 Subjects

Subjects were nine female university lacrosse players (mean age: 20.0 ± 1.3 years, height: 159.6 ± 5.8 cm, weight: 55.1 ± 6.3 kg, lacrosse playing experience: 24 ± 3.4 months) who were members of the women's lacrosse league at one university. Players who had a lower limb injury that could affect the performance of motions tested in this study were excluded. All subjects were given an explanation of the purpose and methods of the study as well as ethical considerations, and their consent was obtained. This study was conducted with the approval of the human research ethics committee of Waseda University.

18.2.2 Test Protocol

The tested motion was a two-legged DVJ from a bench. Subjects were instructed to stand with both feet on a 30 cm high bench, jump down with both feet when signaled, and immediately perform a maximal vertical jump upon landing. They were also instructed to land with their right foot on a force plate both when jumping down from the bench and when landing after the vertical jump. DVJs were performed under three conditions. The first was a two-legged DVJ with nothing held in the hands (not holding a stick: NH) in which subjects were instructed to raise both

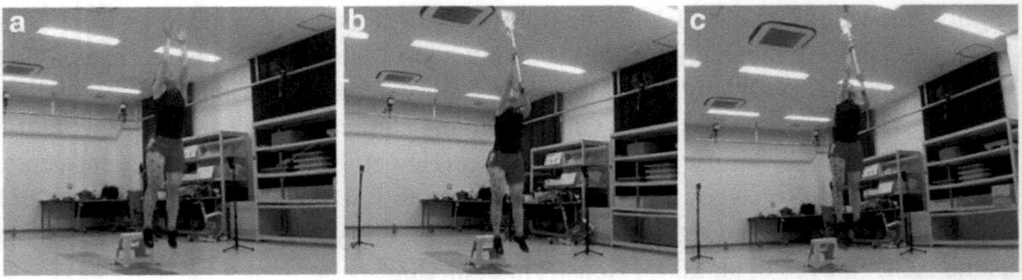

Fig. 18.1 Drop vertical jumps. *NH* not holding a stick (**a**), *RH* holding a stick with right hand (**b**), *RHT* holding a stick with right hand with a target (**c**)

hands toward the ceiling while jumping (Fig. 18.1a). The second was a two-legged DVJ performed while holding a lacrosse stick in the right hand (holding a stick with right hand: RH) in which subjects were instructed to raise the tip of the stick toward the ceiling while jumping and to keep both hands on the stick during the movement (Fig. 18.1b). The third was a DVJ performed under the same conditions as RH, but while also trying to hit a ball hung from the ceiling with the tip of the stick (holding a stick with right hand with a target: RHT) The ball was hung at a height where the tip of the stick would not actually hit the ball (Fig. 18.1c). Each DVJ was performed three times and successful jumps were recorded. A repetition was considered invalid when the subject: (1) could not perform a vertical jump upon landing after jumping off the bench, (2) did not land with her right foot on the force plate either when she jumped down from the bench or when she landed after the vertical jump, or (3) did not make a stable landing after the DVJ.

18.2.3 Data Collection

Motion measurements were taken using a 3D optical motion analysis system composed of eight infrared cameras (MotionAnalysis, Kissei Comtec Co., Ltd., Matsumoto, Japan) and a force plate (Kistler Japan Co., Ltd., Tokyo, Japan). Reflective markers were analyzed at a sampling frequency of 240 Hz and ground reaction forces at 2,400 Hz (Fig. 18.2a). The reflective markers were attached to nine points on the skin (the midpoints of and below the left and right anterior superior iliac spine and the left and right superior posterior iliac spine, the right greater trochanter, the medial and lateral epicondyles of the femur, the medial and lateral malleolus, and the second metatarsal head) as well as ten points on the thigh and six points on the calf according to the point cluster method (Andriacchi et al. 1998) (Fig. 18.2b). Three reflective markers were also attached to the heels of the shoes to determine the positions and orientations of foot segments (Fig. 18.2b).

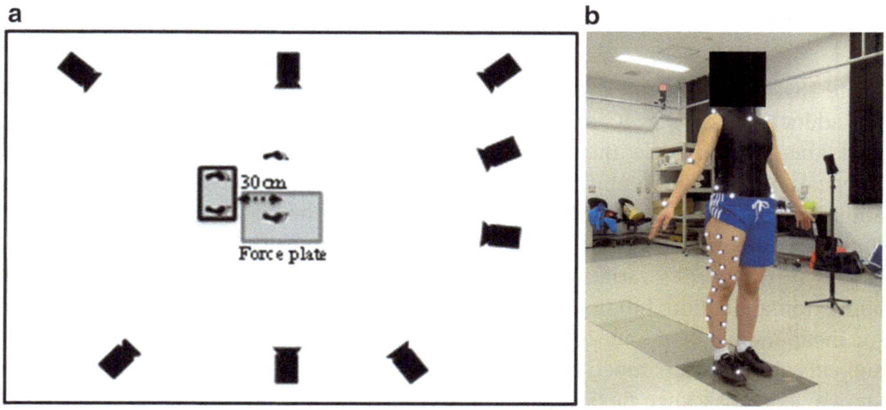

Fig. 18.2 Camera setting (**a**) and marker position (**b**)

18.2.4 Kinematic, Kinetic and Data Analysis

The angles of flexion and abduction for the knee joint as well as for the angles of flexion and adduction for the hip joint were calculated from the obtained 3D coordinate data using a joint coordinate system described by Grood and Suntay (1983). Knee joint abduction moments and hip joint adduction moments were also calculated from ground reaction force data using inverse dynamics. Joint moments were calculated as external moments. Each joint moment was normalized by dividing the moment by the product of the subject's height and weight. The period analyzed was from the point where part of the foot contacted the floor (initial contact: IC) to the point where the tips of the toes left the floor (toe off: TO) based on data obtained from the force plate; this period was designated as the landing phase. The kinematic parameters were peak knee flexion angle, peak knee abduction angle, peak hip flexion angle, and peak hip adduction angle between IC and the 30 % point of the landing phase where the total duration of the landing phase was 100 %. The kinetic parameters were peak knee abduction moment and peak hip adduction moment between IC and the 30 % point of the landing phase. The peak values between IC and the 30 % point of the landing phase were used as parameters because the peak vertical ground reaction force was observed between IC and the 30 % point of the landing phase, in addition to the fact that many reports indicate that real ACL injuries occur near the peak point of maximum ground reaction force seen after IC (Koga et al. 2010; Krosshaug et al. 2007). For statistical analysis, a one-way analysis of variance was used to compare parameters between different jump conditions; significant items were further analyzed utilizing the Bonferroni multiple comparison method. A significance level of <5 % was adopted.

18.3 Results

Table 18.1 depicts the mean angles of knee flexion, knee abduction, hip flexion, and hip adduction between IC and the 30 % point of the landing phase, as well as the peak knee flexion angle, the peak knee abduction angle, the peak hip flexion angle, the peak hip adduction angle, the peak knee abduction moment, and the peak hip adduction moment between IC and the 30 % point of the landing phase as measured during all repetitions of NH, RH, and RHT jumps performed by each subject. In Fig. 18.3, time series variations in the mean angles of knee flexion/extension, knee adduction/abduction, hip flexion/extension, and hip adduction/abduction, as well as the mean moments of knee adduction/abduction and hip adduction/abduction for all subjects during each repetition of each jump condition in the landing phase are shown. The peak knee flexion angle between IC and the 30 % point of the landing phase was significantly lower during RHT as compared with RH ($p<0.05$). No other significant difference was observed in comparisons between the various parameters of the jump conditions.

18.4 Discussion

This study compared the effect of lacrosse-specific DVJs while holding a stick and while aiming for a target on lower limb kinematics and kinetics in female lacrosse players. The results revealed that the peak knee flexion angle while holding a stick and aiming for a target was significantly smaller than when not aiming for a target.

Table 18.1 Knee and hip angle and moment during drop vertical jumps (mean ± SD)

	NH	RH	RHT
Initial contact			
Knee flexion angle (degree)	26.7 ± 11.6	26.4 ± 8.9	23.5 ± 6.2
Knee abduction angle (degree)	−1.5 ± 3.4	−1.5 ± 3.7	−2.3 ± 2.8
Hip flexion angle (degree)	36.8 ± 11.5	36.3 ± 8.8	34.2 ± 6.6
Hip adduction angle (degree)	−5.6 ± 3.2	−7.8 ± 4.4	−7.7 ± 4.5
Landing phase[a]			
Peak knee flexion angle (degree)[b]	113.0 ± 10.6	113.5 ± 9.8	109.7 ± 10.4
Peak knee abduction angle (degree)	−5.1 ± 8.0	−5.8 ± 7.8	−6.3 ± 8.0
Peak hip flexion angle (degree)	92.8 ± 7.4	91.9 ± 9.3	88.6 ± 8.9
Peak hip adduction angle (degree)	2.0 ± 6.2	0.4 ± 7.7	0.8 ± 7.9
Peak knee abduction moment (Nm/(kg*h))	−0.17 ± 0.14	−0.15 ± 0.12	−0.14 ± 0.10
Peak hip adduction moment (Nm/(kg*h))	0.27 ± 0.10	0.30 ± 0.13	0.26 ± 0.11

Negative values indicate positions or moments in case of adduction

NH not holding a lacrosse stick, *RH* holding a lacrosse stick (right hand), *RHT* holding a lacrosse stick (right hand) with a target

[a]The values from IC to 30 % phase in the landing phase

[b]Significant difference between RH and RHT ($p < 0.05$)

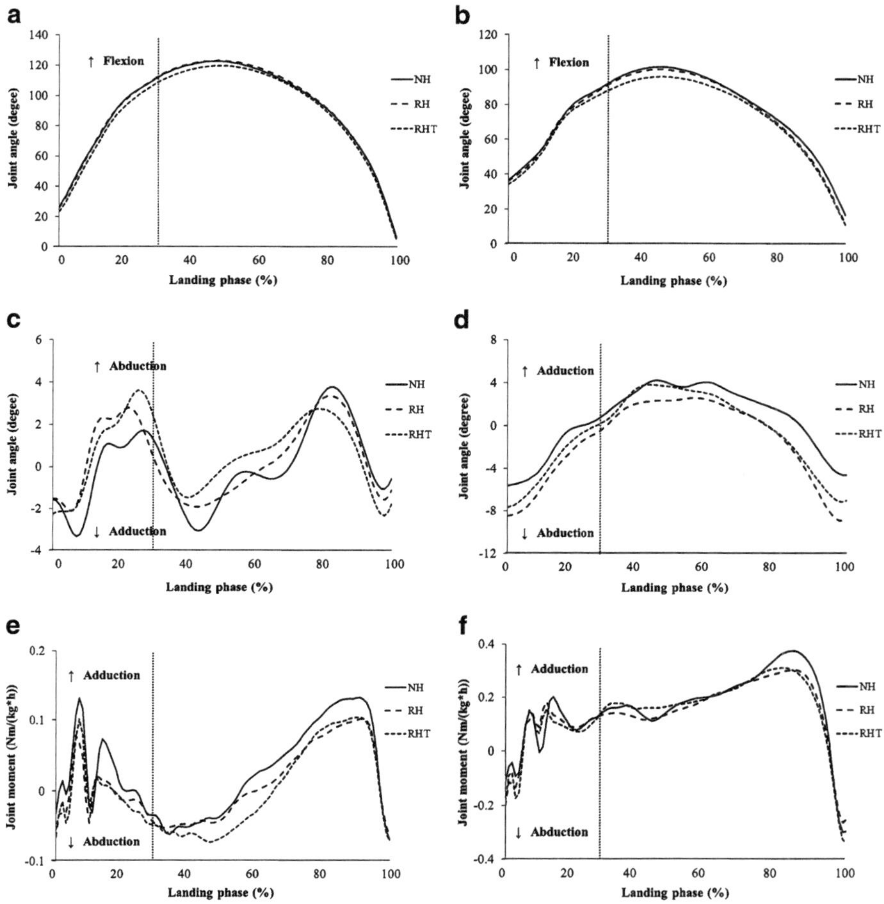

Fig. 18.3 Mean knee and hip joint angle and moment during the landing phase: knee joint flexion/extension angle (**a**), hip joint flexion/extension angle (**b**), knee joint adduction/abduction angle (**c**), hip joint adduction/abduction angle (**d**), knee joint adduction/abduction moment (**e**), hip joint adduction/abduction moment (**f**)

Previous studies have compared lower limb alignment during the DVJ task of female athletes with that of male athletes, and demonstrated that female athletes have greater knee abduction and hip adduction angles, and a smaller knee flexion angle. Many reports state that this dynamic lower limb alignment is a cause of ACL injury in female athletes (Hewett et al. 2006; Renstrom et al. 2008). Furthermore, Decker et al. (2003) reported that during drop landing, female athletes have significantly reduced knee flexion angles as compared to those of male athletes, while Yu et al. (2005) reported that in a stop jump task, the peak female knee flexion angle was significantly smaller. Using data on knee kinematics and kinetics obtained from a DVJ task as baseline values, Hewett et al. (2005) conducted a prospective study comparing an ACL injury group and a non-injury group and demonstrated that the ACL injury group had a significantly greater valgus knee angle than the non-injury group, whereas the peak knee flexion angle as seen from the sagittal plane was significantly smaller (Hewett et al. 2005). Furthermore, Nunley et al. (2003) examined the risk of ACL injury in male and female athletes from an anatomical perspective using X-ray imaging and found that at the same angle of knee flexion, the anterior tibial shear force caused by contraction of the quadriceps femoris muscle was greater in females than males, and thus increased the tensile stress on the ACL.

In a comparison of motions while holding a stick and aiming or not aiming for a target, the motion of holding a stick without aiming for the target can be performed while maintaining an awareness of the execution of a relatively simple motion such as jumping after landing, whereas with inclusion of holding a stick and aiming at a target during the motion task, there is a strong awareness of the need to jump high to reach the target and to maintain a higher line of sight when jumping after landing; thus, the athlete is less aware of the need to increase the lower limb—particularly the knee—flexion angle when jumping. This may have led to the decreased knee flexion angle. Although no significant difference was noted, the hip flexion angle tended to be smaller during RHT than during RH and NH (Fig. 18.3b, Table 18.1). According to data from a previous study analyzing successive video images taken of ACL injury that occurred during an actual competition, slight flexion of the knee was seen continuously (Boden et al. 2009). These previous studies have demonstrated that one cause of ACL injury is a small knee flexion angle during landing, and our study also showed that the mean knee flexion angle from IC until reaching the peak knee flexion angle tended to be smaller during RHT than during RH and NH (Fig. 18.3a). The values calculated in the present study indicated that the difference in peak mean knee flexion angle between RHT and RH was 3.8°; however, all earlier reports utilizing comparisons between ACL groups and control groups indicate that a small knee flexion angle while running is a risk factor for ACL injury. Situations in which athletes jump while holding a stick in both hands in order to catch a ball high above them are often seen in actual competitions. In the present experiment, a ball hanging from the ceiling was set as the target and our results suggested that the act of jumping as high as possible to reach a target while holding the stick may be a good task for evaluating a lower limb alignment specific to lacrosse. In this experiment, we were able to observe how holding a stick and aiming or not aiming for a target impacts the knee in the sagittal plane. Screening to prevent ACL injury includes the commonly used DVJ task, as well as having subjects perform motions specific to lacrosse, i.e. holding a stick with both hands and aiming for a target, which is an index for evaluating dynamic lower limb alignment. This index may help examine the risk of lower limb injury such as to those of the ACL.

In recent years, various programs to prevent lower limb injury such as to those of the ACL have been introduced and have yielded positive outcomes. Thus, the implementation of such programs is recommended (Bien 2011). The content of such programs focuses on knee motions and provides instruction on how to prevent the knee from turning inward (preventing "knee-in") as seen from the front, how to position the knee in line with the toes (knee over the toes position), and how to land with the knees bent deeply (to absorb the impact) when landing after jumping or when changing direction (Myer et al. 2004). Several reports have described the impact that tilting the trunk has on knee abduction while changing direction or when landing after jumping (Hewett and Myer 2011). These reports also emphasize the importance of obtaining guidance on how to maintain a stable posture. Based on the contents of this study, when issuing precautions from the perspective of preventing ACL injury, it is important to emphasize that the knees should be bent deeply, especially when landing after jumping. Instructors should also teach that abduction and valgus positions of the knee should be avoided and that the trunk should be maintained in a stable posture.

18.5 Conclusion

We compared the effect of holding and manipulating a lacrosse stick on lower limb kinematics and kinetics during drop vertical jumps in female lacrosse players and found that the peak knee flexion angle seen between IC and the 30 % point of the landing phase was significantly lower while holding a stick and aiming for a target than when not aiming for a target. When screening to prevent ACL injury in female lacrosse players, the drop vertical jump task while holding a stick and aiming for a target may be a useful means of predicting the risk of lower limb injury such as an ACL tear. Furthermore, when teaching movements to prevent ACL injury, it is important to emphasize landing with the knees deeply bent, especially when landing after a jump.

References

Agel J, Arendt EA et al (2005) Anterior cruciate ligament injury in national collegiate athletic association basketball and soccer: a 13-year review. Am J Sports Med 33:524–530

Andriacchi TP, Alexander EJ et al (1998) A point cluster method for in vivo motion analysis: applied to a study of knee kinematics. J Biomech Eng 120:743–749

Bien DP (2011) Rationale and implementation of anterior cruciate ligament injury prevention warm-up programs in female athletes. J Strength Cond Res 25:271–285

Boden BP, Dean GS et al (2000) Mechanisms of anterior cruciate ligament injury. Orthopedics 23:573–578

Boden BP, Torg JS et al (2009) Video analysis of anterior cruciate ligament injury: abnormalities in hip and ankle kinematics. Am J Sports Med 37:252–259

Carter EA, Westerman BJ et al (2010) Common game injury scenarios in men's and women's lacrosse. Int J Inj Contr Saf Promot 17:111–118

Caswell SV, Lincoln AE et al (2012) Video incident analysis of head injuries in high school girls' lacrosse. Am J Sports Med 40:756–762

Decker MJ, Torry MR et al (2003) Gender differences in lower extremity kinematics, kinetics and energy absorption during landing. Clin Biomech 18:662–669

Diamond PT, Gale SD (2001) Head injuries in men's and women's lacrosse: a 10 year analysis of the NEISS database. National Electronic Injury Surveillance System. Brain Inj 15:537–544
Dick R, Lincoln AE et al (2007) Descriptive epidemiology of collegiate women's lacrosse injuries: National Collegiate Athletic Association Injury Surveillance System, 1988–1989 through 2003–2004. J Athl Train 42:262–269
Goldenberg MS, Hossler PH (1995) Head and facial injuries in interscholastic women's lacrosse. J Athl Train 30:37–39
Grood ES, Suntay WJ (1983) A joint coordinate system for the clinical description of three-dimensional motions: application to the knee. J Biomech Eng 105:136–144
Hewett TE, Myer GD (2011) The mechanistic connection between the trunk, hip, knee, and anterior cruciate ligament injury. Exerc Sport Sci Rev 39:161–166
Hewett TE, Myer GD et al (2005) Biomechanical measures of neuromuscular control and valgus loading of the knee predict anterior cruciate ligament injury risk in female athletes: a prospective study. Am J Sports Med 33:492–501
Hewett TE, Myer GD et al (2006) Anterior cruciate ligament injuries in female athletes: part 1, mechanisms and risk factors. Am J Sports Med 34:299–311
Hinton RY, Lincoln AE et al (2005) Epidemiology of lacrosse injuries in high school-aged girls and boys: a 3-year prospective study. Am J Sports Med 33:1305–1314
Hootman JM, Dick R et al (2007) Epidemiology of collegiate injuries for 15 sports: summary and recommendations for injury prevention initiatives. J Athl Train 42:311–319
Lincoln AE, Hinton RY et al (2007) Head, face, and eye injuries in scholastic and collegiate lacrosse: a 4-year prospective study. Am J Sports Med 35:207–215
Marar M, Mcllvain NM et al (2012) Epidemiology of concussions among United States high school athletes in 20 sports. Am J Sports Med 40:747–755
Matz SO, Nibbelink G (2004) Injuries in intercollegiate women's lacrosse. Am J Sports Med 32:608–611
Mihata LC, Beutler AI et al (2006) Comparing the incidence of anterior cruciate ligament injury in collegiate lacrosse, soccer, and basketball players: implications for anterior cruciate ligament mechanism and prevention. Am J Sports Med 34:899–904
Myer GD, Ford KR et al (2004) Rationale and clinical techniques for anterior cruciate ligament injury prevention among female athletes. J Athl Train 39:352–364
Myklebust G, Maehlum S et al (1998) A prospective cohort study of anterior cruciate ligament injuries in elite Norwegian team handball. Scand J Med Sci 8:149–153
Noyes FR, Barber-Westin SD et al (2005) Drop-jump screening test: difference in lower limb control by gender and effect of neuromuscular training in female athletes. Am J Sports Med 33:197–207
Nunley RM, Wright D et al (2003) Gender comparison of patellar tendon tibial shaft angle with weight bearing. Res Sports Med 11:173–185
Otago L, Adamcewicz E et al (2007) The epidemiology of head, face and eye injuries to female lacrosse players in Australia. Int J Inj Contr Saf Promot 14:259–261
Pollard CD, Davis IM et al (1994) Influence of gender on hip and knee mechanics during a randomly cued cutting maneuver. Clin Biomech 19:1022–1031
Renstrom P, Ljungqvist A et al (2008) Non-contact ACL injuries in female athletes: an international olympic committee current concepts statement. Br J Sports Med 42:394–412
Sanomura M, Hosokawa Y et al (2013) Collegiate women's lacrosse injuries: a 2-year prospective surveillance study. Jpn J Phys Fitness Sports Med 62:399–411
US Lacrosse. 2011 US lacrosse participation survey. http://www.uslacrosse.org/LinkClick.aspx?fileticket=tIqhe4P9DsM%3D&tabid=14645. Accessed 10 Jan 2013
Waicus KM, Smith BW (2002) Eye injuries in women's lacrosse players. Clin J Sport Med 12:24–29
Yu B, Lin CF et al (2005) Lower extremity biomechanics during the landing of a stop-jump task. Clin Biomech 21:297–305

Chapter 19
ACL Injury Mechanism in Badminton: Survey of Injury Situation and Motion Analysis Study

Eiichi Tsuda, Yuka Kimura, and Yasuyuki Ishibashi

Abstract An interviewer-administered survey to investigate playing situations of ACL injury and a laboratory study which included a motion analysis of frequently performed movements was conducted to demonstrate the ACL injury mechanism in badminton. The interview with 21 ACL-injured badminton players showed that 2 typical injury situations related to the sports-specific movement. One is ACL injury opposite to the racket-hand side during a single-leg-landing after an overhead stroke in the backhand side of the court, and the other is that of the racket-hand side in the course of a plant-and-cut maneuver during side or backward stepping in the forehand side of the court. The kinematic and kinetic data in motion analysis study demonstrated that larger knee valgus angle and moment during single-leg-landing following back-steps to the backhand-side rear compared with those following back-steps to the forehand-side rear. These biomechanical features of knee motion in badminton might cause the high incidence of ACL injuries which occurred during single-leg-landing after an overhead stroke in the backhand-side rear court. A better understanding of ACL injury pattern in badminton helps to avoid the players from a risk of ACL injury and develop the efficient prevention program.

Keywords Anterior cruciate ligament • Injury mechanism • Non-contact injury • Badminton • Motion analysis

19.1 Introduction

In badminton, overuse injuries and acute injuries in the lower extremities are generally the most frequent (Jørgensen and Winge 1987; Krøner et al. 1990; Shariff et al. 2009). In a 2007 report (Uchiyama 2007), a high incidence of anterior

Electronic supplementary material The online version of this chapter (doi: 10.1007/978-4-431-55318-2_19) contains supplementary material, which is available to authorized users.

E. Tsuda, MD (✉) • Y. Kimura, MD • Y. Ishibashi, MD
Department of Orthopaedic Surgery, Hirosaki University
Graduate School of Medicine, Hirosaki, Japan
e-mail: eiichi@cc.hirosaki-u.ac.jp

K. Kanosue et al. (eds.), *Sports Injuries and Prevention*,
DOI 10.1007/978-4-431-55318-2_19

cruciate ligament (ACL) injuries in badminton was reported and accounted for 37 % of all injuries which required surgical treatment. ACL injuries in badminton are purely non-contact, since the players are separated from each other by a net and physical interaction between the opponents is avoided. Competitive badminton requires frequent quick starts, stops and changes in direction. Players hold a racket in their dominant hand, which limits their arm position and leads to an asymmetric posture. There is no doubt that a better understanding of the sport-specific risk factors for ACL injury is important for the establishment of an efficient injury prevention strategy. We conducted a survey of ACL injury in badminton players (Kimura et al. 2010) and a laboratory study which included a motion analysis of frequently performed movements by players engaged in this sport (Kimura et al. 2012). We utilized this information to develop an ACL injury prevention program for badminton players.

19.2 Survey of ACL Injury

19.2.1 Materials and Methods

Twenty-one patients who had suffered ACL injury while playing badminton participated in this survey. Information on injury mechanism was collected from medical records and interviews obtained within 1 month of the injury.

Age, gender and subject characteristics, such as years of badminton experience, dominant hand (racket hand), the event during which the injury occurred (practice or game), and the injured side (right or left), were collected from medical records. The players were interviewed by a physician using a standardized questionnaire. They were each asked to describe the circumstances (playing situation, motion, handling of the shuttle) and court position when the injury occurred. Court position was specified by both row (front, middle or rear) and side (forehand or backhand). Based on the results obtained from questionnaires filled out by the patients, the event inciting ACL injury was placed into three groups. They were: (1) An overhead stroke (smash or clear) involving a single-leg-landing (Video 19.1). (2) A plant-and-cut maneuver that occurred during stepping (Video 19.2). (3) Other motions that produced the injury.

19.2.2 Results

The mean age of the 21 injured players was 21.9 ± 7.9 (mean $\pm$ SD) years (13–38); 18.6 ± 5.3 years (15–34) for the 15 females and 27.4 ± 9.6 years (18–38) for the 6 males. Twenty players were right-hand dominant, and 1 player was left-hand dominant. Twelve players injured knees that were on the side opposite to that of the dominant hand while nine players injured knees that were on the same side as the

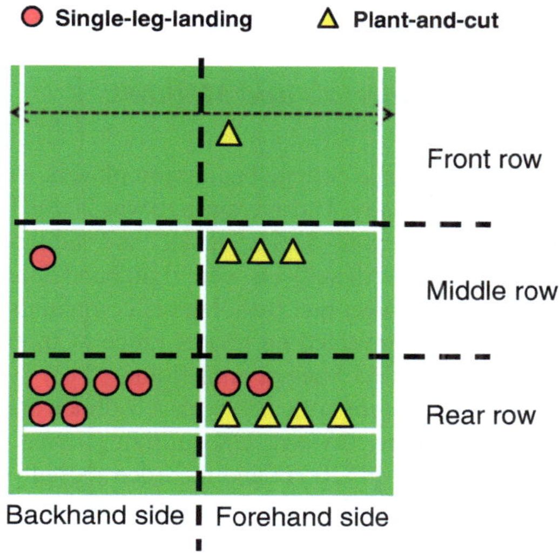

Fig. 19.1 Court position where the injury occurred in a left-handed player who converted to right-handed play. *Red circles* show the injury during the single-leg-landing and *yellow triangles* show that during plant-and-cut (Cited from Kimura et al. (2010))

dominant hand. All players except one participated at various levels in badminton clubs. The exception was a player who was injured in a gym class. The mean playing experience of injured participants was 5.8 ± 5.8 (mean ± SD) years. The range was 0.4–21 years. Fourteen players of 21 (67 %) received ACL injury during games.

The most common movement sequence preceding injury was a single-leg-landing after an overhead stroke (smash or clear shot) following a backward step. This scenario described the immediate cause of injury for 10 players (47.6 %). Six injuries occurred in the backhand side of the rear corner; 2 injuries in the forehand side of the rear corner; 1 injury in the backhand side of the middle court and 1 injury in an unknown court position (Fig. 19.1). Nine of the ten players injured the knee opposite to the racket-hand side. The second most common movement sequence (8 players; 38.1 %) was plant-and-cut; 4 injuries occurred during a backward step made to turn the body towards the rear corner in order to execute a forehand stroke. Five injuries occurred during a side-step to receive the shuttle on the forehand side of the court. Four of these were in the middle and one was in the front of the court. All eight players injured the knee on the racket-hand side. In these situations, the players land the leg of the racket-hand side by transferring their body weight while taking the shot, and then push off on the leg of the racket-hand side for recovery and also to return the body to the center of the court. This sequence needs to be put into effect very rapidly in order to prepare for the next shot. One player hyperextended and injured the knee opposite to the racket-hand side when receiving the shuttle from a forehand stroke at the front of the court. One player who was injured on the forehand side could not remember the event. Another player could not remember the inciting event or the court location at the time of the injury. This player ruptured her ACL in a gym class and had little experience in competitive badminton.

19.3 Motion Analysis

19.3.1 Materials and Methods

Seventeen female college badminton players with no history of knee injury participated in this study. The average age was 20.3 ± 1.8 years, height was 158.4 ± 5.4 cm, body weight was 51.7 ± 5.8 kg and athletic experience of badminton was 5.3 ± 2.1 years. All subjects were right hand dominant, based on racket holding preference. No subject had pathological complaints. Administered orthopedic physical examinations revealed no loss of range of motion or muscular weakness in trunk, lumbar spine, hip, knee or ankle joint.

Trial data were collected with a 3-dimensional motion analysis system (Vicon; Oxford Metrics, Oxford, UK) utilizing seven infrared cameras. Kinematics data were sampled at 120 Hz and recorded digitally on a Dual Pentium III 1 GHz personal computer. Cameras were positioned so that each retroreflective marker was detected by at least two cameras throughout the task. Ground reaction force was collected at 120 Hz using a calibrated and leveled force plate (model OR6-6-1; AMTI, Watertown, Massachusetts, USA) embedded in the floor and synchronized with a Vicon system for simultaneous collection. Each subject was barefooted and wore black shorts during the testing. Sixteen retroreflective markers (25 mm) were placed on specific anatomical landmarks (anterior superior iliac spine, posterior superior iliac spine, mid thigh, lateral condyle of the femur, mid lower leg, lateral malleolus of ankle, heel and second metatarsalis) to calculate motion of the hips, knees and ankles in the sagittal, frontal and transverse planes (Kadaba et al. 1990). Three dimensional marker trajectories were recorded and kinematic and kinetic variables were calculated using Plug-in Gait in Vicon Workstation (version 4.6; Oxford Metrics, Oxford, UK). Inverse dynamics analyses were used to calculate external joint moments from kinematic data and force plate data with Plug-in Gait. Marker trajectory data were filtered using a Woltering quintic spline filter with a predicted mean square error of 20 mm (Woltering 1986). The force plate data were filtered through a low-pass Butterworth digital filter at a cut-off frequency of 12 Hz. All kinetic data were normalized to body weight (kg) and height (m).

Before participation, the testing procedures were explained to each subject. Anthropometric measurements including height, weight, segmental lengths and diameters of the ankles and knees, length of the feet, and pelvic width were recorded for each subject. The footwork and overhead stroke task were then demonstrated to each of the subjects by a badminton coach who had approximately 15 years of experience as a competitive player. Subjects were asked to perform two different testing tasks: a right back-step task and a left back-step task. In the right back-step task, the subjects stepped back three steps 45° diagonally to the right from the starting position. This simulated the back-step towards the forehand-side rear of the badminton court. They then made an overhead stroke with a racket in their right hand, immediately landed with their left leg on the force plate and then returned to the starting position (Video 19.3). In the left back-step task, the subjects stepped back three steps 45° diagonally to the left from the starting position. This simulated the back-step towards the backhand-side rear of the badminton court. This was followed by an overhead stroke. They then landed on the force plate and returned to the starting position in the same way as in the right back-step task (Video 19.3). Subjects were allowed to practice the tasks several times, and then performed three to five consecutive trials. To avoid any coaching effect on the subjects' natural performance of the tasks, no further instructions relating to stepping, landing or stroking techniques were provided.

The data analysis was performed using a trial in which the subject successfully landed within the force plate without overrunning it. In general, right-handed badminton players land on the left leg after an overhead stroke (Video 19.1). Therefore all data were analyzed for the left leg. Because non-contact ACL injuries generally occur in the initial phase of landing (Koga et al. 2010), the data analysis was only focused on the impact phase after an overhead stroke. The impact phase was defined as between the time of initial contact (IC) of the left toe and the floor after an overhead stroke and maximum knee flexion (MKF). All kinematic data were time normalized to the impact phase. The values thus ranging from 0 (equal to IC) to 100 % (equal to MKF). From each trial, the flexion/extension, adduction/abduction and internal/external rotation angles of the left hip, the flexion/extension and varus/valgus of the left knee, and the dorsi/plantar flexion, inversion/eversion and adduction/abduction angles of the left ankle at IC and MKF were determined. The peak values of the external knee valgus moment were extracted from IC to MKF, and normalized to height and body weight. The mean and SD of the relevant measures were calculated for all trials. To evaluate the significance of the comparison between the right and left back-step tasks for kinematic and kinetic data during single-leg-landing, paired t tests were utilized. A significance level of 0.05 was adopted. Statistical analyses were conducted using SPSS Version 16.0 software (SPSS Inc, Chicago, Illinois, USA).

19.3.2 Results

Figures 19.2, 19.3, and 19.4 show the time course of the hip, knee and ankle kinematics during the impact phase. The hip flexion angle at IC was not significantly different between the right and the left back-step, whereas the hip flexion angle at MKF after the right back-step was 22.7° ± 6.0°, and significantly smaller compared with the 34.4° ± 11.5° seen after the left back-step ($p = 0.0002$) (Fig. 19.2a). The hip adduction/abduction angle at IC was 9.8° ± 6.4° of abduction after the right back-step and 17.0° ± 7.6° of abduction after the left back-step. This represents a significant difference ($p = 0.0004$) (Fig. 19.2b). This significant difference disappeared at MKF. There was no significant difference in the hip internal/external rotation angle at IC or MKF (Fig. 19.2c). The knee flexion angle at IC was not significantly different between the right and the left back-step, whereas those at MKF were 50.4° ± 8.4° and 55.6° ± 10.6°, which represented a significant difference ($p = 0.0304$) (Fig. 19.3a).

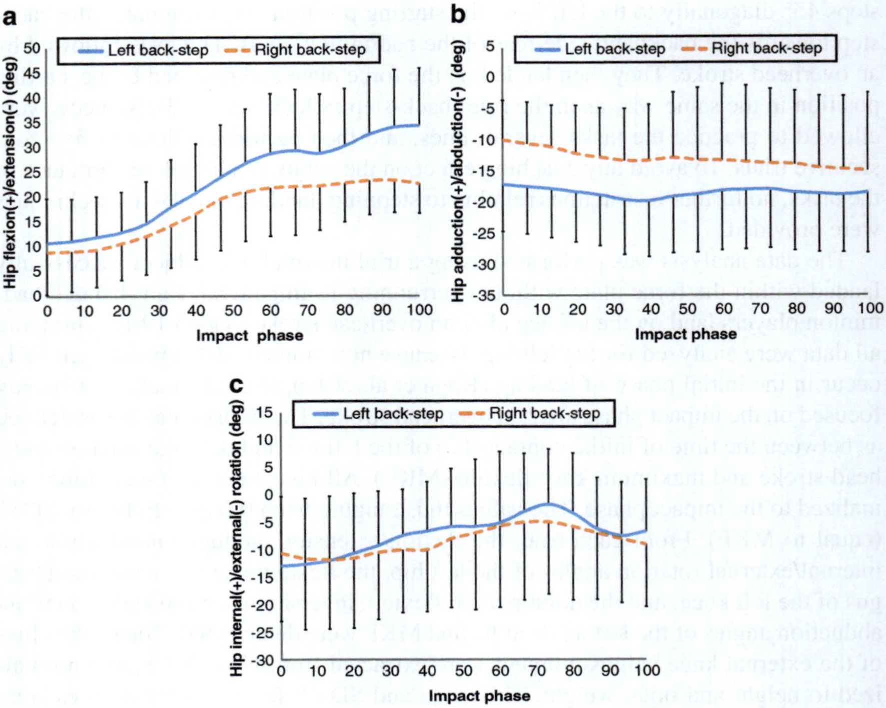

Fig. 19.2 Values are mean ± SD during impact phase. (**a**) Hip flexion/ extension angle; (**b**) hip abduction/adduction angles and (**c**) hip internal/external rotation angle (Cited from Kimura et al. (2012))

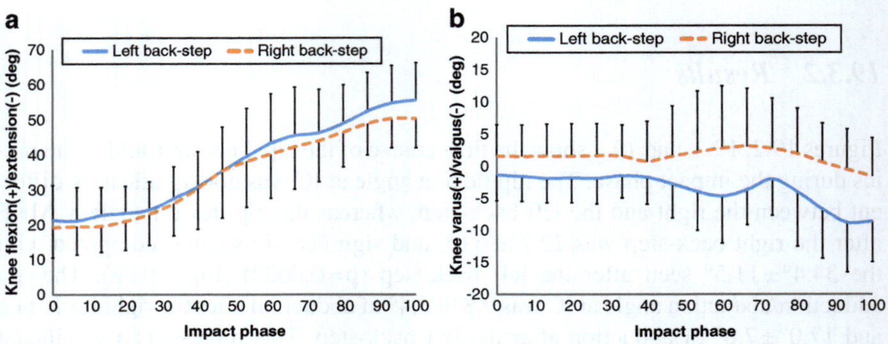

Fig. 19.3 Values are mean ± SD. (**a**) Knee flexion/extension angle and (**b**) knee varus/valgus angle (Cited from Kimura et al. (2012))

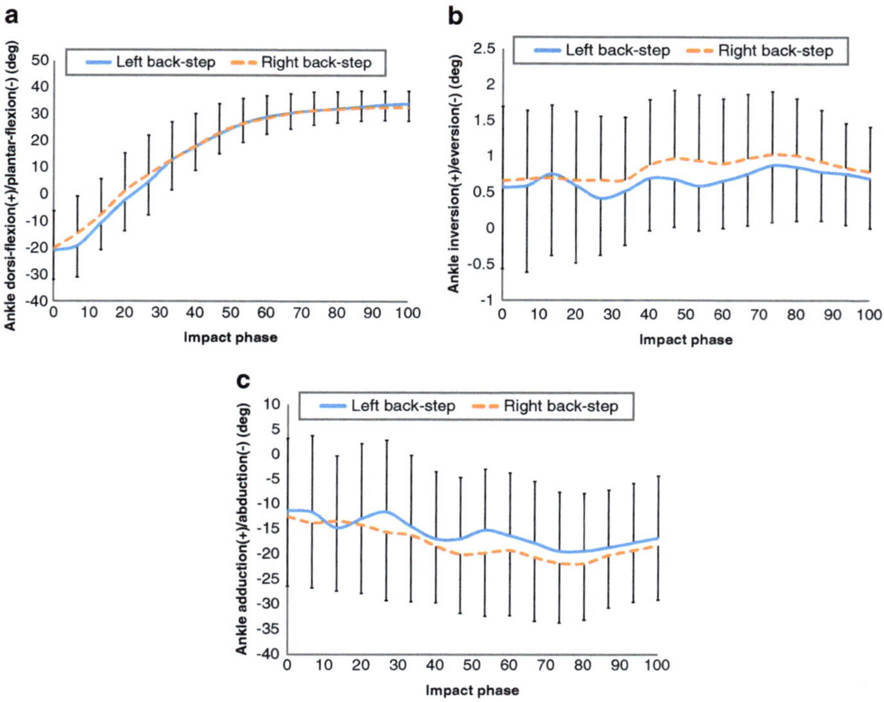

Fig. 19.4 Values are mean ± SD. (**a**) Ankle dorsiflexion/plantar flexion angle; (**b**) ankle inversion/eversion angle and (**c**) ankle adduction/abduction rotation angle (Cited from Kimura et al. (2012))

The knee varus/valgus angle at IC was 1.6° ± 4.9° of varus after the right back-step and 1.3° ± 4.1° of valgus after the left back-step and those at MKF were 1.6° ± 6.0° of valgus and 8.6° ± 6.3° of valgus. Both comparisons were significantly different ($p = 0.0214$ and $p = 0.0294$) (Fig. 19.3b). There was no significant difference in the ankle kinematics at IC or MKF between the right and the left back-step (Fig. 19.4a–c). The maximum knee valgus moment was 0.42 ± 0.26 Nm/kg/m after the left back-step and significantly higher than the value of 0.28 ± 0.25 Nm/kg/m seen after the right back-step ($p = 0.0039$) (Fig. 19.5).

19.4 Conclusion

The survey of ACL injury in badminton showed that most injuries are the result of two injury mechanisms. In one case, the knee opposite to the racket-hand side is injured during a single-leg-landing after an overhead stroke and it usually occurs in the backhand side of the court. In the other case the knee of the racket-hand side is injured in the course of a plant-and-cut maneuver during side or backward stepping

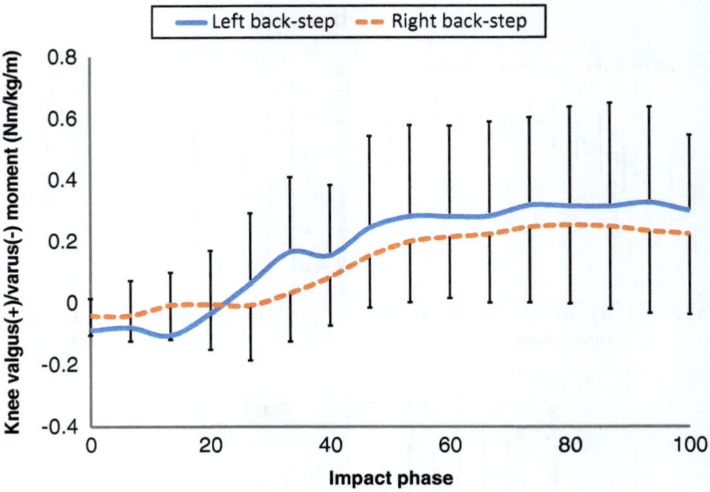

Fig. 19.5 Values are mean ± SD of the knee valgus moment (Cited from Kimura et al. (2012))

in the forehand side of the court. Following an overhead stroke, players transfer their main support from the back to the front, and land on the leg opposite to their racket-hand side. This allows them to balance and push their body from the back to the base position. In an overhead stroke on the backhand side it is more difficult to keep the body in balance than it is on the forehand side, because the players have to bend their trunk to their backhand side. This occurs laterally as their arm comes through. In addition, quick footwork in a diagonal direction may also result in imbalance. Therefore, many players were injured by a single-leg-landing after an overhead stroke in the rear court. This injury usually involved the knee opposite to the racket-hand side.

The motion analysis portion of the study demonstrated that larger knee valgus angle and moment during single-leg-landing following back-steps to the backhand-side rear compared with those following back-steps to the forehand-side rear. This might be related to the high incidence of ACL injuries which occurred during single-leg-landing after an overhead stroke in the backhand-side rear court. For the prevention of ACL injury in badminton players, it is suggested that an understanding of the 'at risk' posture and adaptation of training methods to teach landing techniques considering badminton-specific movements is needed.

References

Jørgensen U, Winge S (1987) Epidemiology of badminton injuries. Int J Sports Med 8:379–382

Kadaba MP, Ramakrishnan HK, Wootten ME (1990) Measurement of lower extremity kinematics during level walking. J Orthop Res 8:383–392

Kimura Y, Ishibashi Y, Tsuda E et al (2010) Mechanisms for anterior cruciate ligament injuries in badminton. Br J Sports Med 44:1124–1127

Kimura Y, Ishibashi Y, Tsuda E et al (2012) Increased knee valgus alignment and moment during single-leg-landing after overhead stroke as a potential risk factor of anterior cruciate ligament injury in badminton. Br J Sports Med 46:207–213

Koga H, Nakamae A, Shima Y et al (2010) Mechanisms for noncontact anterior cruciate ligament injuries: knee joint kinematics in 10 injury situations from female team handball and basketball. Am J Sports Med 38:2218–2225

Krøner K, Schmidt SA, Nielsen AB et al (1990) Badminton injuries. Br J Sports Med 24:169–172

Shariff AH, George J, Ramlan AA (2009) Musculoskeletal injuries among Malaysian badminton players. Singapore Med J 50:1095–1097

Uchiyama E (2007) Badminton. In: Hayashi M, Iwasaki Y (eds) Sports injuries in different athletic events, authored by national team physicians. Nankodo, Tokyo, pp 180–185

Woltering HJ (1986) A fortran package for generalized, cross-validatory spline smoothing and differentiation. Adv Eng Softw 8:104–113

Part IV
ACL Injury: Post-operative Rehabilitation, Recovery of Function, and Tendon Regeneration

Chapter 20
Functional Hop Tests Contribute to Safe Return to Sports After Anterior Cruciate Ligament Reconstruction

Yumi Nomura and Toru Fukubayashi

Abstract In this chapter we examine ways to evaluate the potential for injury free return to competitive sports by evaluating athletes who had undergone ACL reconstruction and already returned to previous sport activities. We sought to identify the relationship among various hop tests scores, muscle strength, and self-reported outcomes. The subjects were 15 collegiate athletes (8 men, 7 women; age, 20.5 years ± 1.2) who had successfully undergone primary ACL reconstruction with a hamstring autograft 9–12 months (10.0 ± 1.6) previously. Paticipants completed a self-report questionnaire and underwent one-legged hop tests and isokinetic strength measurements. A paired t-test was used to test for side-to-side differences in each of the hop tests, as well as for isokinetic strength. A Pearson's product–moment correlation coefficient was used to examine the relationship among all measurements. The subjects in our study had a mean clinical outcome score of 90.1, which is considered to be indicative of normal knee function. The hop distance in the operative limb was significantly shorter than that in the nonoperative limb ($p < 0.01$). The extention torque in the operative limb was also significantly lower than that of the nonoperative limb ($p < 0.01$). We found a significant correlation between extention torque and hop distance ($p < 0.01$). This study demonstrates that deficits in hop-distance and quadriceps strength persist at 9–12 months after ACL reconstruction, even after a return to sports activities. To optimize functional and clinical outcomes and a safe return to sports, we propose to use evidence based medicine (EBM) assessments with functional practice, hop tests, and isokinetic strength tests, with an emphasis on evaluating side-to-side assymmetry.

Keywords Anterior cruciate ligament • Hop tests • Strength • KOOS • Return

---

Y. Nomura (✉)
Graduate School of Sport Sciences, Waseda University, 2-579-15 Mikajima, Tokorozawa, Saitama 359-1192, Japan
e-mail: yumi.nmr@gmail.com

T. Fukubayashi
Faculty of Sport Sciences, Waseda University, Saitama, Japan
e-mail: fukubayashi@waseda.jp

K. Kanosue et al. (eds.), *Sports Injuries and Prevention*,
DOI 10.1007/978-4-431-55318-2_20

20.1 Introduction

Anterior cruciate ligament (ACL) injuries are common among the young, active population, particularly for competitive athletes who participate in sports that involve cutting, jumping, and pivoting (Griffin et al. 2000). The aim of ACL reconstruction is to restore knee stability to prevent further intra-articular injury and to return the patients to their preinjury level of performance. Ardern et al. (2011a, b) used a systematic review combined with a meta-analysis to evaluate postoperative return-to-sport outcomes. They found that while 63 % of the previously injured athletes had returned to a preinjury level of participation, only 44 % had returned to competitive sports. Unfortunately, ACL reinjury occurs in 6–13 % of operative knees (Myklebust et al. 2003; Salmon et al. 2005; Walden et al. 2006) and 2–6 % sustain a contralateral ACL injury (Salmon et al. 2005; Wright et al. 2007; Leys et al. 2012). A second ACL injury can add physical and psychic loads, and unfortunately may have a negative impact on the athlete's career and future life.

Accordingly, standardized outcome measures that are appropriate for assessing patients who have undergone ACL reconstruction are required for evaluating individual patient progress. Recoverd lower extremity muscle function, for example, knee extensor and flexor muscle strength and single-legged jumping ability, is regarded as important after an ACL injury in order to make a safe return to sports or physical activity. Among the various outcome assessments commonly used to decide upon the proper time for return to sports, the most popular are hop tests, which are functional tests that simulate athletic activities and provide objective information about neuromascular control, functional power and postual stability (Petschnig et al. 1998; Logerstedt et al. 2012; Xergia et al. 2013). Hop tests are composed of a single hop for distance, a 6-m timed hop, a triple hop for distance, and a crossover hop for distance (Noyes et al. 1991). This series of hop tests includes a variety of movement principles such as direction change, speed, acceleration-deceleration, and jumping. These movements resemble the demands of dynamic knee stability during sports activities and are suggested to provide a good evaluation for whether the patient should return to such activities. Several studies have reported persistent symmetry deficits in hop tests following ACL rconstruction (Reid et al. 2007; Øiestad et al. 2011; Xergia et al. 2013).

Salmon et al. (2005) stated that the risk of reinjury is higher during the first 12 months postoperatively, and that after this point, the risk falls to a value which is equivalent to the rate of ruputure of the contralateral ACL. On these grounds athletes are typically permitted to return to sport between 6 and 12 months postoperatively (Cascio et al. 2004), and most are expected to return to sport within 12 months (Myklebust and Bahr 2005). Competitive sports which include pivoting, pivoting with contact, and jumping are have a higher incedence of reinjury (Laboute et al. 2010). Younger athletes have demonstrated the highest reinjury rate (Shelbourne et al. 2009) and may also have an increased risk of a contralateral injury (Leys et al. 2012) as compared with older athletes. Therefore, it is important to have information regarding the efficacy of ACL reconstruction in terms of achieving the desired aim of returning an athlete to sports participation at 12 months. However, to our knowledge, no study has comprehensively investigated hop tests, muscule strength and self-reported outcome in a homogeneous group of young atheletes who return to competitive sports, and who may have a high risk of ACL reinjury.

Although there is no unified veiw regarding gender defferences for second injury rate (Laboute et al. 2010; Bourke et al. 2012; Brophy et al. 2012), it is common knowledge that the primary ACL injury rate is two to eight times higher for females than for males (Boden et al. 2000; Griffin et al. 2000; Myklebust et al. 2003; Agel et al. 2005). Therefore, a further analysis regarding gender differences in the outcome evaluations is necessary to develop criteria for a safe return of female athletes to sports.

The purpose of this study was to assess hop test performance and muscle strength in competitive athelets who had undergone ACL reconstruction within the past 9–12 months and had already returned to previous sport activities. We also sought to identify the relationship among hop test scores, muscle strength, and self-report outcomes. Our hypothesis was that muscle strength and hop test asymmetries would exist between the operated limb and the uninjured limb. A secondary hypothesis was that hop test scores are affected by knee extention strength, but cannot be detected by self-report outcome scores.

20.2 Materials and Methods

20.2.1 Subjects

This study involved 15 collegiate athletes (8 men, 7 women; mean age, 20.5 years ± 1.2) who successfully underwent primary ACL reconstruction. The subjects underwent arthroscopically assisted reconstruction with an autogenous quadrupled ipsilateral semitendinosus tendon as was first described by Rosenberg et al. in 1997. The mean postoperative time was 10.0 ± 1.6 months (range, 9–12 months). The inclusion criterion were (1) The subjects were 9–12 months post-surgical and (2) Had returned to the sport activity that subjects had participated in before the ACL injury. All subjects participated in sports activities classified as level 1 or 2. Level 1 sports involve jumping, pivoting, and hard cutting, while level 2 sports also involve lateral motion but with less jumping or hard cutting (Daniel et al. 1994). Descriptive data for all patients are presented in Table 20.1. The study was approved by the human research ethics committee of the School of Sport Sciences of Waseda University and is consistent with their requirements for human experimentation (approval no. 2012–284). This study conforms to the Declaration of Helsinki. Written informed consent statements were obtained after participants read the volunteer information sheet and answered questions related to the study.

Table 20.1 Descriptive characteristics of subjects' age, height, weight, postoperative time and sports activity

All patients	Age (years)	Height (cm)	Weight (kg)	Postoperative time (months)	Sports
All subjects	20.5 ± 1.2	165.3 ± 8.6	63.8 ± 11.3	10.0 ± 1.6	–
Males	20.8 ± 1.1	170.9 ± 6.0	71.6 ± 6.0	9.8 ± 1.3	Soccer, rugby, lacrosse, handball
Females	20.1 ± 1.3	158.0 ± 5.2	53.9 ± 8.2	10.7 ± 2.3	Soccer, cheerleading, volleyball, handball

Values are mean ± SD

20.2.2 Rehabilitation

The same rehabilitation program was used for each patient. After the acute phase, the patients were permitted to exercise to promote increased range of motion. They used isotonic knee extension and flexion exercises. The patients were allowed partial weight bearing at 2 postoperative weeks and progressed to full weight bearing at postoperative week 3. Rehabilitation exercises involved stationary bicycling and exercises with partial weight bearing at 3 weeks after surgery. Isotonic strengthening exercises for the quadriceps and hamstring muscle were initiated at week 4. Until week 8, a restriction in the range of motion was set for the knee extension exercise. Jogging and full-speed running were permitted after 12 and 16 postoperative weeks, and progressed to agility and sports-specific drills. Return to unlimited sports activity was allowed at 6–8 postoperative months.

20.2.3 Self-Report Measures

Subjects answered the questions of a standard questionnaire to provide a Knee Injury and Osteoarthritis Outcome Score (KOOS). The questionnaire's answers provide an internationally used orthopaedic patient self-assessed outcome score. The KOOS supplies a global assesment (symptoms/stiffness, pain, functional impact on every day activities and sports activities, and quality of life) of post-traumatic or degenerative knee conditions. Each subscale score of 36 questions is calculated independently. The mean score of the individual items of each subscale are calculated and divided by 4. Scores for the summed (overall KOOS) and individual subdomains are transformed to a scale of 0–100, with 0 representing extreme knee problems and 100 representing no knee problems. For a detailed explanation, consult Roos et al. (1998).

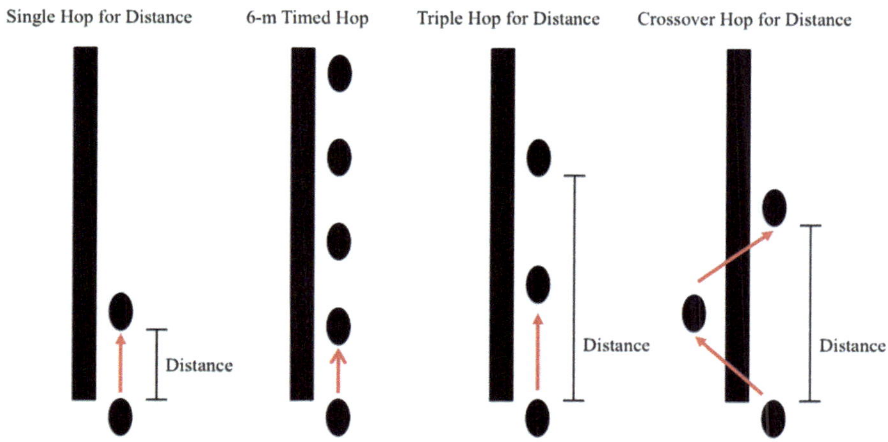

Fig. 20.1 Diagrammatic representation of the series of four hop tests

20.2.4 Hop Test Procedure

We evaluated 4 single-leg function tests: (1) single hop for distance; (2) 6-m timed hop; (3) triple hop for distance; and (4) cross-over hop for distance (Fig. 20.1) (Noyes et al. 1991). For each hop test, the subjects performed one practice trial for each limb, followed by two measured and recorded trials. The final landing on the hop limb had to be stabilized and held for 2 s to be recorded as a successful trial. Subjects were instructed to begin with the nonoperative leg, with their lead toe behind a marked starting line. For single hop for distance, the subjects stood on the leg to be tested. The athlete was instructed to initiate the hop by swinging the arms forward, simultaneously extending at the hip and knee, and to hop forward as far as possible while being able to land safely on the same limb. In the triple hop for distance, the subjects stood on one leg and performed three consecutive hops going as far as possible, using the same limb. In the crossover hop for distance, the subjects were instructed to land safely on the same foot, but on the opposite side of the line medial to the stance limb. Athletes were instructed to immediately redirect into two subsequent forward directed hops, crossing over the midline with each hop. For the 6-m timed hop, the subjects stood on one leg, and then hopped as fast as possible over a marked distance of 6 m. The fastest time measured using the Speed Trap timing system (Brower Timing Systems, Utah, USA) was used for data analysis. The hop test scores were calculated as the average of two recorded trials for each limb. We used the hop test scores of the four tasks to calculate a limb symmetry index (LSI) as per the following formula: (operated lower extremity/nonoperated lower extremity) × 100 %.

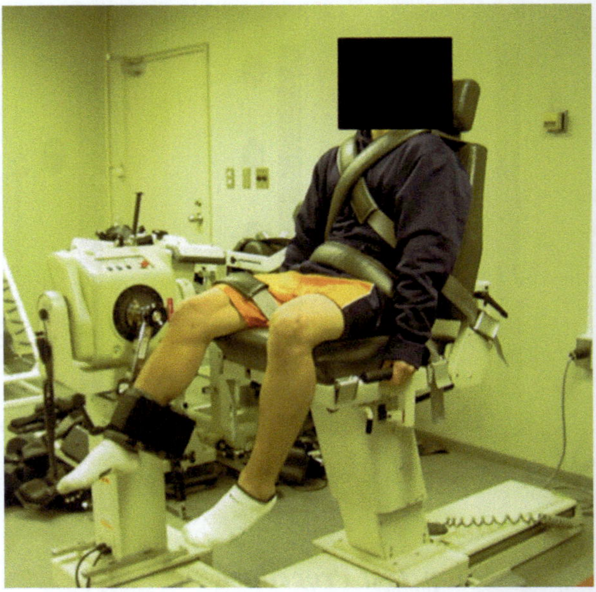

Fig. 20.2 Testing position of the isokinetic torque measurment

20.2.5 Isokinetic Strength Evaluation

Isokinetic concentric strength was measured in the both limbs using a dynamometer (Biodex System III, Biodex Co., New York, USA) (Fig. 20.2). Before evaluation, compensation was performed to exclude the effect of gravity on the measurement of torque. The subject was first supervised in the performance of the movements, and then allowed several practice trials. Three trials of isokinetic contraction were performed at angular velocities of 60° and 180°/s. The range of motion was set from 90° of knee flexion to full extention. The reconstructed limb was tested before the uninjured limb. During the isokinetic contraction, the values for the peak torque were recorded for analysis and normalized to body weight. We calculated LSI by using the same formula as with the hop testing.

20.2.6 Data Analysis

Means and standard deviations were calculated for all variables of interest. All statistical analyses were conducted using a statistical analysis software program (SPSS ver. 21.0; SPSS Japan, Tokyo, Japan). A paired t-test was used to test for side-to-side differences in each hop test and for isokinetic strength. To compare differences between males and females, a non-paired t-test was performed on the hop tests and strength LSI evaluations. A Pearson's product–moment correlation coefficient (*r*) was used to examine the relationship among hop test LSI, strength LSI and overall KOOS scores. Statistical significance was established at $p < 0.05$.

20.3 Results

20.3.1 Self-Report Measures

Subejects had a mean overall KOOS score of 90.1 (range, 76–100), pain score of 90.6 (range, 72–100), symptoms/stifness score of 83.5 (range, 68–100), activities of daily living score of 97.4 (range, 88–100), sports/recreation score of 82.5 (range, 65–100), and QOL score of 79.3 (range, 44–100) (Table 20.2).

20.3.2 Hop Tests

Table 20.2 shows hop test scores for the three subject groups. The hop distance for the operative limb was significantly less than for that of the nonoperative limb ($p < 0.01$) (Table 20.3). Mean LSI value for hop test was single hop of 89 % (range,

Table 20.2 Overall KOOS and individual KOOS subdomains

Overall KOOS	Pain	Symptoms/stifness	ADL	Sports/recreation	QOL
90.1 ± 8.1	90.6 ± 9.8	83.5 ± 12.6	97.4 ± 3.9	82.5 ± 12.4	79.3 ± 19.2

ADL activities of daily living, *QOL* quality of life

Table 20.3 Hop test scores on the operative limbs and nonoperative limbs

	Single hop (cm)	6-m timed hop (s)	Triple hop (cm)	Crossover hop (cm)
All subjects				
Operative limb	118.7 ± 16.3**	1.94 ± 0.23	458.5 ± 66.1*	434.1 ± 81.3*
Nonoperative limb	133.5 ± 20.7	1.92 ± 0.25	492.6 ± 89.9	461.3 ± 89.7
Males				
Operative limb	124.0 ± 17.5**	1.80 ± 0.13	500.5 ± 34.3**	490.0 ± 52.5
Nonoperative limb	142.6 ± 21.0	1.76 ± 0.12	559.4 ± 38.1	525.0 ± 56.2
Females				
Operative limb	111.8 ± 12.6**	2.12 ± 0.20	404.5 ± 41.3**	362.1 ± 46.2
Nonoperative limb	121.9 ± 14.3	2.12 ± 0.23	406.6 ± 54.3	379.3 ± 44.7

$^{*}p < 0.05$; $^{**}p < 0.01$, compared with the nonoperative limb

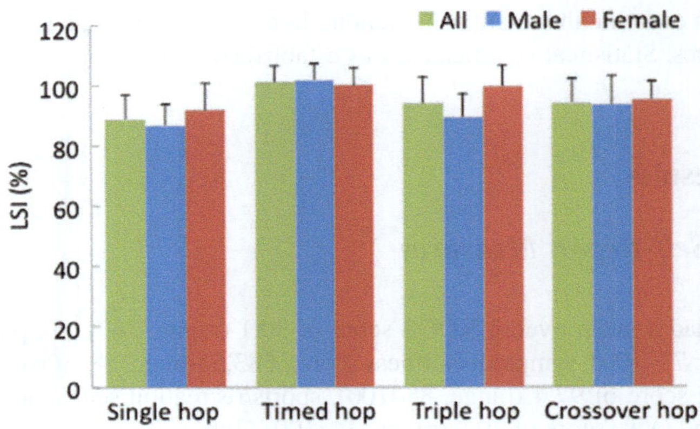

Fig. 20.3 Limb symmetry index for hop tests

77–102 %), 6-m timed hop of 101 % (range, 94–111 %), triple hop of 94 % (range, 75–105), and crossover hop of 95 % (range, 72–104). There was no statistical difference between the operative limb and the nonoperative limb in the 6-m timed hop. There were no significant differences in the LSI of the hop tests between males and females (Fig. 20.3).

20.3.3 Isokinetic Strength

Table 20.3 shows knee extention and flexion isokinetic assessment for the three subject groups. For all subjects, the extention torque in the operative limb was significantly lower than that in the nonoperative limb at 60°/s (2.48 ± 0.49 Nm/kg) and 180°/s (1.75 ± 0.28 Nm/kg), respectively ($p < 0.01$); however, the difference was not significant in flexion torque at 60°/s (1.48 ± 0.32 Nm/kg) and 180°/s (1.16 ± 0.28 Nm/kg) (Table 20.4). Mean LSI value of extention torque was 88 % (range, 77–107 %) at 60°/s, 92 % (range, 81–104 %) at 180°/s (Fig. 20.4). Mean LSI value of flexion torque was 95 % (range, 78–120 %) at 60°/s, 97 % (range, 74–107 %) at 180°/s (Fig. 20.5). There were no significant differences in the strength LSI between males and females.

20.3.4 Relationship Between Hop Test Scores and Other Measurements

We found a significant correlation between knee extention torque at 60°/s and single hop ($r = 0.449$, $p = 0.040$), triple hop ($r = 0.493$, $p = 0.026$), and crossover hop ($r = 0.799$, $p = 0.000$) (Table 20.5). Knee extention torque at 180°/s was significantly

Table 20.4 Isokinetic peak torque of the operative limbs and nonoperative limbs

	60°/s (Nm/kg)	180°/s (Nm/kg)
All subjects		
Extention, operative limb	2.48±0.49**	1.75±0.28**
Extention, nonoperative limb	2.82±0.45	1.91±0.23
Flexion, operative limb	1.48±0.32	1.16±0.28
Flexion, nonoperative limb	1.57±0.32	1.20±0.28
Males		
Extention, operative limb	2.59±0.61*	1.87±0.26
Extention, nonoperative limb	2.99±0.49	2.02±0.18
Flexion, operative limb	1.68±0.28	1.30±0.28
Flexion, nonoperative limb	1.75±0.29	1.38±0.24
Females		
Extention, operative limb	2.33±0.23**	1.59±0.21**
Extention, nonoperative limb	2.60±0.31	1.77±0.21
Flexion, operative limb	1.22±0.15	0.97±0.15
Flexion, nonoperative limb	1.33±0.16	0.97±0.13

*$p < 0.05$; **$p < 0.01$, compared with the nonoperative limb

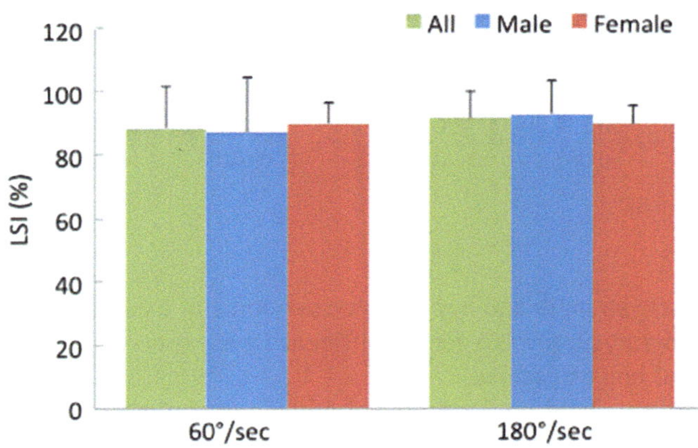

Fig. 20.4 Limb symmetry index for peak torque of isokinetic knee extention

correlated with crossover hop ($r=0.680$, $p=0.002$). There was no relationdship between hop test scores and overall KOOS scores.

20.4 Discussion

In the present study, we evaluated self-reported outcome, various hop tests, and knee flexion/extension isokinetic strength in collegiate athletes who had returned to sports activity following ACL reconstruction. Our main finding was that both

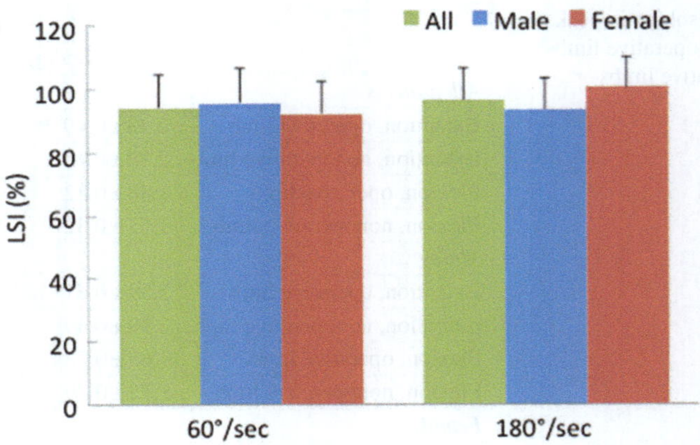

Fig. 20.5 Limb symmetry index for peak torque of isokinetic knee flexion

Table 20.5 Correlation coefficients (*r*) among LSI of the hop testing and overall KOOS score and, LSI of the isokinetic torque

	Overall KOOS	Extention at 60°/s	Extention at 180°/s	Flexion at 60°/s	Flexion at 180°/s
Single hop	0.071	0.449*	0.145	0.111	0.339
6-m timed hop	0.190	−0.135	−0.137	−0.032	−0.133
Triple hop	0.154	0.493*	0.294	0.212	0.093
Crossover hop	0.087	0.799**	0.680**	0.124	0.257

*$p < 0.05$; **$p < 0.01$

extension muscle strength and hop distance assymmetry existed 9–12 months after hamstring autograft surgey. No gender difference was observed in recovery of muscle strength and hop test scores.

The KOOS is a reliable and valid self-reported measurement of symptoms, function, and sports activity, and is certified for use in patients who have undergone ACL reconstruction. The subjects in our study had a mean overall score of 90.1, which considered to be indicative of normal knee function. As previously stated, our study suggests that there is a discrepancy between self-reported results for knee symptoms/function and the results of hop tests. This could mean that the reported outcome measures that are commonly used are neither demanding nor sensitive enough to identify functional disability and do not correlate well with patients' sports-specific performance.

Single-leg hop tests demonstrated potentially significant functional deficits in athletes that had returned to full sports activities. The ability to generate and maintain isolated single-limb power is important during single-limb cutting movements in various sports. In addition, the LSI is a useful tool for deciding whether side-to-side differences are normal or abnormal, and the LSI can be utilized to decide when to return to sports. The acceptable threshold of the LSI for hop tests for safely progressing to more intense sports specific activity after ACL reconstruction is 90 % (Myer et al. 2011; Thomeé et al. 2012). Although, in the present study, 44 % of subjects had scores of greater than 90 % in the LSI for all 4 tests, around two-third of the subjects did not attain a 90 % LSI score. These findings are similar to those of previous research on athlete's level of recovery when returning to sports following ACL reconstruction (Myer et al. 2011; Xergia et al. 2013). Lower extremity asymmetries during athletic tasks may be potential risk factors for lower extremity injury, particularly a second ACL injury, and should be minimized prior to return to sports. Athletes who demonstrate biomechanical assymmetries during a drop vertical jump are at increased risk of ACL injury when compared to those with more symmetrical lower extremity biomechanics (Hewett et al. 2005). Paterno et al. (2010) reported that for athletes who returned to sport post ACL reconstruction, altered neuromuscular control of the hip and side-to-side asymmetry in knee joint recruitment during drop vertical jump are predictors of a second ACL injury. Sports-specific activities are more challenging than landing from a planned hop in a controlled environment, and thus the deficits seen in single-leg hop performance may be magnified in sports-related activities, potentially causing the operated or contralateral knee to sustain injury. Before and after return to sports following ACL reconstruction, it is important to evaluate and minimize these asymmetries, not only for muscle strength but in the performance of athletic maneuvers.

We also evaluated isokinetic strength. If 90 % LSI in all 4 procedure is considered to be an acceptable standard, only 36 % of the subjects in the present study had reached this standard. In particular, extention torque in the operative limb was significantly lower than that in the nonoperative limb at 60°/s (mean, 88 % LSI) and 180°/s (mean, 92 % LSI). With regard to the relationship between strength and the hop tests, extention torque at 60°/s showed a positive correlation with hop-distance in three tasks. Thus, quadriceps weakness may contribute to low hop performance and should be identified and corrected.

The influence of quadriceps strength on postsurgical function after ACL reconstruction is well established (de Jong et al. 2007; Eitzen et al. 2010). Expectedly, recovery of quadriceps strength is important in restoring normal knee function, as persistent quadriceps weakness may have a bad influece on sport-specific function because of their primary role as force controller and generator in knee mechanics. Muscle strength deficits following ACL reconstruction may in some cases accrue to insufficient post-rehabilitation protocols. However, recent rehabilitation protocols have evolved greatly, shifting from a conservative method involving prolonged immobilization and delayed strengthening to current paradigms that recommend immediate weightbearing, early motion and progressive strengthening, and neuromuscular training. In spite of these efforts, as our results indicate, muscle weakness and assymetrical movement still occur after ACL reconstruction. Hewett and his colleague note that neuromuscular training, in various forms, has been effectively used in the prevention of ACL injuries, enhancing function and movement behaviors early after the injury, and improving function and movement behaviors after ACL reconstruction. These investigators also introduced sports symmetry training and preventitive multiplane dynamic movement tasks that are commonly used as preventative training for minimizing primary ACL injury. The protocols of Hewett et al. include exercises such as single-leg progressions, deep knee flexion exercises, balance and propriocepition training, as well as trunk and hip control.

In the present study, even though men obtained higher scores than women in the strength and hop periformance tests, both genders followed the same general LSI pattern. Both genders, thus, reached the same level, at the 9–12 months follow-up, on their operated limbs as compared to their contralateral limb. Nevertheless, females are exhibit lower postsurgical activity levels (Dunn and Spindler 2010) and are associated with a decreased likelihood of returning to sport within 1 year postoperatively (Ardern et al. 2011b) as compared to their male counterparts. While some important factors (such as gender) may be relatively difficult to address with rehabilitation programs and may contribute to second injury risk and incidence, further reseach on the various factors of the risk profile for ACL patients as well as each risk factor's amenability to rehabilitation would be of great value.

There are several limitations to the present study that should be acknowledged. First, since the number of subjects was small, our findings cannot be generalized to all ACL reconstructed patients. Another limitation is that the state of the reconstructed graft, such as knee laixty, was not included in this study. This knowledge would have provided important information about the disabilities seen during the functional tests. Despite these limitations, the findings of the present study demonstrate that deficits in hop-test performance and quadriceps strength persist 9–12 months after ACL reconstruction in athletes that have returned to sports. The asymmetrical charactristics revealed by our investigation may cause future knee injuries, most likely a second rupture of the ACL. In conclusion, to optimize functional and clinical outcomes after ACL reconstruction and to prevent a second knee injury, an evidence based medicine (EBM) approach is proposed in this study; when the criteria for return to sports have not been achieved, then more practice and functional interventions should be performed as necessary.

20.5 Conclusion

This study demonstrates that deficits in hop-distance and quadriceps strength persist 9–12 months after ACL reconstruction, even though the subjucts had returned to their sports activities. To optimize functional and clinical outcomes and to prevent a second injury after ACL reconstruction, we propose to use an EBM assessment with functional practice tests as the criteria for return. Such tests would include various hop tests and isokinetic strength tests, and there would be a focus on the degree of side-to-side assymmetry. The deficits observed in the current study illustrate the need for focused sports symmetry training and highlight their potential contribution to a safe return to high-risk activity by the prevention of future knee injuries.

References

Agel J, Arendt EA, Bershadsky B (2005) Anterior cruciate ligament injury in national collegiate athletic association basketball and soccer: a 13-year review. Am J Sports Med 33:524–530

Ardern CL, Webster KE, Taylor NF, Feller JA (2011a) Return to sport following anterior cruciate ligament reconstruction surgery: a systematic review and meta-analysis of the state of play. Br J Sports Med 45:596–606

Ardern CL, Webster KE, Taylor NF, Feller JA (2011b) Return to the preinjury level of competitive sport after anterior cruciate ligament reconstruction surgery: two-thirds of patients have not returned by 12 months after surgery. Am J Sports Med 39:538–543

Boden BP, Dean GS, Feagin JA Jr, Garrett WE Jr (2000) Mechanisms of anterior cruciate ligament injury. Orthopedics 23:573–578

Bourke HE, Salmon LJ, Waller A, Patterson V, Pinczewski LA (2012) Survival of the anterior cruciate ligament graft and the contralateral ACL at a minimum of 15 years. Am J Sports Med 40:1985–1992

Brophy RH, Schmitz L, Wright RW, Dunn WR, Parker RD, Andrish JT, McCarty EC, Spindler KP (2012) Return to play and future ACL injury risk after ACL reconstruction in soccer athletes from the Multicenter Orthopaedic Outcomes Network (MOON) group. Am J Sports Med 40:2517–2522

Cascio BM, Culp L, Cosgarea AJ (2004) Return to play after anterior cruciate ligament reconstruction. Clin Sports Med 23:395–408

Daniel DM, Stone ML, Dobson BE, Fithian DC, Rossman DJ, Kaufman KR (1994) Fate of the ACL-injured patient: a prospective outcome study. Am J Sports Med 22:632–644

de Jong SN, van Caspel DR, van Haeff MJ, Saris DB (2007) Functional assessment and muscle strength before and after reconstruction of chronic anterior cruciate ligament lesions. Arthroscopy 23:21–28

Dunn WR, Spindler KP (2010) Predictors of activity level 2 years after anterior cruciate ligament reconstruction (ACLR): a Multicenter Orthopaedic Outcomes Network (MOON) ACLR cohort study. Am J Sports Med 38:2040–2050

Eitzen I, Moksnes H, Snyder-Mackler L, Risberg MA (2010) A progressive 5-week exercise therapy program leads to significant improvement in knee function early after anterior cruciate ligament injury. J Orthop Sports Phys Ther 40:705–721

Griffin LY, Agel J, Albohm MJ et al (2000) Noncontact anterior cruciate ligament injuries: risk factors and prevention strategies. J Am Acad Orthop Surg 8:141–150

Hewett TE, Myer GD, Ford KR, Heidt RS, Colosimo AJ, McLean SG, Bogert AJ, Paterno MV, Succop P (2005) Biomechanical measures of neuromuscular control and valgus loading of the knee predict anterior cruciate ligament injury risk in female athletes: a prospective study. Am J Sports Med 33:492–501

Laboute E, Savalli L, Puig P, Trouve P, Sabot G, Monnier G, Dubroca B (2010) Analysis of return to competition and repeat rupture for 298 anterior cruciate ligament reconstructions with patellar or hamstring tendon autograft in sports people. Ann Phys Rehabli Med 53:598–614

Leys T, Salmon L, Waller A, Linklater J, Pinczewski L (2012) Clinical results and risk factors for reinjury 15 years after anterior cruciate ligament reconstruction: a prospective study of hamstring and patellar tendon grafts. Am J Sports Med 40:595–605

Logerstedt D, Lynch A, Eitzen I, Engebretsen L, Risberg MA, Axe MJ (2012) Single-legged hop tests as predictors of self-reported knee function after anterior cruciate ligament reconstruction : the Delaware-Oslo ACL cohort study. Am J Sports Med 40:2348–2356

Myer GD, Schmitt LC, Brent JL, Ford KR, Barber Foss KD, Scherer BJ, Heidt RS Jr, Divine JG, Hewett TE (2011) Utilization of modified NFL combine testing to identify functional deficits in athletes following ACL reconstruction. J Orthop Sports Phys Ther 41:377–387

Myklebust G, Bahr R (2005) Return to play guidelines after anterior cruciate ligament surgery. Br J Sports Med 39:127–131

Myklebust G, Holm I, Maehlum S, Engebretsen L, Bahr R (2003) Clinical, functional, and radiologic outcome in team handball players 6–11 years after anterior cruciate ligament injury: a follow-up study. Am J Sports Med 3:981–989

Noyes FR, Barber SD, Mangine RE (1991) Abnormal lower limb symmetry determined by function hop tests after anterior cruciate ligament rupture. Am J Sports Med 19:513–518

Øiestad BE, Holm I, Engebretsen L, Risberg MA (2011) The association between radiographic knee osteoarthritis and knee symptoms, function and quality of life 10–15 years after anterior cruciate ligament reconstruction. Br J Sports Med 45:583–588

Paterno MV, Schmitt LC, Ford KR, Rauh MJ, Myer GD, Huang B (2010) Biomechanical measures during landing and postural stability predict second anterior cruciate ligament injury after anterior cruciate ligament reconstruction and return to sport. Am J Sports Med 38:1968–1978

Petschnig R, Baron R, Albrecht M (1998) The relationship between isokinetic quadriceps strength test and hop tests for distance and 1-legged vertical jump test following anterior cruciate ligament reconstruction. J Orthop Sports Phys Ther 28:23–31

Reid A, Birmingham TB, Stratford PW, Alcock GK, Giffin JR (2007) Hop testing provides a reliable and valid outcome measure during rehabilitation after anterior cruciate ligament reconstruction. Phys Ther 87:337–349

Roos EM, Roos HP, Lohmander LS, Ekdahl C, Beynnon BD (1998) Knee Injury and Osteoarthritis Outcome Score (KOOS): development of a selfadministered outcome measure. J Orthop Sports Phys Ther 28:88–96

Rosenberg TD, Brown GC, Deffner KT (1997) Anterior cruciate ligament reconstruction with a quadrupled semitendinosus autograph. Sports Med Arthrosc Rev 5:51–58

Salmon L, Russell V, Musgrove T, Pinczewski L, Refshauge K (2005) Incidence and risk factors for graft rupture and contralateral rupture after anterior cruciate ligament reconstruction. Arthroscopy 21:948–957

Shelbourne KD, Gray T, Haro M (2009) Incidence of subsequent injury to either knee within 5 years after anterior cruciate ligament reconstruction with patellar tendon autograft. Am J Sports Med 37:246–251

Thomeé R, Neeter C, Gustavsson A, Thomee P, Augustsson J, Eriksson B, Karlsson J (2012) Variability in leg muscle power and hop performance after anterior cruciate ligament reconstruction. Knee Surg Sports Traumatol Arthrosc 20:1143–1151

Walden M, Hagglund M, Ekstrand J (2006) High risk of new knee injury in elite footballers with previous anterior cruciate ligament injury. Br J Sports Med 40:158–162

Wright RW, Dunn WR, Amendola A, Andrish JT, Bergfeld J, Kaeding CC, Marx RG, McCarty EC, Parker RD, Wolcott M, Wolf BR, Spindler KP (2007) Risk of tearing the intact anterior cruciate ligament in the contralateral knee and rupturing the anterior cruciate ligament graft during the first 2 years after anterior cruciate ligament reconstruction: a prospective MOON cohort study. Am J Sports Med 35:1131–1134

Xergia SA, Pappas E, Zampelif F, Georgiou S, Georgoulis SD (2013) Asymmetries in functional hop tests, lower extremity kinematics, and isokinetic strength persist 6 to 9 months following anterior cruciate ligament reconstruction. J Orthop Sports Phys Ther 43:154–162

Chapter 21
Biomechanical Adaptations in Subjects After Anterior Cruciate Ligament Reconstruction: Preventing Secondary Injury

Takuma Hoshiba and Toru Fukubayashi

Abstract A secondary anterior cruciate ligament (ACL) injury in a surgically reconstructed limb is of critical concern after ACL reconstruction. Although risk factors for a secondary ACL injury have been recently identified, the risk factors actually caused by ACL reconstruction and their associated behavioral characteristics remain unclear. In this chapter I introduce methods for evaluating biomechanical adaptations during dynamic exercises and summarize current concepts for preventing secondary ACL injury after reconstruction. In earlier studies, the altered loading strategy and joint movement patterns of the reconstructed limb on landing during dynamic exercises were present; thus, to prevent secondary ACL injury, the altered biomechanical function needs to be addressed and eliminated before subjects return to sports participation.

Keywords Knee • Secondary ACL injury • Lower limb kinematics and kinetics

21.1 Introduction

Anterior cruciate ligament (ACL) reconstruction is the most common surgical reconstruction after ACL injury (Lyman et al. 2009). More than 100,000 reconstructions are estimated to be carried out in the United States every year (Lyman et al. 2009). Most orthopedic surgeons advocate surgical reconstruction if ACL-injured athletes expect to return to sports participation (Marx et al. 2003) since surgical reconstruction enables athletes to return to their pre-injury level of activity and prevents torsional overloading on the menisci (Ajuied et al. 2013). One study reported that an average of 63 % of patients return to their pre-injury level of sport

T. Hoshiba (✉)
Graduate School of Sport Sciences, Waseda University, Saitama, Japan
e-mail: takuma9@asagi.waseda.jp

T. Fukubayashi
Faculty of Sport Sciences, Waseda University, Saitama, Japan
e-mail: fukubayashi@waseda.jp

K. Kanosue et al. (eds.), *Sports Injuries and Prevention*,
DOI 10.1007/978-4-431-55318-2_21

participation after ACL reconstruction, even though some of these do not participate in sports at a competitive level (Ardern et al. 2010). Current guidelines for a return to sports require that the reconstructed limb be within 80–90 % of that of the contralateral limb during specific clinical tests that focus on the asymmetry between limbs (Barber-Westin and Noyes 2011). The amount of asymmetry is considered one of the crucial determinants of whether ACL-reconstructed subjects can resume sports participation (Hewett et al. 2012; Paterno et al. 2010).

Although ACL reconstruction is successfully performed and restores joint function by allowing athletes to regain joint stability, the reconstructed and contralateral knee are both at risk of a secondary ACL injury (Salmon et al. 2005; Wright et al. 2007). While the rate of primary ACL injury is 1 in 3,500 for the general population (Beynnon et al. 2005), the rate of secondary ACL injury (either re-injury or contralateral initial injury) after return to sports participation following ACL reconstruction is as high as 1 in 4 (Paterno et al. 2010), indicating that the risk of a secondary ACL injury is much higher than that of primary ACL injury. One of the major risk factors involved in a secondary ACL injury involves biomechanical adaptations that may be affected by surgical technique, time since surgery, age, sex, and activity level (Di Stasi et al. 2013). It is pertinent that the mechanism of a secondary ACL injury is more complicated than the mechanism of a primary ACL injury since both risk factors related to surgical reconstruction and biomechanical adaptations leading to primary ACL injury need to be simultaneously taken into account.

Biomechanical adaptations after ACL reconstruction that affect the movement patterns of lower limb joints during exercises have been commonly reported (Dai et al. 2012, 2013; Decker et al. 2002; Delahunt et al. 2012; Gokeler et al. 2009; Hart et al. 2010; Ortiz et al. 2007; Orishimo et al. 2010; Paterno et al. 2010; Stearns and Pollard 2013). The protective mechanism, which is referred to as "movement patterns compensated by the hip and ankle joints" in these previous studies, is often observed during dynamic exercises and functions to avoid overloading the reconstructed knee joint. This possibly alters the force transfer to the lower limb joints and causes abnormal alignment that results in asymmetry between the reconstructed and contralateral limbs. To understand the biomechanical adaptations that occur after ACL reconstruction, it is necessary to comprehensively assess the joint movement patterns in ACL-reconstructed subjects in terms of a secondary ACL injury.

Although the knee might be altered by ACL injury and/or surgical reconstruction, what is actually caused by these risk factors and what appears as behavioral characteristics remains unclear. A limited number of studies have reported on the risk factors for and measures of preventing secondary ACL injury after ACL reconstruction. This stands in contrast to the large number of studies concerning the prevention of primary ACL injury.

In this chapter I review the dynamic exercises (drop vertical jump/landing, side cutting maneuver, and single hop maneuver) that have been frequently utilized in laboratory settings and described in earlier studies. I evaluate how these exercises, and others, are typically used to provide an understanding of knee mechanics during the recovery from ACL reconstruction surgery. I compare the usual clinical assessments with those done in laboratory situations. My focus will be on evaluating the potential risk of an injury recurrence. My main emphasis will be on the biomechanical, physical, and nervous alterations of the repaired limb that are adopted by athletes recovering from ACL surgery. An awareness of such adjustments is important, since these alterations can impair the viability of an athlete returning to sports

21.2 Study Selection

An online Pubmed search using the following terms and phrases was performed in this chapter: "anterior cruciate ligament reconstruction and drop vertical jump" (n = 7), "anterior cruciate ligament reconstruction and side cutting" (n = 5), and "anterior cruciate ligament reconstruction and single hop" (n = 117). The studies written in English including ACL-reconstructed subjects compared to a contralateral limb and/or healthy subjects were extracted in this chapter.

21.3 Tasks to Identify Biomechanical Adaptations After ACL Reconstruction

Biomechanical adaptations in ACL-reconstructed subjects have been evaluated by using a three-dimensional **motion** analysis system and ground reaction forces to calculate kinematics and kinetics at the lower limb joints (hip, knee, and ankle) during dynamic exercises (Dai et al. 2012, 2013; Decker et al. 2002; Delahunt et al. 2012; Gokeler et al. 2009; Ortiz et al. 2007; Orishimo et al. 2010; Paterno et al. 2010; Stearns and Pollard 2013). In these dynamic exercises, the lower-limb stabilization and/or postural control are typically assessed, along with how the loading forces are transferred and distributed between the lower limb joints. After ACL reconstruction, biomechanical adaptations in the reconstructed knee are commonly reported, which may cause asymmetry between the reconstructed and contralateral limbs (Hewett et al. 2012; Paterno et al. 2010). This can be one of the crucial predictors of a secondary ACL injury risk; the altered biomechanical adaptations need to be analyzed and can be effectively addressed with rehabilitation that includes prevention training before the return to sports participation in terms of preventing a secondary ACL injury (Di Stasi et al. 2013; Hewett et al. 2012).

In the laboratory setting in earlier studies, three dynamic exercises performed by ACL-reconstructed subjects to evaluate their biomechanical adaptations are introduced: a drop vertical jump/drop landing (Decker et al. 2002; Delahunt et al. 2012; Paterno et al. 2010), side cutting maneuver (Dai et al. 2012; Decker et al. 2002; Stearns and Pollard 2013), and single hop maneuver (Gokeler et al. 2009; Orishimo et al. 2010).

1. Drop vertical jump/drop landing (Fig. 21.1)
   The drop vertical jump/drop landing is composed of dropping off a box onto the ground (drop vertical jump/drop landing) and performing a vertical jump imme-

diately after landing (drop vertical jump). In this task, subjects are instructed to stand on the box and drop off the box without jumping or stepping off (drop vertical jump/drop landing). As soon as they land on the ground, they are required to perform a vertical jump with maximal effort (drop vertical jump).

2. Side-cutting maneuver (Fig. 21.2)
   The side-cutting task consists of an approaching run followed by a right or left landing on the ground and then changing direction to either the right or left at a designated angle. Prior to this task, subjects are usually instructed as to which direction they should execute the side-cutting maneuver with the reconstructed and contralateral limbs, unless the anticipated condition is necessary. The required approach speed is usually fixed at maximum speed to perform the task.
3. Single hop maneuver (Fig. 21.3)
   The single hop maneuver consists of single-limb jumping and landing on the ground with the same limb. Subjects are instructed to stand on a single leg at the starting position and jump as far as possible. They then land on the same limb and maintain a stable lower limb. In general, the subjects are not instructed to perform any upper limb movements (such as swinging the arms) during this task.

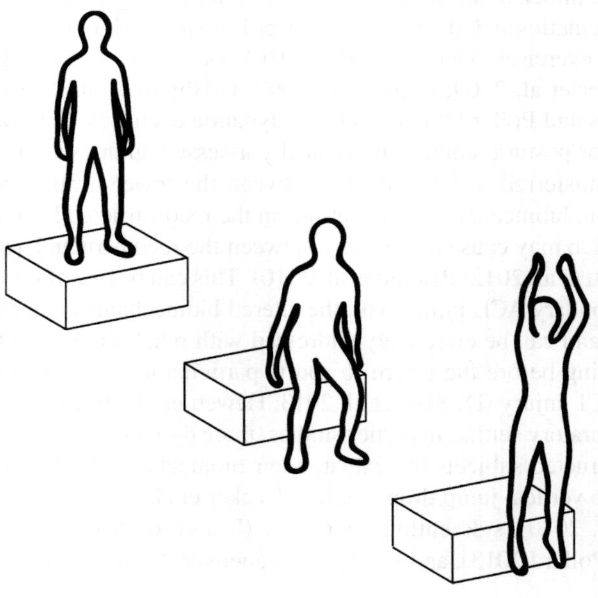

Fig. 21.1 The drop vertical jump

Fig. 21.2 The side-cutting maneuver

Fig. 21.3 The single hop maneuver

21.4 Kinematics and Kinetics Representing Biomechanical Adaptations

Joint mechanics include kinematics and kinetics that are calculated using the three-dimensional coordinate system and inverse dynamics approach.

With regard to landing patterns, the altered loading strategy and joint movements on the reconstructed limb are typically present, and reveal that the hip and ankle joints compensate upon the loading absorption of the initial contact (Decker et al. 2002). Asymmetries between limbs in sagittal plane knee moments are also present (Paterno et al. 2010). The time-course changes in the lower kinematics upon landing indicate that the hip is likely to be more extended and the ankle is more plantar-flexed during initial contact (Decker et al. 2002). Following initial contact, the ACL-reconstructed subjects tend to exhibit a more adducted hip position and a less adducted and flexed knee position (Delahunt et al. 2012).

In the side-cutting maneuver, in which the athlete is required to first make a stable landing and then execute a side cut, increased knee abduction angles occur knee adductor moments on the reconstructed limb were also observed during the first 20 % of the stance phase defined as the early deceleration phase (Stearns and Pollard 2013). Decreased initial knee flexion velocity on the reconstructed limb was also observed upon landing (Dai et al. 2012). It has also been demonstrated that asymmetries in sagittal plane knee moment are associated with asymmetries in the vertical ground reaction force. Thus, compensation strategies in side cutting may exist in ACL reconstructed subjects (Dai et al. 2013).

In the single hop maneuver, which requires trunk and lower limb stabilization for takeoff and landing, decreased range of motion at the hip, knee, and ankle were observed, and the knee flexion angle was less on the reconstructed limb at the initial posture takeoff posture (Gokeler et al. 2009) and at landing (Orishimo et al. 2010). A low peak-knee extension moment and peak power on the reconstructed limb, and high peak hip and ankle extension moments and power were observed at takeoff (Orishimo et al. 2010). Peak power, considered as force absorption during landing, on the reconstructed limb was also low at the knee and high at the hip and ankle at landing (Orishimo et al. 2010). A low knee extension moment tends to be primarily compensated by the ankle joint (Gokeler et al. 2009).

Altered sagittal and frontal plane of lower limb joint mechanics are commonly observed after ACL reconstruction during introduced dynamic exercises, and these alterations have been considered critical risk factors for secondary ACL injury. Sagittal and frontal plane kinematics and kinetics as well as postural stability are key factors in the evaluation of the physical characteristics of ACL-reconstructed subjects and important predictors of a secondary ACL injury after ACL reconstruction (Hewett et al. 2012; Paterno et al. 2010).

21.5 Current Views in ACL-Reconstructed Subjects

Altered lower limb joint mechanics on the reconstructed limb during dynamic exercises are commonly observed after ACL reconstruction. A combination of deceleration and acceleration with high velocity is a frequent requirement in many sports. These movements have a major effect on loading strategies and joint movement patterns and increase the risk of secondary ACL injury. To prevent a secondary ACL injury, rehabilitation programs after ACL reconstruction need to include restoration of the proper lower limb movement patterns, especially at landing, before the subject returns to sports participation. Proper lower limb alignment enables the encountered forces to be properly transferred to the joints, which should be one of the goals after ACL reconstruction, both to enhance performance and to prevent a secondary ACL injury.

In the clinical setting, recovery tends to be focused more on clinical performance tests which evaluate the strength of the muscles around the knee joint as well as agility variables which are measured in time and distance (Czuppon et al. 2013; Thomee et al. 2011). Although conducting these tests does not require special measurement equipment and the tests are easily performed in clinical settings, outcomes from these tests depict only performance results; compensatory adaptations that allow a good performance but increase the risk of a secondary ACL injury may be masked. The relationship between clinical outcomes and laboratory-based outcomes should be taken into account when evaluating ACL-reconstructed subjects who have biomechanical impairments. In most cases such assessments should take place before the athlete returns to sports participation.

In addition to biomechanical adaptations, adaptations in muscle activity, particularly in the onset times and in peak activity, are also commonly seen during dynamic exercises after ACL reconstruction (Coats-Thomas et al. 2013; DeMont et al. 1999; Gokeler et al. 2009; Swanik et al. 1999). In dynamic exercises, the primary function of muscle is to stabilize the joints by generating force (Gokeler et al. 2009; Solomonow and Krogsgaard 2001). Deficits in neuromuscular function and dynamic strength as well as the inhibition of muscular exertion, however, may create a situation where the muscles cannot absorb the required loading and subsequently prepare the joint for ground contact that provides for a proper lower extremity alignment. This may lead to compensatory movement patterns in lower limb joints on the reconstructed limb that result in altered kinematics and kinetics.

In this chapter I have evaluated the use of dynamic exercises to assess biomechanical adaptations and described the biomechanical characteristics observed in ACL-reconstructed subjects after ACL reconstruction. Although it is unknown which risk factors are strongly associated with a secondary ACL injury, adequate intervention strategies are needed to eliminate risk factors, restore optimal function, and enhance performance. Further studies are needed to simultaneously investigate both biomechanical and neuromuscular adaptations that increase our understanding of the mechanisms that underly secondary ACL injuries.

21.6 Limitations

A limited number of studies, often with a small sample size, were available for the review performed in this chapter. In addition, the review involved a summary of ACL-reconstructed subjects who were not differentiated according to surgical technique, time since surgery, age, or sex.

References

Ajuied A, Wong F, Smith C, Norris M, Earnshaw P, Back D, Davies A (2013) Anterior cruciate ligament injury and radiologic progression of knee osteoarthritis: a systematic review and meta-analysis. Am J Sports Med. doi:10.1177/0363546513508376

Ardern CL, Webster KE, Taylor NF, Feller JA (2010) Return to sport following anterior cruciate ligament reconstruction surgery: a systematic review and meta-analysis of the state play. Br J Sports Med 45:596–606. doi:10.1136/bjsm.2010.076364

Barber-Westin SD, Noyes FR (2011) Factors used to determine return to unrestricted aports activities after anterior cruciate ligament reconstruction. Arthroscopy 27:1697–1705. doi:10.1016/j.arthro.2011.09.009

Beynnon BD, Johnson RJ, Abate JA, Fleming BC, Nicholas CE (2005) Treatment of anterior cruciate ligament injuries, partI. Am J Sports Med 33:1579–1602. doi:10.1177/0363546505279913

Coats-Thomas MS, Miranda DL, Badger GJ, Fleming BC (2013) Effects of ACL reconstruction surgery on muscle activity of the lower limb during a jump-cut maneuver in males and females. J Orthop Res 31:1890–1896. doi:10.1002/jor.22470

Czuppon S, Racette BA, Harris-Hayes M (2013) Variables associated with return to sport following anterior cruciate ligament reconstruction: a systematic review. Br J Sports Med. doi:10.1136/bjsports-2012-091786

Dai B, Butler RJ, Garrett WE, Queen RM (2012) Anterior cruciate ligament reconstruction in adolescent patients: limb asymmetry and functional knee bracing. Am J Sports Med 40:2756–2763. doi:10.1177/0363546512460837

Dai B, Butler RJ, Garrett WE, Queen RM (2013) Using ground reaction force to predict knee kinetic asymmetry following anterior cruciate ligament reconstruction. Scand J Med Sci Sports. doi:10.1111/sms.12118

Decker MJ, Torry MR, Noonan TJ, Riviere A, Sterett WI (2002) Landing adaptations after ACL reconstruction. Med Sci Sports Exerc 34:1408–1413. doi:10.1249/01.MSS.0000027627.82650.1F

Delahunt E, Sweeny L, Chawke M, Kelleher J, Murphy K, Patterson M, Prendiville A (2012) Lower limb kinematic alternations during drop vertical jumps in female athletes who have undergone anterior cruciate ligament reconstruction. J Orthop Res 30:72–78. doi:10.1002/jor.21504

DeMont RG, Lephart SM, Giraldo JL, Swanik CB, Fu FH (1999) Muscle preactivity of anterior cruciate ligament-deficient and-reconstructed females during functional activities. J Athl Train 34:115–120

Di Stasi SL, Myer GD, Hewett TE (2013) Neuromuscular training to target deficits associated with second anterior cruciate ligament injury. J Orthop Sports Phys Ther 43:777–792. doi:10.2519/jospt.2013.4693

Gokeler A, Hof AL, Arnold MP, Dijkstra PU, Postema K, Otten E (2009) Abnormal landing strategies after ACL reconstruction. Scand J Med Sci Sports 20:e12–e19. doi:10.1111/j.1600-0838.2008.00873.x

Hart JM, Ko JW, Konold T, Pietrosimone B (2010) Sagittal plane knee joint moments following anterior cruciate ligament injury and reconstruction: a systematic review. Clin Biomech (Bristol, Avon) 25:277–283. doi:10.1016/j.clinbiomech.2009.12.004

Hewett TE, Di Stasi SL, Myer GD (2012) Current concepts for injury prevention in athletes after anterior cruciate ligament reconstruction. Am J Sports Med 41:216–224. doi:10.1177/0363546512459638

Lyman S, Koulouvaris P, Sherman S, Do H, Mandl LA, Marx RG (2009) Epidemiology of anterior cruciate ligament reconstruction. J Bone Joint Surg Am 91:2321–2328. doi:10.2106/JBJS.H.00539

Marx RG, Jones EC, Angel M, Wickiewicz TL, Warren RF (2003) Beliefs and attitudes of members of the American Academy of Orthopaedic Surgeons regarding the treatment of anterior cruciate ligament injury. Arthroscopy 19:762–770. doi:10.1016/S0749-8063(03)00398–0

Orishimo KF, Kremenic IJ, Mullaney MJ, McHugh MP, Nicholas SJ (2010) Adaptations in single-leg hop biomechanics following anterior cruciate ligament reconstruction. Knee Surg Sports Traumatol Arthrosc 18:1587–1593. doi:10.1007/s00167-010-1185-2

Ortiz A, Olson S, Libby CL, Trudelle-Jackson E, Kwon YH, Etnyre B, Bartlett W (2007) Landing mechanics between noninjured women and women with anterior cruciate ligament reconstruction during 2 jump tasks. Am J Sports Med 36:149–157. doi:10.1177/0363546507307758

Paterno MV, Schmitt LC, Ford KR, Rauh MJ, Myer GD, Huang B, Hewett TE (2010) Biomechanical measuresduring landing and postural stability predict second anterior cruciate ligament injury after anterior cruciate ligament reconstruction and return to sport. Am J Sports Med 38:1968–1978. doi:10.1177/0363546513476853

Salmon L, Russell V, Musgrove T, Pinczewski L, Refshauge K (2005) Incidence and risk factors for graft rupture and contralateral rupture after anterior cruciate ligament reconstruction. Arthroscopy 21:948–957. doi:10.1016/j.arthro.2005.04.110

Solomonow M, Krogsgaard M (2001) Sensorimotor control of knee stability. A review. Scand J Med Sci Sports 11:64–80. doi:10.1034/j.1600-0838.2001.011002064.x

Stearns KM, Pollard CD (2013) Abnormal frontal plane knee mechanics during sidestep cutting in female soccer athletes after anterior cruciate ligament reconstruction and return to sport. Am J Sports Med 41:918–923. doi:10.1177/0363546513476853

Swanik CB, Lephart SM, Giraldo JL, DeMont RG, Fu FH (1999) Reactive muscle firing of anterior cruciate ligament-injured females during functional activities. J Athl Train 34:121–129

Thomee R, Kaplan Y, Kvist J, Myklebust G, Risberg MA, Theisen D, Tsepis E, Werner S, Wondrasch B, Witvrouw E (2011) Muscle strength and hop performance criteria prior to return to sports after ACL reconstruction. Knee Surg Sports Traumatol Arthrosc 19:1798–1805. doi:10.1007/s00167-011-1669-8

Wright RW, Dunn WR, Amendola A, Andrish JT, Bergfeld J, Kaeding CC, Marx RG, McCarty EC, Parker RD, Wolcott M, Wolf BR, Spindler KP (2007) Risk of tearing the intact anterior cruciate ligament in the contralateral knee and rupturing the anterior cruciate ligament graft during the first 2 years after anterior cruciate ligament reconstruction. Am J Sports Med 35:1131–1134. doi:10.1177/0363546507301318

Chapter 22
Morphology and Function of the Semitendinosus Muscle-Tendon Complex After Harvesting Its Tendon for Anterior Cruciate Ligament Reconstruction

Akie Nishino and Toru Fukubayashi

Abstract A considerable decrease in knee flexion strength during deep knee flexion is reported after harvesting the semitendinosus (ST) tendon for anterior cruciate ligament (ACL) reconstruction, although the peak knee flexion strength of the ST tendon-harvested limb can eventually recover almost completely. Strength deficits in deep knee flexion may influence athletic performance in sports activities that require more strength at deep knee flexion. Here we describe morphological changes in the ST as a muscle-tendon complex that may affect knee flexion strength after harvesting the ST tendon for ACL reconstruction. In conclusion, deficits in knee-flexion torque at deep knee flexion are associated with the atrophy and shortening of the ST after harvesting the ST tendon for ACL reconstruction. Maintenance of the ST morphology as a muscle-tendon complex is necessary to ensure postoperative recovery of knee-flexion torque at both shallow and deep knee flexion. We propose that patients should undergo ACL reconstructive surgery and postoperative rehabilitation appropriately to enable regeneration of the ST tendon-like structure while maintaining the morphology of the ST.

Keywords Anterior cruciate ligament reconstruction • Semitendinosus • Tendon graft harvest • Tendon regeneration • Deep knee flexion

A. Nishino (✉)
Institute of Sport Science, ASICS Corporation, Hyogo, Japan
e-mail: akie.nishino@asics.com

T. Fukubayashi
Faculty of Sport Sciences, Waseda University, Saitama, Japan
e-mail: fukubayashi@waseda.jp

K. Kanosue et al. (eds.), *Sports Injuries and Prevention*,
DOI 10.1007/978-4-431-55318-2_22

22.1 Introduction

The semitendinosus (ST) tendon is commonly used as a replacement graft during anterior cruciate ligament (ACL) reconstruction. There are many advantages of using the ST tendon, including ease of harvest during ACL reconstructive surgery, suitable morphology for use as an ACL graft, and lower donor-site morbidity (Rosenberg et al. 1997). Despite these advantages, a considerable decrease in knee flexion strength during deep knee flexion has been reported after harvesting the ST tendon for ACL reconstruction (Nakajima et al. 1996; Ohkoshi et al. 1998; Tadokoro et al. 2004; Tashiro et al. 2003), although the peak knee flexion torque of the ST tendon-harvested limb can eventually recover to at least 90 % of normal (Aglietti et al. 1994, 1997; Anderson et al. 1994, 2001; Cooley et al. 2001; Cross et al. 1992; Ferretti et al. 2002; Gobbi et al. 2003; Irie and Tomatsu 2002; Karlson et al. 1994; Lipscomb et al. 1982; Lipscomb and Anderson 1986; Maeda et al. 1996; Siegel and Barber-Westin 1998; Simonian et al. 1997; Yasuda et al. 1995). Strength deficits in deep knee flexion can influence athletic performance in sports activities that require more strength at deep knee flexion. For example, judo athletes cannot perform effectively standing techniques using the leg, and gymnasts and ballet dancers cannot pose in maximum knee flexion if they have strength deficits in deep knee flexion.

In most cases, the ST tendon can regenerate with a morphology similar to the native tendon after being harvested for use as an ACL graft (Cross et al. 1992; Eriksson et al. 1999, 2001a, b; Ferretti et al. 2002; Hioki et al. 2003; Nakamae et al. 2005; Nakamura et al. 2004; Papandrea et al. 2000; Rispoli et al. 2001; Simonian et al. 1997; Tadokoro et al. 2004; Williams et al. 2004). More recently, morphological changes including atrophy (Eriksson et al. 1999, 2001a; Irie and Tomatsu 2002; Williams et al. 2004) and shortening (Hioki et al. 2003; Nakamae et al. 2005; Williams et al. 2004) of the ST muscle belly have been confirmed in patients with ACL reconstruction using the ST tendon. The ST is one of the agonist muscles in knee flexion. Therefore, these morphological changes in the ST as a muscle-tendon complex could be reasonably assumed to be a factor causing strength deficits in deep knee flexion after harvesting the ST tendon. However, little information is available on the relationship between the morphology of the ST muscle-tendon complex and deep knee flexion strength in the ST tendon-harvested limb.

In this chapter we utilize our research findings to describe morphological changes in the ST as a muscle-tendon complex that may affect knee flexion strength after harvesting the ST tendon for ACL reconstruction. An earlier form of this section first appeared in the *Medicine & Science in Sports & Exercise* Vol. 38, No. 11, pp. 1895–1900, 2006 (Nishino et al. 2006).

Table 22.1 Muscle volume and length of the ST (Nishino et al. 2006)

	Contralateral limb	ACL-reconstructed limb
Muscle volume of the ST (cm^3)	152.5±65.5	111.4±55.7[a]
Muscle length of the ST (cm)	28.1±3.1	24.3±4.3[a]

Values are mean±SD
[a]$p < 0.0001$ compared with the contralateral limb

22.2 Morphology of the Semitendinosus Muscle-Tendon Complex After Harvesting Its Tendon

For 23 patients (10 males, 13 females, mean age±SD: 22±4 years) with unilateral ACL reconstruction (average: 23 months after surgery; range: 12–43 months) with an autogenous quadrupled ipsilateral ST tendon, the muscle volume and length of the ST in both limbs, and the presence of the regenerated ST tendon in ACL-reconstructed limb were evaluated by using magnetic resonance imaging (MRI) scans.

Evaluation results of the muscle volume and length of the ST are shown in Table 22.1. The volume of the ST in the ACL-reconstructed limb (111.4±55.7 cm^3) was significantly smaller than in the contralateral limb (152.5±65.5 cm^3; $P < 0.0001$) (Table 22.1). Thus, atrophy of the ST in the reconstructed limb was confirmed.

The muscle length of the ST in the reconstructed limb (24.3±4.3 cm) was significantly shorter than in the contralateral limb (28.1±3.1 cm; $P < 0.0001$) (Table 22.1). In 19 of the 23 patients, the shorter muscle length of the ST was attributable to a proximal shift of the distal musculotendinous junction in the ACL-reconstructed limb (Fig. 22.1a). In the remaining four patients, the length of the ST in the reconstructed limb was the same as in the contralateral limb (Fig. 22.1b).

In 21 of the 23 patients, regeneration of an ST tendon-like structure was confirmed (Fig. 22.2a). In these patients, the entire regenerated tendon-like structure passed the knee joint and inserted into the distal structures. In the remaining two patients, a tendon-like structure was not identified (Fig. 22.2b).

22.3 Relationship Between Knee Flexion Strength and Morphology of the Semitendinosus Muscle-Tendon Complex After Harvesting Its Tendon

Based on the morphological changes in the ST muscle-tendon complex after harvesting its tendon, patients were divided into three groups. In four patients, regeneration of an ST tendon-like structure was confirmed, and the ST muscle length of

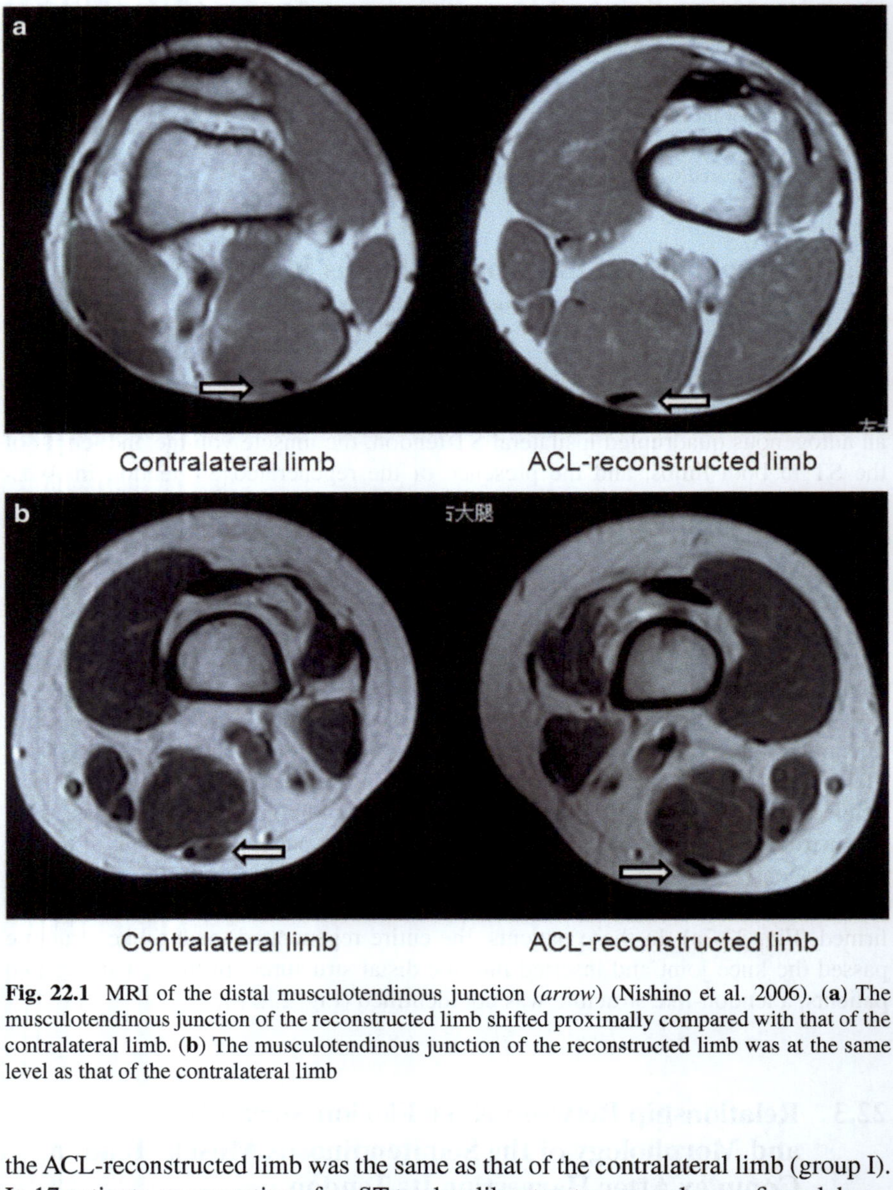

Fig. 22.1 MRI of the distal musculotendinous junction (*arrow*) (Nishino et al. 2006). (**a**) The musculotendinous junction of the reconstructed limb shifted proximally compared with that of the contralateral limb. (**b**) The musculotendinous junction of the reconstructed limb was at the same level as that of the contralateral limb

the ACL-reconstructed limb was the same as that of the contralateral limb (group I). In 17 patients, regeneration of an ST tendon-like structure was also confirmed; however, the ST muscle of the ACL-reconstructed limb was shorter than that of the contralateral limb (group II). In the remaining two patients, an ST tendon-like structure could not be identified, and the length of the ST muscle of the ACL-reconstructed limb was shorter than that of the contralateral limb (group III).

To evaluate the relationship between knee flexion strength and morphological changes in the ST muscle-tendon complex after harvesting the ST tendon for ACL

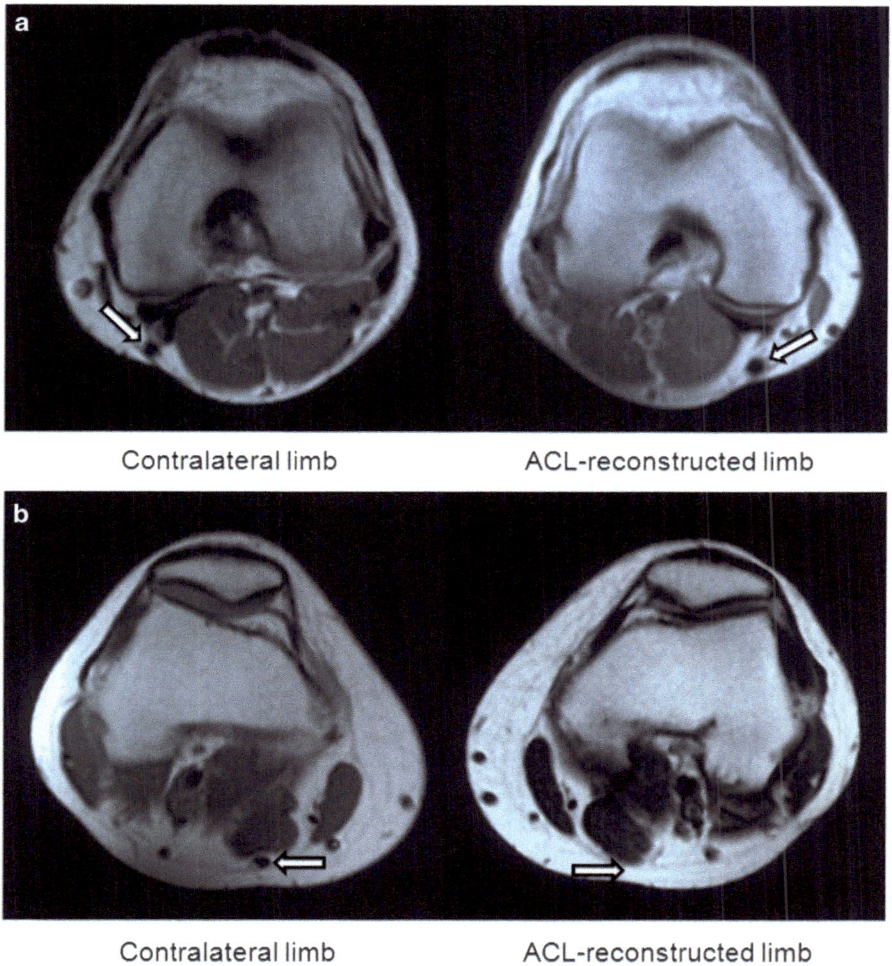

Fig. 22.2 MRI of the ST tendon (*arrow*) (Nishino et al. 2006). (**a**) Tendon-like structure of the ST was confirmed at the ST tendon harvest site. (**b**) Tendon-like structure of the ST could not be identified at the ST tendon harvest site

reconstruction, the percentage of muscle volume and length of the ST and knee flexion torque in the ACL-reconstructed limb relative to those in the contralateral limb (% contralateral) were summarized in Table 22.2 for each of the three groups mentioned above. In the patients in group I, the isometric knee flexion torque of the reconstructed limb was similar to that of the contralateral limb at both 45° and 90° of knee flexion. In group II, knee flexion torque of the reconstructed limbs was similar to the contralateral limb at 45°; however, the torque value at 90° tended to be lower than that of the contralateral limb. In group III, knee flexion torque of the reconstructed limbs tended to be considerably lower than that of the contralateral limbs at both 45° and 90°.

Table 22.2 Summarization of the percentages of the measured variables in the ACL reconstructed limb to those in the contralateral limb for the three patient's groups (% contralateral) (Nishino et al. 2006)

	Group I (n = 4)	Group II (n = 17)	Group III (n = 2)
Tendon-like structure of the ST	Regenerated	Regenerated	Unidentified
Muscle length of the ST (%)	100	86.5 ± 6.0	58.2 ± 8.2
Muscle volume of the ST (%)	85.6 ± 18.0	74.8 ± 13.9	37.4 ± 10.8
Knee flexion torque at 45° (%)	101.3 ± 7.7	96.7 ± 13.6	57.4 ± 0.4
Knee flexion torque at 90° (%)	98.0 ± 37.3	71.9 ± 21.6	44.4 ± 8.5

Values are mean ± SD

Isometric knee flexion torque with maximum voluntary effort were measured at 45° and 90°, representing shallow and deep angles of knee flexion, respectively

Group I, the regeneration of the ST tendon-like structure was confirmed and the ST muscle length of the ACL- reconstructed limb was the same as that of the contralateral limb; Group II, the regeneration of the ST tendon-like structure was also confirmed; however the ST muscle of the ACL-reconstructed limb was shorter than that of the contralateral limb; Group III, the ST tendon-like structure could not be identified and the length of the ST muscle of the ACL-reconstructed limb was shorter than that of the contralateral limb

22.4 Function of the Semitendinosus Muscle-Tendon Complex After Harvesting Its Tendon

The findings described in the previous subsection indicate that changes in knee flexion strength are associated with the degree of morphological change in the ST as a muscle-tendon complex after harvesting its tendon for ACL reconstruction. The deficits in knee flexion strength of the ST tendon-harvested limb are partly explained by the morphological changes in the ST muscle-tendon complex.

For the patients with regeneration of the ST tendon-like structure at the ST tendon-harvested site (groups I and II), knee flexion torque of the ST tendon-harvested limb was recovered at shallow knee flexion angles. In all cases with regeneration, the regenerated tendon-like structure passed the knee joint and inserted into the structures distal to the knee joint. On the other hand, patients without regeneration of the tendon-like structure (group III) demonstrated deficits of knee flexion torque at both shallow and deep knee flexion, most likely because of the lack of a functional tendon to transmit forces from the ST to the tibia. Taken together, our results indicate that the regenerated tendon-like structure has a function similar to that of the native ST tendon when contributing to knee flexion. The functionality of the regenerated tendon-like structure is supported by several studies (Eriksson et al. 2001b; Ferretti et al. 2002; Hioki et al. 2003). The histological study of Ferretti et al. (2002) indicated that the regenerated ST tendon seemed to be very similar to a normal tendon. Eriksson et al. (2001b) reported that adequate tension in the regenerated ST tendon was created by voluntary muscle contraction. Hioki et al. (2003) used a novel MRI technique called the tagging snapshot technique and claimed that the ST with a regenerated tendon-like structure moved proximally during active knee flexion.

Knee flexion torque in the ACL-reconstructed limb was recovered almost completely at deep knee flexion in only those patients with ST tendon-like structure regeneration and the same ST length as the contralateral limb (group I). This suggests that the recovery of knee flexion strength at deep knee flexion is attributable to not only regeneration of the ST tendon-like structure but also to the maintenance of ST muscle length. On the other hand, a postoperative decrease in knee flexion torque at deep knee flexion in the patients from groups II and III might have been a result of shortening and atrophy of the ST. To prevent deficits in knee flexion strength after ACL reconstructive surgery, further studies are needed to investigate the operative technique and rehabilitation program, enabling regeneration of the ST tendon-like structure while maintaining the morphology of the ST.

A limitation of our study is that the morphology of the regenerated ST tendon-like structure was not evaluated in detail. Investigations of the tendon morphology show that the insertion site of the regenerated ST tendon is more proximal than the native tendon (Nakamura et al. 2004; Papandrea et al. 2000; Simonian et al. 1997). The insertion sites of the tendons of knee flexor muscles, including the ST, affect the knee flexion moment arm that contributes to the production of knee flexion torque. In patients with ACL reconstruction using the ST tendon, it would be useful to investigate the morphology of the ST muscle-tendon complex in more detail.

22.5 Conclusion

Changes in knee flexion strength were associated with morphological changes in the ST as a muscle-tendon complex after harvesting the ST tendon for ACL reconstruction. Regeneration of the ST tendon-like structure and maintenance of the morphology of the ST are necessary to ensure postoperative recovery of knee flexion strength at both shallow and deep knee flexion.

References

Aglietti P, Buzzi R, Zaccherotti G, De Biase P (1994) Patellar tendon versus doubled semitendinosus and gracilis tendons for anterior cruciate ligament reconstruction. Am J Sports Med 22:211–218

Aglietti P, Zaccherotti G, Buzzi R, De Biase P (1997) A comparison between patellar tendon and doubled semitendinosus/gracilis tendon for anterior cruciate ligament reconstruction. A minimum five-year follow-up. J Sports Traumatol Rel Res 19:57–68

Anderson AF, Snyder RB, Lipscomb AB (1994) Anterior cruciate ligament reconstruction using the semitendinosus and gracilis tendons augmented by the losee iliotibial band tenodesis. A long-term study. Am J Sports Med 22:620–626

Anderson AF, Snyder RB, Lipscomb AB (2001) Anterior cruciate ligament reconstruction. A prospective randomized study of three surgical methods. Am J Sports Med 29:272–279

Cooley VJ, Deffner KT, Rosenberg TD (2001) Quadrupled semitendinosus anterior cruciate ligament reconstruction: 5-year results in patients without meniscus loss. Arthroscopy 17:795–800

Cross MJ, Roger G, Kujawa P, Anderson IF (1992) Regeneration of the semitendinosus and gracilis tendons following their transection for repair of the anterior cruciate ligament. Am J Sports Med 20:221–223

Eriksson K, Larsson H, Wredmark T, Hamberg P (1999) Semitendinosus tendon regeneration after harvesting for ACL reconstruction. A prospective MRI study. Knee Surg Sports Traumatol Arthrosc 7:220–225

Eriksson K, Hamberg P, Jansson E, Larsson L, Shalabi A, Wredmark T (2001a) Semitendinosus muscle in anterior cruciate ligament surgery: morphology and function. Arthroscopy 17:808–817

Eriksson K, Kindblom LG, Hamberg P, Larsson H, Wredmark T (2001b) The semitendinosus tendon regenerates after resection. A morphologic and MRI analysis in 6 patients after resection for anterior cruciate ligament reconstruction. Acta Orthop Scand 72:379–384

Ferretti A, Conteduca F, Morelli F, Masi V (2002) Regeneration of the semitendinosus tendon after its use in anterior cruciate ligament reconstruction. A histologic study of three cases. Am J Sports Med 30:204–207

Gobbi A, Tuy B, Mahajan S, Panuncialman I (2003) Quadrupled bone-semitendinosus anterior cruciate ligament reconstruction: a clinical investigation in a group of athletes. Arthroscopy 19:691–699

Hioki S, Fukubayashi T, Ikeda K, Niitsu M, Ochiai N (2003) Effect of harvesting the hamstrings tendon for anterior cruciate ligament reconstruction on the morphology and movement of the hamstrings muscle: a novel MRI technique. Knee Surg Sports Traumatol Arthrosc 11:223–227

Irie K, Tomatsu T (2002) Atrophy of semitendinosus and gracilis and flexor mechanism function after hamstring tendon harvest for anterior cruciate ligament reconstruction. Orthopedics 25:491–495

Karlson JA, Steiner ME, Brown CH, Johnston J (1994) Anterior cruciate ligament reconstruction using gracilis and semitendinosus tendons. Comparison of through-the-condyle and over-the-top graft placements. Am J Sports Med 22:659–666

Lipscomb AB, Anderson AF (1986) Tears of the anterior cruciate ligament in adolescents. J Bone Joint Surg Am 68:19–28

Lipscomb AB, Johnston RK, Snyder RB, Warburton MJ, Gilbert PR (1982) Evaluation of hamstring strength following use of semitendinosus and gracilis tendons to reconstruct the anterior cruciate ligament. Am J Sports Med 10:340–342

Maeda A, Shino K, Horibe S, Nakata K, Buccafusca G (1996) Anterior cruciate ligament reconstruction with multistranded autogenous semitendinosus tendon. Am J Sports Med 24:504–509

Nakajima R, Maruyama Y, Shitoto K, Yamauchi Y, Nakajima H (1996) Decreased hamstring strength after harvest of semitendinosus and gracilis tendons for anterior cruciate ligament reconstruction (in Japanese). J Jap Clin Sports Med Assoc 6:681–686

Nakamae A, Deie M, Yasumoto M, Adachi N, Kobayashi K, Yasunaga Y, Ochi M (2005) Three-dimensional computed tomography imaging evidence of regeneration of the semitendinosus tendon harvested for anterior cruciate ligament reconstruction. A comparison with hamstring muscle strength. J Comput Assist Tomogr 29:241–245

Nakamura E, Mizuta H, Kadota M, Katahira K, Kudo S, Takagi K (2004) Three-dimensional computed tomography evaluation of semitendinosus harvest after anterior cruciate ligament reconstruction. Arthroscopy 20:360–365

Nishino A, Sanada A, Kanehisa H, Fukubayashi T (2006) Knee-flexion torque and morphology of the semitendinosus after ACL reconstruction. Med Sci Sports Exerc 38:1895–1900

Ohkoshi Y, Inoue C, Yamane S, Hashimoto T, Ishida R (1998) Changes in muscle strength properties caused by harvesting of autogenous semitendinosus tendon for reconstruction of contralateral anterior cruciate ligament. Arthroscopy 14:580–584

Papandrea P, Vulpiani MC, Ferretti A, Conteduca F (2000) Regeneration of the semitendinosus tendon harvested for anterior cruciate ligament reconstruction. Evaluation using ultrasonography. Am J Sports Med 28:556–561

Rispoli DM, Sanders TG, Miller MD, Morrison WB (2001) Magnetic resonance imaging at different time periods following hamstring harvest for anterior cruciate ligament reconstruction. Arthroscopy 17:2–8
Rosenberg TD, Brown GC, Deffner KT (1997) Anterior cruciate ligament reconstruction with a quadrupled semitendinosus autograft. Sports Med Arthrosc Rev 5:51–58
Siegel MG, Barber-Westin SD (1998) Arthroscopic-assisted outpatient anterior cruciate ligament reconstruction using the semitendinosus and gracilis tendons. Arthroscopy 14:268–277
Simonian PT, Harrison SD, Cooley VJ, Escabedo EM, Deneka DM, Larson RV (1997) Assessment of morbidity of semitendinosus and gracilis tendon harvest for ACL reconstruction. Am J Knee Surg 10:54–59
Tadokoro K, Matsui N, Yagi M, Kuroda R, Kurosaka M, Yoshiya S (2004) Evaluation of hamstring strength and tendon regrowth after harvesting for anterior cruciate ligament reconstruction. Am J Sports Med 32:1644–1650
Tashiro T, Kurosawa H, Kawakami A, Hikita A, Fukui N (2003) Influence of medial hamstring tendon harvest on knee flexor strength after anterior cruciate ligament reconstruction. A detailed evaluation with comparison of single- and double-tendon harvest. Am J Sports Med 31:522–529
Williams GN, Snyder-Mackler L, Barrance PJ, Axe MJ, Buchanan TS (2004) Muscle and tendon morphology after reconstruction of the anterior cruciate ligament with autologous semitendinosus-gracilis graft. J Bone Joint Surg Am 86:1936–1946
Yasuda K, Tsujino J, Ohkoshi Y, Tanabe Y, Kaneda K (1995) Graft site morbidity with autogenous semitendinosus and gracilis tendons. Am J Sports Med 23:706–714

Part V
Muscle Functions, Occurrence Mechanisms, and Program to Prevent Strain in the Hamstring Muscles

Chapter 23
Functional Differences Among Hamstring Muscles in Hip Extension and Knee Flexion Exercises

Osamu Yanagisawa

Abstract The hamstring muscles consist of the long (BF-L) and short (BF-S) heads of the biceps femoris, semitendinosus (ST), and semimembranosus (SM). The BF-L, ST, and SM act on both hip extension and knee flexion as biarticular muscle, whereas the BF-S flexes the knee as a uniarticular muscle. The hamstrings are occasionally treated as a muscle group (hip extensors or knee flexors), but previous studies have reported morphological and functional differences among the hamstrings. The hamstrings show individual characteristics in architectural parameters (e.g. fiber length, pennation angle, and physiological cross-sectional area). These architectural differences are likely closely associated with differences in the force-generation capacity of each hamstring muscle. In addition, each muscle's function is affected by the angle of the hip and/or knee joints. The morphology and function of one hamstring muscle cannot be considered representative of the whole muscle group and vice versa. Thus, treating the hamstrings as a single muscle unit by assuming a uniform inter-muscular architecture can result in an inaccurate account of hamstring muscle function.

Keywords Hamstrings • Hip extension • Knee flexion • Morphology • Function

23.1 Introduction

The hamstring muscles are located in the posterior portion of the thigh and consist of the long (BF-L) and short (BF-S) heads of the biceps femoris, semitendinosus (ST), and semimembranosus (SM). The BF-L, ST, and SM lie across both the hip and knee joints as a biarticular muscle, whereas the BF-S crosses only the knee joint. Based on this gross anatomy, the hamstrings are thought to act in concert as

O. Yanagisawa (✉)
Sports and Health Management Program, Faculty of Business and Information Sciences, Jobu University, Gunma, Japan
e-mail: yanagisawa@ic.jobu.ac.jp

K. Kanosue et al. (eds.), *Sports Injuries and Prevention*,
DOI 10.1007/978-4-431-55318-2_23

hip extensors (except for the BF-S) as well as knee flexors. Therefore, the hamstrings are often treated as a one muscle unit, not only in clinical settings but also in the research field.

On the other hand, morphological and functional differences among the hamstrings have been reported. Kellis et al. (2012) revealed that the highest architectural similarity is between BF-L and SM and the lowest between BF-L and ST. These investigators note that the architecture of one hamstring muscle is not representative of the whole muscle group and vice versa. The ST and BF-S have much longer fiber length (FL) than the BF-L and SM, while the BF-L and SM display a relatively larger physiological cross-sectional area (PCSA) (Woodley and Mercer 2005). These architectural similarities and differences must have effects on the force-generation capacity of each hamstring muscle. In addition, the hamstrings are well known to be susceptible to muscle strain injury, but the incidence of muscle strain differs among the hamstrings; the BF-L has the highest risk of muscle strain (Koulouris et al. 2007; Malliaropoulos et al. 2010; Slavotinek et al. 2002). This difference may be also attributed to the morphological and functional differences among these muscles.

The main purpose of this chapter is to describe functional differences among the hamstrings in hip extension and knee flexion. Moreover, this paper refers to the anatomical characteristics of individual hamstring muscle because muscle morphology is closely related to muscle function (Fukunaga et al. 1996). Learning about the intra- and inter-muscular differences in hamstring muscle morphology would lead to a better understanding of the individual functions of each hamstring.

23.2 Morphology and Function of the Hamstring Muscles

Anatomical knowledge of the hamstrings is necessary to better understand the function of each muscle. The hamstrings are located in the posterior compartment of the thigh with the BF-L and BF-S (lateral hamstrings) in the lateral part and with the ST and SM (medial hamstrings) in the medial part of the posterior thigh. The ischial tuberosity is the common origin site of the BF-L, ST, and SM, while the BF-S origin is from the most proximal part of the lateral lip of the femur. The BF-L attaches on the fibular head and the BF-S shares a common distal tendon with the BF-L. The ST and SM attach on the medial surface of the upper part of tibia and on the protruded medial surface of the medial condyle of the tibia, respectively. The BF-L, ST, and SM are biarticular muscles that cross from the hip to the knee joints, whereas the BF-S crosses only the knee joint. The hamstrings have similar anatomical features with each other, but previous studies have reported intra- and inter-muscular architectural variations among these muscles. Considering the close relationship between muscle architecture and function (Fukunaga et al. 1996), the differing anatomical features of individual muscles likely indicate a functional difference among the hamstrings.

The FL, pennation angle (PA), and PCSA of a muscle have been considered important parameters that affect muscle function. The FL is thought to be proportional to maximum muscle excursion (or velocity), and the PCSA proportional to maximum muscle force (Lieber and Friden 2000). A muscle with relatively long fibers (a large number of sarcomeres arranged in series) with a smaller PA is suited for joint excursion, whereas those composed of relatively short fibers (a large number of fibers arranged in parallel) with a larger PA and PCSA are suitable for force generation (Woodley and Mercer 2005). The BF-L and SM are typically classified as pennate muscle due to their structure, while the ST is regarded as a fusiform muscle with a parallel fiber configuration (Chleboun et al. 2001; Kellis et al. 2010, 2012; Woodley and Mercer 2005). Each muscle differs with respect to architectural parameters such as the muscle length, FL, PA, PCSA, and volume (Kellis et al. 2010, 2012; Kumazaki et al. 2012; Woodley and Mercer 2005). Woodley and Mercer (2005) suggested that the BF-L is designed to allow for long excursion since its FL is relatively long and its PCSA is intermediate in size. On the contrary, Kellis et al. (2012) inferred that the BF-L have a higher force generation capacity with shorter pennate fibers and larger PCSA. The SM is designed for force production rather than excursion because this muscle has a relatively short FL but a relatively large PCSA (Kellis et al. 2012; Woodley and Mercer 2005). The ST has a parallel fiber configuration and longer FL coupled with a very long distal tendon (Kellis et al. 2010, 2012; Kumazaki et al. 2012; Woodley and Mercer 2005), indicating that this muscle has a higher excursion capacity. The BF-S possesses a relatively long FL among the hamstrings, but the PCSA is smaller compared with the other hamstrings (Kellis et al. 2012; Woodley and Mercer 2005). Therefore, the magnitude of force exerted by the BF-S is likely to be small.

The BF-L and BF-S form lateral hamstrings running along the lateral side of the thigh, whereas the ST and SM run medially within the thigh as medial hamstrings. The medial hamstrings have more distal attachments than the lateral ones. Therefore, functional differences between medial and lateral groups in knee flexion have been suggested. For instance, the ST has an almost double moment arm (MA) as compared to the BF-L, suggesting that the ST is superior in the production of knee flexion torque (Herzog and Read 1993). On the other hand, Kellis et al. (2012) reported that the pairs of lateral (BF-L and BF-S) and medial (ST and SM) muscles show a low similarity in their muscle architectures. According to their results, the BF-L and SM have shorter pennate fibers and larger PCSA among the hamstrings, whereas the ST has the longest FL coupled with a very long distal tendon. The BF-S has a longer FL, but possesses the smallest PCSA among the hamstrings. Therefore, they suggest that there are noticeable differences between medial or lateral muscles such that each pair has one muscle which is designed for excursion (BF-S or ST) and another muscle which is suitable for force production (BF-L or SM).

There are also intramuscular differences in the architecture of each hamstring muscle. Kellis et al. (2010) revealed that the BF-L has greater PA and longer FL at the proximal side than at the distal one, whereas the ST exhibits a significantly smaller PA and shorter FL proximally than distally. The SM consists of three distinct regions (proximal and intermediate unipennate, and distal bipennate regions) on the basis of fascicular orientation, and the distal region has the largest PCSA (Woodley and Mercer 2005). These data suggest that the architecture of each muscle is not uniform. Therefore, modeling each muscle by assuming a uniform architecture along muscle length can yield an inaccurate representation of hamstring muscle function (Kellis et al. 2010). Moreover, it is possible that intramuscular architectural variation produces higher tension in some regions of the muscle, while it allows for greater efficiency of force transfer in line with the tendon in other regions (Kellis et al. 2010; Scott et al. 1993).

Measured torque value at any joint is the sum of torques produced by several muscles. The amount of torque produced by a muscle depends on the number of motor units activated, the sarcomere length (based on a length–tension relationship), and the MA of the muscle during force generation in addition to muscle size. Moreover, the non-contractile component of the muscle (the endomysium, perimysium, epimysium, and tendon) can be an important factor in joint torque production because stretched connective tissues produce passive tension that is added to the active tension generated by muscle contraction (Magnusson 1998). Therefore, the connective tissues of the hamstrings are likely more stretched with a flexed hip and/or an extended knee, which may contribute to total joint torque production (Lunnen et al. 1981; Mohamed et al. 2002).

Each hamstring muscle possesses relatively long tendons at both the proximal and distal sides (except for the proximal side of the BF-S). The long proximal and distal tendons of each hamstring muscle extend into the muscle bellies, thereby forming elongated musculotendinous junctions. The BF-L and SM have proximal and distal tendons which overlap to some extent within the belly of the muscle, and the distal tendon of the BF-L receives fascicles from the BF-S (Woodley and Mercer 2005). The ST possesses a much longer tendon on the distal side than on the proximal side (Kellis et al. 2012; Woodley and Mercer 2005), and unlike other muscles, has a tendinous inscription within the muscle belly that divides it into two partitions, with each region receiving innervation from one muscle nerve, or from a primary branch of the nerve (Woodley and Mercer 2005). Considering that hamstring strain injuries often occur at the musculotendinous junction (Koulouris et al. 2007; Slavotinek et al. 2002), the structures of tendons and musculotendinous junctions may be key factors. In addition, Kumazaki et al. (2012) revealed that muscle fibers of the BF-L and SM with unipennate architecture are elongated more than those of the ST and BF-S during knee extension. These authors suggest that BF-L and SM have a higher risk of muscle strain during extension of the knee joint. Moreover, there is a gender difference. Blackburn et al. (2009) showed that musculotendinous stiffness of the BF-L is greater in males than in females, but that the elastic modulus does not differ significantly across the sexes. They also indicated that the male BF-L has a greater capacity for resisting changes in length imposed via joint motion from a structural viewpoint. However, this property is at least partially attributable to a greater muscle size in males, and is probably similar between genders from a material viewpoint.

Many previous studies used cadaveric lower limbs to study the architecture of the hamstrings (Chleboun et al. 2001; Kellis et al. 2010, 2012; Kumazaki et al. 2012;

Woodley and Mercer 2005). However, the shrinkage of muscle fibers, which can occur during or after the embalming process, could potentially affect the fascicular dimensions. Also, elderly specimens may have smaller FL and volume than younger ones. Although these limitations might have led to some inconsistent findings among earlier studies, understanding the detailed anatomy of the hamstrings is important to develop biomechanical models of relationship between morphology and function for each hamstring muscle. Moreover, these morphological data should be useful for discussing injury mechanisms as well as for proposing injury prevention and treatment procedures for individual hamstring muscles.

23.3 Function of the Hamstring Muscles at the Hip and Knee Joints

23.3.1 Hip Extension

The hamstrings (except for the BF-S) have the role of hip extension. However, hip extension strength has received relatively less attention compared with knee flexion strength. Cahalan et al. (1989) reported that as the velocity of exercise increased, the hip extension torque decreased. They also showed that isometric hip extension torque was greater at 90° hip flexion than at 45° hip flexion, with the knee approximately 90° flexed, regardless of age or gender. Similarly, Worrell et al. (2001) revealed that isometric hip extension torque was greatest at 90° of hip flexion with the knee 90° flexed and gradually decreased as the hip angle was extended. These findings imply that isometric hip extension strength is greater when the agonist muscles (BF-L, ST, and SM) are in a lengthened position. Skeletal muscle fibers have an optimal length for producing the largest contraction force (Gordon et al. 1966). Chleboun et al. (Chleboun et al. 2001) indicated that when both the hip and knee joints are flexed at 90°, the sarcomere length of the BF-L is optimal. Moreover, it was demonstrated that the effect of changing hip angle on the length of the BF-L is much larger than that of changing the knee angle because the BF-L has a larger MA at the hip joint than at the knee joint (Visser et al. 1990). With an increase in MA, there is a greater excursion of the muscle and a greater change in the FL. The MA of a muscle per se is an important factor that affects joint function. The MA of the BF-L at the hip joint slightly increases with increasing hip flexion angle (Visser et al. 1990). Considering that the ST and SM have essentially the same origin on the ischial tuberosity as the BF-L, both muscles may also show similar increases in MA when the hip flexion angle is increased. This increased MA of the hamstrings may be partly associated with the finding that a greater hip extension force is recorded with increased hip flexion angle than with a decreased flexion angle (Cahalan et al. 1989; Nemeth et al. 1983). On the other hand, Jacobs et al. (1996) showed using Hill-based muscle models that the relative work contribution of the hamstrings (as an average of the BF-L and ST) done in hip extension was 7 % in the one leg jump push-off phase and 11 % in the sprint push-off phase. Their findings suggest that the relative contribution of the BF-L and ST to hip extension during jumping and sprinting push-off phases is relatively low.

We should also consider the contribution of the gluteus muscles to hip extension, especially the gluteus maximus (GM), which has the largest muscle volume. The GM primarily acts as a powerful extensor of the hip joint. The GM arises from the posterior gluteal line of the ilium and the posterior surface of the sacrum and coccyx, and is directed downward and outward into the iliotibial tract and the gluteal tuberosity of the femur. Thus, the involved muscle fibers are assumed to run parallel to the direction of muscle pull, probably leading to optimized muscle activation. Kang et al. (2013) found that electromyography (EMG) activity of the GM was greatest at 30° of hip abduction, followed by 15° and then 0° hip during prone hip extension with 90° knee flexion, whereas activity of the hamstrings decreased with increasing hip abduction angle. These findings indicate that hip abduction angle affects the relative contribution of the GM and hamstrings to hip extension. The activity of the hamstrings possibly becomes weak with increasing hip abduction angle. Additionally, Worrell et al. (2001) demonstrated that isometric hip extension torque gradually decreased as the hip angle was extended, but that EMG activity of the GM at a hip angle of 0° was greater than at 30°, 60°, or 90°, and was lower at 90° than at all other angles. Moreover, considering that the MA of the GM decreases with increasing hip flexion angle (Nemeth and Ohlsen 1985), the relative contribution of the GM to hip extension may become smaller with increasing the hip flexion angle.

23.3.2 Knee Flexion

The peak torque of knee flexion becomes greater with increasing hip flexion angle (Bohannon et al. 1986; Deighan et al. 2012; Guex et al. 2012; Lunnen et al. 1981; Worrell et al. 1989), which is not dependent on muscle contraction mode or velocity (Deighan et al. 2012; Guex et al. 2012; Worrell et al. 1989). These findings are presumably related to the length-tension curve, which describes an optimal length at which a muscle can develop maximal force (Gordon et al. 1966). The hamstrings except for the BF-S are lengthened by hip flexion and a more optimal actin-myosin cross-bridging may occur in the flexed hip position compared with the extended one. When both the hip and knee joints are flexed at 90°, the BF-L shows an optimal sarcomere length (Chleboun et al. 2001). Guex et al. (2012) found that eccentric peak torque of the hamstrings during knee flexion was higher in a flexed hip position than in an extended one for both investigated angular velocities (60°/s and 150°/s), but the amount of hip flexion did not influence EMG activities in the medial or lateral hamstrings. Lunnen et al. (1981) reported that their subjects showed significantly less isometric knee flexion torque at 0° hip position with a 60° knee angle than at 90° and 135° hip positions, and found that during maximal isometric knee flexion, an increase in EMG activity of the BF-L and a decrease in knee flexion torque occurred as the muscle was shortened; the opposite occurred when the muscle was at lengthened positions. In short, their findings imply that when the MA of muscle at knee joint is held constant, the EMG activity of the BF-L decreases as the muscle is lengthened, but the muscle develops a greater amount of torque in the lengthened state than in the shortened position.

Isometric knee flexion torque tends to be larger as the knee flexion angle decreases (Kumazaki et al. 2012; Mohamed et al. 2002; Onishi et al. 2002; Worrell et al. 2001). Concentric and eccentric knee flexion forces are also greater when obtained at more extended knee positions (Aagaard et al. 1998). Onishi et al. (Onishi et al. 2002) demonstrated that isometric knee flexion torque at 60° knee flexion was greater than that at 90° and that EMG activities of the ST and SM were significantly lower at 60° knee flexion than at 90° knee flexion, whereas the EMG of the BF-L at 60° flexion was significantly greater than the value at 90° knee flexion. No significant difference in the EMG of the BF-S was observed between the knee flexion angles of 60 and 90° in their study. Moreover, they found that the peak EMG activity of the BF-L was observed at a knee angle between 15 and 30° at which the peak torque occurred, while the other muscles (ST, SM, and BF-S) showed the largest activity at a knee angle between 90 and 105°, and suggested that the BF-L participates strongly in knee flexion torque at the early stages of knee flexion, whereas the other three muscles greatly contribute to flexion torque at a knee angle of nearly 90°.

Some researchers (Aagaard et al. 1998; Baratta et al. 1988; Croce and Miller 2006; Miller et al. 2000) have postulated that the co-activation (antagonist activity) of the hamstrings during knee extension assists the anterior cruciate ligament in maintaining knee joint stability by exerting an opposing force to anterior tibial translation. Active contraction of the quadriceps may create significant anterior tibial translation or shear, particularly at high contraction forces and with the knee toward full extension (Hirokawa et al. 1992; More et al. 1993). The co-activation of the hamstrings may contribute to a counterbalancing of this tibial shear (Baratta et al. 1988). Croce and Miller (2006) indicated that when acting as an antagonist in knee extension, the BF-L exhibits greater EMG activity than the medial hamstrings as the knee approaches terminal extension (0–20° knee flexion). Moreover, several researchers (Croce and Miller 2003, 2006) noted an increase in the antagonistic hamstrings' activity with increasing velocity of knee extension, while Miller et al. (2000) observed velocity dependence in the BF-L but not in the medial hamstrings. This could be associated with eccentric muscle contraction of the hamstrings as antagonists to control the extension forces produced by the quadriceps during the final stage of knee extension.

The FL of the hamstrings except for the BF-S is affected by the positions of the hip joint as well as the knee joint. As mentioned above, the FL has a great influence on the force production capacity of each muscle (Gordon et al. 1966). According to Chleboun et al. study (Chleboun et al. 2001), the sarcomere length of the BF-L is optimal for potential force production with both the hip and knee flexed at 90°, while the length is shortest in the 0° hip and 90° knee position and longest in the 90° hip and 0° knee position. They also revealed that the FL of the BF-L is more sensitive to changes in hip position with knee position constant compared to changes in knee position with hip position constant. In fact, Mohamed et al. (2002) compared maximum isometric knee flexion force at 6 joint positions of varying the hip (0°, 90°) and knee (0°, 45°, 90°) angles, and recorded minimum torque with the hip 0° and knee 90° position. However, in their study, the 90° hip and 90° knee position did not show maximum torque among 6 joint positions. Considering that several physiological, biomechanical, and morphological factors (except for the FL of the BF-L) will contribute to knee flexion torque, these findings are not surprising.

23.4 Perspective

In biarticular muscles, the architectural parameters (e.g. FL, PA, and MA) are affected by the angle change of each joint. Therefore, we need to think of the individual functions of the BF-L, ST, and SM, and consider the combination of the hip and knee angles as well. Moreover, the variation of architectural parameters with changing hip and/or knee angles does not remain uniform among the hamstrings. Some previous studies did not describe the specific names of the hamstrings in their methods section (i.e., hamstrings, lateral or medial hamstrings), and apparently regarded the hamstrings as a unitary muscle. In this chapter we made the case that the assumption of uniform inter-muscular architecture leads to a less accurate estimation of hamstring muscle function. Therefore, further research that focuses on the individual function of each of the hamstrings would be needed. It would also be of value to evaluate each function during the complex movement of the hip and knee joints. These would lead to a better understanding of the injury mechanisms, aid in the planning of training and rehabilitation programs, and help establish treatment strategies for each hamstring muscle.

References

Aagaard P, Simonsen EB, Magnusson SP, Larsson B, Dyhre-Poulsen P (1998) A new concept for isokinetic hamstring: quadriceps muscle strength ratio. Am J Sports Med 26(2):231–237

Baratta R, Solomonow M, Zhou BH, Letson D, Chuinard R, Dambrosia R (1988) Muscular coactivation – the role of the antagonist musculature in maintaining knee stability. Am J Sports Med 16(2):113–122

Blackburn JT, Bell DR, Norcross MF, Hudson JD, Kimsey MH (2009) Sex comparison of hamstring structural and material properties. Clin Biomech (Bristol, Avon) 24(1):65–70

Bohannon RW, Gajdosik RL, LeVeau BF (1986) Isokinetic knee flexion and extension torque in the upright sitting and semireclined sitting positions. Phys Ther 66(7):1083–1086

Cahalan TD, Johnson ME, Liu S, Chao EY (1989) Quantitative measurements of hip strength in different age groups. Clin Orthop Relat Res (246):136–145

Chleboun GS, France AR, Crill MT, Braddock HK, Howell JN (2001) In vivo measurement of fascicle length and pennation angle of the human biceps femoris muscle. Cells Tissues Organs 169(4):401–409

Croce RV, Miller JP (2003) The effect of movement velocity and movement pattern on the reciprocal co-activation of the hamstrings. Electromyogr Clin Neurophysiol 43(8):451–458

Croce RV, Miller JP (2006) Coactivation patterns of the medial and lateral hamstrings based on joint position and movement velocity during isokinetic movements. Electromyogr Clin Neurophysiol 46(2):113–122

Deighan MA, Serpell BG, Bitcon MJ, De Ste Croix M (2012) Knee joint strength ratios and effects of hip position in rugby players. J Strength Cond Res 26(7):1959–1966

Fukunaga T, Roy RR, Shellock FG, Hodgson JA, Edgerton VR (1996) Specific tension of human plantar flexors and dorsiflexors. J Appl Physiol 80(1):158–165

Gordon AM, Huxley AF, Julian FJ (1966) Variation in isometric tension with sarcomere length in vertebrate muscle fibres. J Physiol 184(1):170–192

Guex K, Gojanovic B, Millet GP (2012) Influence of hip-flexion angle on hamstrings isokinetic activity in sprinters. J Athl Train 47(4):390–395

Herzog W, Read LJ (1993) Lines of action and moment arms of the major force-carrying structures crossing the human knee joint. J Anat 182(Pt 2):213–230

Hirokawa S, Solomonow M, Lu Y, Lou ZP, Dambrosia R (1992) Anterior-posterior and rotational displacement of the tibia elicited by quadriceps contraction. Am J Sports Med 20(3):299–306

Jacobs R, Bobbert MF, van Ingen Schenau GJ (1996) Mechanical output from individual muscles during explosive leg extensions: the role of biarticular muscles. J Biomech 29(4):513–523

Kang SY, Jeon HS, Kwon O, Cynn HS, Choi B (2013) Activation of the gluteus maximus and hamstring muscles during prone hip extension with knee flexion in three hip abduction positions. Man Ther. doi:10.1016/j.math.2012.11.006

Kellis E, Galanis N, Natsis K, Kapetanos G (2010) Muscle architecture variations along the human semitendinosus and biceps femoris (long head) length. J Electromyogr Kinesiol 20(6):1237–1243

Kellis E, Galanis N, Kapetanos G, Natsis K (2012) Architectural differences between the hamstring muscles. J Electromyogr Kinesiol 22(4):520–526

Koulouris G, Connell DA, Brukner P, Schneider-Kolsky M (2007) Magnetic resonance imaging parameters for assessing risk of recurrent hamstring injuries in elite athletes. Am J Sports Med 35(9):1500–1506

Kumazaki T, Ehara Y, Sakai T (2012) Anatomy and physiology of hamstring injury. Int J Sports Med 33(12):950–954

Lieber RL, Friden J (2000) Functional and clinical significance of skeletal muscle architecture. Muscle Nerve 23(11):1647–1666

Lunnen JD, Yack J, LeVeau BF (1981) Relationship between muscle length, muscle activity, and torque of the hamstring muscles. Phys Ther 61(2):190–195

Magnusson SP (1998) Passive properties of human skeletal muscle during stretch maneuvers. A review. Scand J Med Sci Sports 8(2):65–77

Malliaropoulos N, Papacostas E, Kiritsi O, Papalada A, Gougoulias N, Maffulli N (2010) Posterior thigh muscle injuries in elite track and field athletes. Am J Sports Med 38(9):1813–1819

Miller JP, Croce RV, Hutchins R (2000) Reciprocal coactivation patterns of the medial and lateral quadriceps and hamstrings during slow, medium and high speed isokinetic movements. J Electromyogr Kinesiol 10(4):233–239

Mohamed O, Perry J, Hislop H (2002) Relationship between wire EMG activity, muscle length, and torque of the hamstrings. Clin Biomech (Bristol, Avon) 17(8):569–579

More RC, Karras BT, Neiman R, Fritschy D, Woo SLY, Daniel DM (1993) Hamstrings – an anterior cruciate ligament protagonist – an invitro study. Am J Sports Med 21(2):231–237

Nemeth G, Ohlsen H (1985) In vivo moment arm lengths for hip extensor muscles at different angles of hip flexion. J Biomech 18(2):129–140

Nemeth G, Ekholm J, Arborelius UP, Harms-Ringdahl K, Schuldt K (1983) Influence of knee flexion on isometric hip extensor strength. Scand J Rehabil Med 15(2):97–101

Onishi H, Yagi R, Oyama M, Akasaka K, Ihashi K, Handa Y (2002) EMG-angle relationship of the hamstring muscles during maximum knee flexion. J Electromyogr Kinesiol 12(5):399–406

Scott SH, Engstrom CM, Loeb GE (1993) Morphometry of human thigh muscles – determination of fascicle architecture by magnetic-resonance-imaging. J Anat 182:249–257

Slavotinek JP, Verrall GM, Fon GT (2002) Hamstring injury in athletes: using MR imaging measurements to compare extent of muscle injury with amount of time lost from competition. AJR Am J Roentgenol 179(6):1621–1628

Visser JJ, Hoogkamer JE, Bobbert MF, Huijing PA (1990) Length and moment arm of human leg muscles as a function of knee and hip-joint angles. Eur J Appl Physiol Occup Physiol 61(5–6):453–460

Woodley SJ, Mercer SR (2005) Hamstring muscles: architecture and innervation. Cells Tissues Organs 179(3):125–141

Worrell TW, Perrin DH, Denegar CR (1989) The influence of hip position on quadriceps and hamstring peak torque and reciprocal muscle group ratio values. J Orthop Sports Phys Ther 11(3):104–107

Worrell TW, Karst G, Adamczyk D et al (2001) Influence of joint position on electromyographic and torque generation during maximal voluntary isometric contractions of the hamstrings and gluteus maximus muscles. J Orthop Sports Phys Ther 31(12):730–740

Chapter 24
Anatomical and Functional Differences of Hamstrings

Yukiko Makihara

Abstract A decrease of deep knee flexion torque after anterior cruciate ligament (ACL) reconstruction using semitendinosus (ST) tendon has been reported. We hypothesized that architectural and functional differences of the hamstrings influence this weakness. Fiber length of the ST, gracilis (G), semimembranosus (SM), and biceps femoris (BF) were measured in six human cadavers. Electromyography (EMG) of the hamstrings were measured in both limbs of 16 patients after ACL reconstruction. Magnetic resonance imaging (MRI) was taken over both thighs of those patients to examine muscle volume and ST tendon regeneration. The fiber length of the ST and G were longer than that of the SM and BF. EMG values for the SM and BF of both limbs, and the ST of the ACL reconstructed limb became significantly reduced as the knee flexion angle increased. The volume of the ST in the reconstructed limb was significantly smaller than in the normal limb. Regeneration of the ST tendon was confirmed in all subjects, and the musculotendinous junction of the reconstructed limb shifted proximally when compared to that of the normal limb. The atrophy and shortening of the ST after its tendon is harvested and the lack of compensation from the SM and BF due to architectural differences among the hamstrings could cause the weakness of deep knee flexion torque after ACL reconstruction.

Keywords ACL reconstruction • Flexion torque • Architecture

24.1 Introduction

The semitendinosus (ST) or the ST and gracilis (ST/G) tendon is commonly used as a replacement graft during anterior cruciate ligament (ACL) reconstruction. Many advantages of using this graft have been reported including superior initial fixation strength (Rowden et al. 1997), a reduction in the incidence of anterior knee pain

Y. Makihara (✉)
Department of Life Sciences, Graduate School of Arts and Science,
University of Tokyo, Tokyo, Japan
e-mail: yukikomakihara68@gmail.com

K. Kanosue et al. (eds.), *Sports Injuries and Prevention*,
DOI 10.1007/978-4-431-55318-2_24

(Corry et al. 1999; Eriksson et al. 2001; Shaieb et al. 2002; Yasuda et al. 1995), and recovery of quadriceps muscle strength and extension loss (Frank and Jackson 1997; Rosenberg and Deffner 1997). The potential of the ST tendon to regenerate after being harvested for use as an ACL graft has also been shown (Cross et al. 1992; Eriksson et al. 1999; Papandrea et al. 2000; Rispoli et al. 2001), although the function of the regenerated tendon has not been investigated. Some drawbacks to using the ST tendon as an ACL replacement have been reported including a weakness in internal rotation of the tibia (Armour et al. 2004; Viola et al. 2000).

As for the function of knee flexion after the ST tendon is harvested for ACL reconstruction, many studies reported that the peak knee flexion torque that occurs between 15° and 45° of knee flexion recovers to at least 90 % of that of the normal limb (Aglietti et al. 1994; Carter and Edinger 1999; Cross et al. 1992; Harter et al. 1988; Karlson et al. 1994; Lipscomb et al. 1982; Maeda et al. 1996; Sachs et al. 1989; Shino et al. 1993; Simonian et al. 1997). However, a decrease of knee flexion strength during knee flexion of over 70° has also been reported (Nakajima et al. 1996; Nakamura et al. 2002; Ohkoshi et al. 1998; Tashiro et al. 2003), and the cause of this weakness remains controversial.

Several investigators have shown architectural differences among the knee flexor muscles, including muscle weight, muscle volume, fiber length, and pennation angle (Akima et al. 1995; Friederich and Brand 1990; Herzog 1993; Wickiewicz et al. 1983; Winters and Woo 1990). In this chapter we evaluate our hypothesis that those architectural differences in the knee flexor muscles cause a different function in each muscle during deep knee flexion, resulting in weakness of knee flexion torque at this position. To validate this hypothesis, in-vitro and in-vivo analyses were performed.

24.2 Materials and Methods

24.2.1 In-Vitro Anatomical Study

Muscle belly length (the length from muscle origin to the distal musculotendinous junction), fiber arrangement, pennation angle and fiber length of the ST, the G, the semimembranosus (SM), and the long head of the biceps femoris (BF) were examined directly in six fresh-frozen human cadavers (5 males, 1 female, age range: 36–63 years). The causes of death included lung cancer (n = 2), breast cancer (n = 1), pancreatic cancer (n = 2), and cerebellitis (n = 1). The specimens were thawed overnight at room temperature for 24 h before dissection (Sakane et al. 1997; Woo et al. 1986). After removing the skin, the muscles were extracted from the specimens to measure muscle length, and then incised longitudinally to visualize the fiber arrangement. Bundles of the muscle fibers were selected from several different regions of each muscle, and the pennation angle formed by the individual muscle bundle with the deep aponeurosis was approximated with a protractor, and the length of each bundle was measured to the nearest millimeter using a caliper. It has been documented that muscle fibers have the same length as the bundle in which they are packed (Trotter 1993), therefore the bundle length was reasonably assumed to approximate the fiber length. Measurement of the pennation angle and the fiber length were performed five times in each specimen. The five measurements were averaged and used as a representative score for that specimen. Mean values of the six specimens were calculated.

24.2.2 *In-Vivo Clinical Study*

Sixteen patients (3 males, 13 females, age: 23 ± 5 years) who had undergone an arthroscopically assisted ACL reconstruction with an autogenous quadrupled ipsilateral ST tendon (12 patients) or ST/G tendon (4 patients), according to the technique described by Rosenberg et al. (1997; Rosenberg and Graf 1994), participated in the study. The rehabilitation program used for those patients included (1) placing the knee in a hinged knee brace and immobilizing it for 1 week; isometric quadriceps and hamstring exercises were encouraged 3 days after the surgery, followed by active range of motion exercise and isotonic knee extension and flexion excises, (2) allowing partial weight bearing at 2 weeks, and progression to full weight bearing at 3 weeks, followed by half squat exercises, (3) performing bicycle ergometer and resistance exercises (leg press and leg curl) which began at 3 weeks, (4) permitting jogging and full speed running at 3 and 4 months respectively, (5) agility drills including jumping exercises that began at 5 months, and (6) permission to return to sports activity at 6–8 months. All subjects were either recreational or competitive athletes. At the time of participation in this study (average: 26 months, range: 12–43 months post-operation), the subjects had returned to their previous sports without any pain or restriction, although their pre-injury performance levels were not always attained.

Knee flexion torque was measured in the normal limbs and the ACL reconstructed limbs using a Biodex System III (Biodex Co, New York) with the subject in a prone position. The axis of the subject's knee was adjusted to coincide with the rotating axis of the dynamometer. Before evaluation, compensation was performed to exclude the effect of gravity on the measurement of torque. After several practice trials, three trials of isokinetic contraction were performed at an angular velocity of 60°/s followed by two trials of maximum voluntary isometric contraction (MVC) for five seconds at 45° and 90° of knee flexion. During the isokinetic contraction, the values for peak torque, peak torque angle, and the torque at 15°, 30°, 45°, 60°, 75°, 90°, and 105° were recorded. The trials that produced the highest (or higher) peak torque were chosen for analysis. The torque values were normalized to the subjects' body weight.

After the torque measurement, the electromyographic (EMG) activity of the bilateral ST, SM, and BF during an isometric 50 % MVC at 45° and 90° of knee flexion in the prone position were recorded at a sampling rate of 1 kHz. Surface

EMG activity was recorded using bipolar electrodes (ME3750, MEGA) with an interelectrode distance of 30 mm. The amplified and full-wave rectified EMG signals were integrated (i.e., I-EMG) with respect to a 1-s period during steady state of 50 % MVC at each flexion angle.

Magnetic resonance imaging (MRI) images were taken (FLEXART, TOSHIBA) to measure muscle volume of the ST, SM, and BF, and to examine the ST tendon regeneration and position of the ST musculotendinous junction after harvesting for ACL reconstruction. T1-weighted spin-echo, transaxial sequences were performed over the bilateral thighs perpendicularly to the femoral shaft with the following parameters: a repetition time (TR) of 1,650 ms, an echo time (TE) of 20 ms, a slice thickness of 10 mm, and an interslice gap of 2 mm. The subjects were in a supine position with the knee in full extension. The images were taken from the ischial tuberosity to the knee joint space. To obtain muscle volume, an anatomical cross-sectional area of the ST, SM, and BF from each image was calculated using Scion Image (Scion Corporation, Maryland). The presence of a regenerated ST tendon and the position of the ST musculotendinous junction were examined in the sequence of MRI images starting from the knee joint space level. Once the image in which the musculotendinosus junction first appeared was confirmed, the difference between the normal and ACL reconstructed limbs in the number of images from the knee joint space to the musculotendinosus junction was calculated to evaluate shortening of the ST.

A paired *t*-test was used to test for differences in knee flexion torque, I-EMG values and muscle volume. Statistical significance was set at $p<0.05$. Results are reported as the mean ± one standard deviation.

24.3 Results

24.3.1 In-vitro Anatomical Study

The architectural features of the ST, G, SM, and BF are shown in Table 24.1 and Fig. 24.1. The length of the muscle bellies did not vary significantly among the muscles, however, the fiber length of the ST (23.8 ± 1.8 cm) and G (24.3 ± 1.9 cm) were three to four times longer than that of the SM (6.0 ± 0.8 cm) and BF (7.3 ± 1.3 cm). Furthermore, the ST and G were found to be composed of parallel

Table 24.1 Architectural measurements of knee flexor muscles (mean ± S.D.)

Muscle	Muscle belly length (cm)	Fiber length (cm)	Pennation angle (deg)	Fiber arrangement
ST	26.8 ± 3.6	23.8 ± 1.8	0	Parallel fibered muscle
G	24.9 ± 2.4	24.3 ± 1.9	0	
SM	28.5 ± 2.6	6.0 ± 0.8	31 ± 5	Unipennate muscle
BF	31.2 ± 5.2	7.3 ± 1.3	28 ± 4	

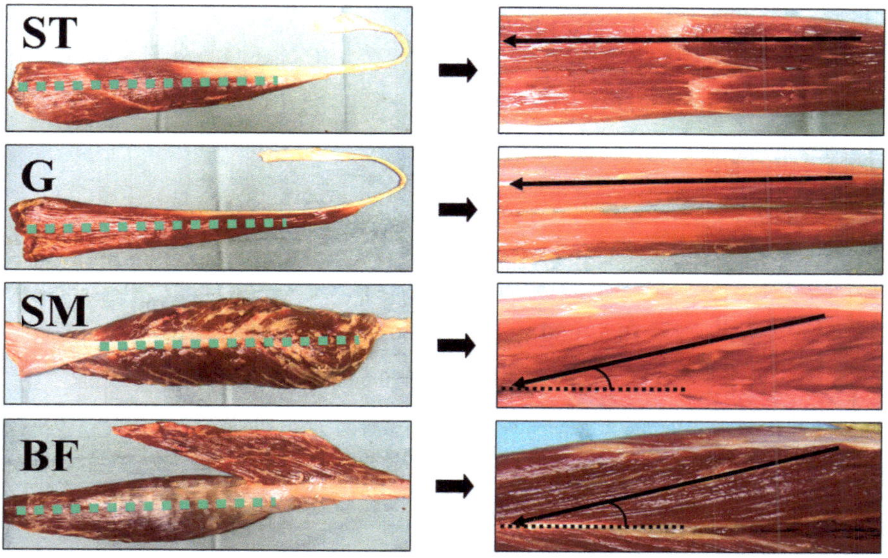

Fig. 24.1 Typical examples of knee flexor muscles from a human cadaver. The pictures on the right show the fiber arrangement visualized after incision along the *dotted lines* in the left

fibered muscles, whereas the SM and BF were made up of unipennate muscles with pennation angles of 31° ± 5° and 28° ± 4°, respectively.

24.3.2 In-vivo Clinical Study

No significant differences were measured in the peak isokinetic flexion torque or the peak torque angle between the normal and the ACL reconstructed limbs. However, differences in isokinetic torque appeared as the knee flexion angle increased. The torque of the ACL reconstructed limb was significantly lower than that of the normal limb at 60°, 75°, 90°, and 105° of knee flexion ($p < 0.05$ at 60°, $p < 0.01$ at 75°, 90°, and 105°). The differences were not significant at 15°, 30°, or 45° (Fig. 24.2). The isometric torque of the ACL reconstructed limb at 45° (130 ± 23 % of body weight: %BW) was not significantly different from that of the normal limb (123 ± 22 %BW), however at 90°, the torque of the ACL reconstructed limb (55 ± 19 %BW) was significantly lower than that of the normal limb (88 ± 28 %BW, $p < 0.01$).

During the 50 % MVC, the I-EMG value for the SM and BF of both limbs was significantly reduced as the knee flexion angle increased ($p < 0.01$) (Fig. 24.3a, b). For the ST, the I-EMG value for the normal limb did not change while the value for the ACL reconstructed limb significantly decreased from 45° to 90° of knee flexion ($p < 0.01$) (Fig. 24.3c). The I-EMG value at 90° for each muscle and for each limb

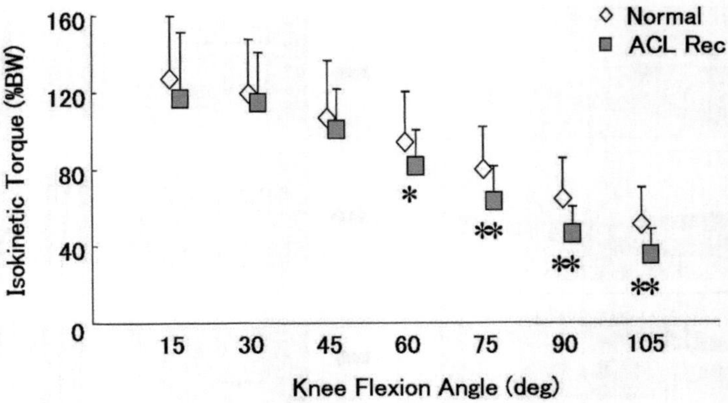

Fig. 24.2 Isokinetic torque at sequential 15° increases in knee flexion angle *: $p<0.05$, **: $p<0.01$; a comparison between normal and ACL reconstructed limbs

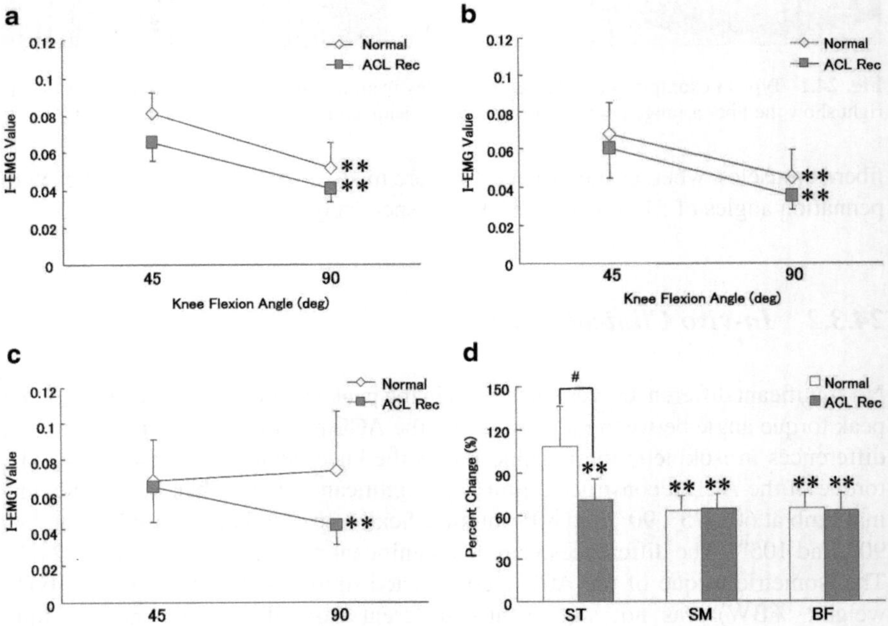

Fig. 24.3 I-EMG value of the SM (**a**), BF (**b**), and ST (**c**) **: $p<0.01$ compared with the value at 45°. (**d**) I-EMG value at 90° calculated as a percentage of the I-EMG value at 45°. #: $p<0.05$, **: $p<0.01$ compared with the value at 45°

is calculated as a percent of the I-EMG value at 45° and shown in Fig. 24.3d. The I-EMG values for the SM and BF of both limbs at 90° were significantly reduced to approximately 65 % of the values at 45° ($p<0.01$). In addition, no significant differences were found between the normal and ACL reconstructed limbs in the SM and BF. However, for the ST a significant difference was found between the limbs ($p<0.05$). The I-EMG value at 90° in the normal limb was 108.5±56.0 % of the value at 45° ($p>0.05$), while in the ACL reconstructed limb, the I-EMG value significantly decreased to 70.4±29.6 % of the value at 45° ($p<0.01$).

The volume of the ST calculated from the MRI images was significantly smaller in the ACL reconstructed limb (68.7±17.4 %) than in the normal limb ($p<0.01$), whereas the volumes of the SM and BF were not different between the limbs. In all subjects, regeneration of the ST tendon was confirmed. Since the regenerated tendon passed the knee joint and inserted into a structure distal to the knee, it could function as a knee flexor. The ST musculotendinous junction in the reconstructed limb in almost all subjects was located more proximally from the knee joint space than that in the normal limb. In only one subject did the musculotendinous junction of the reconstructed limb appear in the same image as the normal limb. The difference in location between the limbs ranged from 0 to 5 images, which corresponds to a distance of 0–6 cm.

24.4 Discussion

In agreement with studies by Lipscomb et al. (1982) and Maeda et al. (1996), we found that after ACL reconstruction using the ST or STG tendon, the peak knee flexion torque of the ACL reconstructed limb that was measured at approximately 20° of knee flexion had recovered to 96 % of that of the normal limb. However, significant differences in knee flexor strength between the normal and ACL reconstructed limbs both in isokinetic and isometric contraction were found at angles of knee flexion greater than 60°. This result is similar to the finding of Tashiro et al. (2003).

From the anatomical study, considerable differences in the muscle architecture of each knee flexor muscle were shown. The ST and G were parallel fibered muscles with long fiber lengths. Thus, the number of sarcomeres in series is large for these muscles which illustrates an eminent potential for the ST and G to shorten at long distances. On the other hand, the SM and BF were unipennate muscles, characterized by short fiber lengths and a pennation angle that increases with increasing knee flexion angles. These features suggest that the SM and BF are insufficient to produce knee flexion torque at deeper angles. As results from our EMG analysis indicate, these architectural differences seem to reflect functional differences between the ST and the SM and BF. Therefore, it appears that the SM and BF cannot compensate for the function of the ST, particularly at deep flexion angles. We believe that the ST plays an important role at over 60° of knee flexion, while the SM and BF work together to produce the peak knee flexion torque occurring at less angles, since no significant difference of the peak torque and peak torque angle between the normal and reconstructed limbs were measured in the present study. Furthermore, since the G is a parallel fibered muscle with a long fiber length, it may compensate for the function of the harvested ST and should be preserved in ACL reconstruction. In fact, Tashiro et al. (2003) have reported that the reduction of the flexion torque at 70° of knee flexion and beyond was more often observed in patients with ACL reconstruction using the ST and G tendon than in patients where only the ST was used for reconstruction.

In this study, regeneration of the ST tendon was confirmed in all subjects. Previously, Eriksson et al. (1999) and Rispoli et al. (2001) confirmed regeneration of the ST tendon in MRI images taken at the level of the tibial plateau and superior pole of the patella, respectively. In addition to the regenerated tendon, we were able to measure the location of the ST musculotendinous junction by analyzing the MRI images in sequence from the knee joint space level and found that the musculotendinous junction of the ACL reconstructed limb shifted proximally as compared to that of the normal limb. Similarly, Papandrea et al. used ultrasonography (Papandrea et al. 2000) to show that the insertion of the regenerated ST tendon was proximal to that of the normal tendon. Our results also showed that the volume of the ST after its tendon was harvested for ACL reconstruction was significantly smaller than the normal ST. However, the volume of the SM and BF did not change significantly. As in the results from our EMG study, these architectural changes, such as atrophy and shortening of the ST in the ACL constructed limb, seem to have an effect on knee function at deep knee angles. Although the ST tendon had regenerated in all subjects, the regenerated ST tendon probably did not function in the same manner as in the intact (i.e., native) tendon.

In conclusion, we suggest that the lack of compensation from the SM and BF as well as the atrophy and shortening of the ST have an influence on the decrease of knee flexion torque which occurs at over 60° of knee flexion in patients with ACL reconstruction using the ST tendon.

References

Aglietti P, Buzzi R, Zaccherotti G et al (1994) Patellar tendon versus doubled semitendinosus and gracilis tendons for anterior cruciate ligament reconstruction. Am J Sports Med 22:211–218

Akima H, Kuno S, Fukunaga T et al (1995) Architectural properties and specific tension of human knee extensor and flexor muscles based on magnetic resonance imaging. Jpn J Phys Fitness Med 44:267–278

Armour T, Forwell L, Litchfield R et al (2004) Isokinetic evaluation of internal/external tibial rotation strength after the use of hamstring tendons for anterior cruciate ligament reconstruction. Am J Sports Med 32:1639–1643

Carter TR, Edinger S (1999) Isokinetic evaluation of anterior cruciate ligament reconstruction: hamstring versus patellar tendon. Arthroscopy 15:169–172

Corry IS, Webb JM, Clingeleffer AJ et al (1999) Arthroscopic reconstruction of the anterior cruciate ligament. A comparison of patellar tendon autograft and four-standard hamstring tendon autograft. Am J Sports Med 27:444–454

Cross MJ, Roger G, Kujawa P et al (1992) Regeneration of the semitendinosus and gracilis tendons following their transection for repair of the anterior cruciate ligament. Am J Sports Med 20:221–223

Eriksson K, Larsson H, Wredmark T et al (1999) Semitendinosus tendon regeneration after harvesting for ACL reconstruction. A prospective MRI study. Knee Surg Sports Traumatol Arthrosc 7:220–225

Eriksson K, Anderberg P, Hamberg P et al (2001) A comparison of quadrupled semitendinosus and patellar tendon grafts in reconstruction of the anterior cruciate ligament. J Bone Joint Surg 83B:348–354

Frank CB, Jackson DW (1997) The science of reconstruction of the anterior cruciate ligament. J Bone Joint Surg 79A:1556–1576

Friederich JA, Brand RA (1990) Muscle fiber architecture in the human lower limb. J Biomech 23:91–95

Harter RA, Osternig LR, Singer KM et al (1988) Long-term evaluation of knee stability and function following surgical reconstruction for anterior cruciate ligament insufficiency. Am J Sports Med 16:434–443

Herzog W (1993) Lines of action and moment arms of the major force-carrying structures crossing the human knee joint. J Anat 182:213–230

Karlson JA, Steiner ME, Brown CH et al (1994) Anterior cruciate ligament reconstruction using gracilis and semitendinosus tendons. Comparison of through-the-condyle and over-the-top graft placement. Am J Sports Med 22:659–666

Lipscomb AB, Johnston RK, Snyder RB et al (1982) Evaluation of hamstring strength following use of semitendinosus and glacilis tendons to reconstruct the anterior cruciate ligament. Am J Sports Med 10:340–342

Maeda A, Shino K, Horibe S et al (1996) Anterior cruciate ligament reconstruction with multistranded autogenous semitendinosus tendon. Am J Sports Med 24:504–508

Nakajima R, Maruyama Y, Ichiao K et al (1996) Decreased hamstring strength after harvest of semitendinosus and gracilis tendons for anterior cruciate ligament reconstruction. J Jpn Clin Sports Med Assoc 6:681–686

Nakamura N, Horibe S, Sasaki S et al (2002) Evaluation of active knee flection and hamstring strength after anterior cruciate ligament reconstruction using hamstring tendons. Arthroscopy 18:598–602

Ohkoshi Y, Inoue C, Yamane S et al (1998) Changes in muscle strengthproperties caused by harvesting of autogenous semitendinosus tendon for reconstruction of contralateral anterior cruciate ligament. Arthroscopy 14:580–584

Papandrea P, Vulpiani MC, Ferretti A et al (2000) Regeneration of the semitendinosus tendon harvested for anterior cruciate ligament reconstruction. Evaluation using ultrasonography. Am J Sports Med 28:556–561

Rispoli DM, Sanders TG, Miller MD et al (2001) Magnetic resonance imaging at different time periods following hamstring harvest for anterior cruciate ligament reconstruction. Arthroscopy 17:2–8

Rosenberg TD, Deffner KT (1997) ACL reconstruction: semitendinosus tendon is the graft of choice. Orthopedics 20:396–398

Rosenberg TD, Graf B (1994) Techniques for ACL reconstruction with Multi-Trac drill guide. Acufex Microsurgucal, Mansfield

Rosenberg TD, Brown GC, Deffner KT (1997) Anterior cruciate ligament reconstruction with a quadrupled semitendinosus autograft. Sports Med Arthrosc Rev 5:51–58

Rowden NJ, Sher D, Rogers GJ et al (1997) Anterior cruciate ligament graft fixation. Initial comparison of patellar tendon and semitendinosus autografts in young fresh cadavers. Am J Sports Med 25:472–478

Sachs RA, Daniel DM, Stone ML et al (1989) Patellofemoral problems after anterior cruciate ligament reconstruction. Am J Sports Med 17:760–765

Sakane M, Fox RJ, Woo SL-Y et al (1997) In situ forces in the anterior cruciate ligament and its bundles in response to anterior tibial loads. J Orthop Res 15:285–293

Shaieb MD, Kan DM, Chang SK et al (2002) A prospective randomized comparison of patellar tendon versus semitendinosus and gracilis tendon sutografte for anterior cruciate ligament reconstruction. Am J Sports Med 30:214–220

Shino K, Nakata K, Horibe S et al (1993) Quantitative evaluation after arthroscopic anterior cruciate ligament reconstruction: allograft and autograft. Am J Sports Med 21:609–616

Simonian PT, Harrison SD, Cooley VJ et al (1997) Assessment of morbidity of semitendinosus and gracilis tendon harvest for ACL reconstruction. Am J Knee Surg 10:54–59

Tashiro T, Kurosawa H, Kawakami A et al (2003) Influence of medial hamstring tendon harvest on knee flexor strength after anterior cruciate ligament reconstruction. Am J Sports Med 31:522–529

Trotter JA (1993) Functional morphology of force transmission in skeletal muscle. A brief review. Acta Anat (Basel) 146:205–222

Viola RW, Sterett WI, Newfield D et al (2000) Internal and external tibial rotation strength after anterior cruciate ligament reconstruction using ipsilateral semitendinosus and gracilis tendon autografts. Am J Sports Med 28:552–555

Wickiewicz TL, Roy RR, Powell PL et al (1983) Muscle architecture of the human lower limb. Clin Orthop 179:275–283

Winters JM, Woo SL (1990) Multiple muscle systems. Springer, New York, pp 732–736

Woo SL-Y, Orlando CO, Camp JF et al (1986) Effects of postmortem storage by freezing on ligament tensile behavior. J Biomech 19:399–404

Yasuda K, Tsujino J, Ohkoshi Y et al (1995) Graft site morbidity with autogenous semitendinosus and gracilis tendons. Am J Sports Med 23:706–714

Chapter 25
Differences in Activation Patterns of the Hamstring Muscles During Sprinting

Ayako Higashihara, Takashi Ono, and Toru Fukubayashi

Abstract In this chapter, we evaluate the functional characteristics of the hamstring muscles during sprinting. During near maximal sprinting speed, the respective hamstring muscles exhibited very different characteristics of electromyographic (EMG) activation within the sprinting gait cycle. The activation demand of the BFlh muscles is high before and after foot contact due to their function as hip extensors, while the ST muscle shows high activation primarily during the control of knee extension and hip flexion in the mid-swing phase. The mechanism underlying these activation characteristics may involve architectural differences between the hamstring muscles, which likely reflect each muscle's function during sprinting. A complex neuromuscular coordination pattern of the hamstring muscles during sprinting is accomplished by the utilisation and activation of each muscle's specific functional differences.

Keywords Biceps femoris • Semitendinosus • Sprinting • Electromyography

25.1 Introduction

The biarticular hamstring muscles consist of the long head of the biceps femoris (BFlh) laterally, and the semimembranosus (SM) and semitendinosus muscles (ST) medially. All of these muscles play a role in the common function of knee flexion and hip extension. Several investigators have described these muscle's morphological differences, including variations in weight, pennation angle, volume, physiological cross-sectional area, and fibre length (Friederich and Brand 1990; Wickiewicz et al. 1983; Woodley and Mercer 2005). Studies have reported that the electromyographic (EMG) activity of each biarticular hamstring muscle varies with knee

A. Higashihara (✉)
Research Fellows of the Japan Society for the Promotion of Science,
Niigata University of Health and Welfare, Niigata, Japan
e-mail: ayako23.w15@gmail.com

T. Ono • T. Fukubayashi
Faculty of Sport Sciences, Waseda University, Saitama, Japan

K. Kanosue et al. (eds.), *Sports Injuries and Prevention*,
DOI 10.1007/978-4-431-55318-2_25

flexion (Kubota et al. 2007; Makihara et al. 2006; Ono et al. 2010) and hip extension (Ono et al. 2011), and have suggested that activation differences exist among these muscles due to their morphological and architectural differences.

The ST muscle is a parallel fibred muscle with long fibre lengths that demonstrates an eminent potential to contract over long distances, such that this muscle produces knee flexion torque at deeper angles compared to the BFlh and SM muscles (Makihara et al. 2006). In addition, the ST muscle is selectively recruited during intensive eccentric knee flexion exercise because of its morphological property of effectively handling strain during lengthening contraction (Kubota et al. 2007; Ono et al. 2010). On the other hand, the BFlh and SM muscles are unipennate muscles that are characterised by their short fibre lengths and pennation angles. These muscles have large cross-sectional areas and are more suitable than fusiform muscles for torque production (Koulouris and Connell 2005; Lieber and Friden 2000; Woodley and Mercer 2005). Another study reports that because of their morphological and architectural properties, the BFlh and SM muscles are selectively recruited to manage hip extension exercise, as these movements demand a high muscle torque (Ono et al. 2011). On the basis of these observations, it is likely that activation differences in the hamstring muscles also exist during dynamic movements such as sprinting.

The hamstring muscles are susceptible to strain injury during sports (Bennell and Crossley 1996; Brooks et al. 2005; Woods et al. 2004), and hamstring muscle strain injuries often occur during high speed or high intensity situations such as sprinting (Brooks et al. 2006; Woods et al. 2004). It has been reported that the majority of hamstring muscle strain injuries occur while the athlete is running at maximal or close to maximal speeds (Askling et al. 2007). An understanding of the function of the hamstring muscles during sprinting is therefore required to provide insight into the risk factors and mechanisms of hamstring muscle strain injury.

25.2 Activation Patterns of the Hamstring Muscles During Sprinting

25.2.1 Function of the Hamstring Muscles During Sprinting

During the late-swing phase of sprinting, the function of the biarticular hamstring muscles is that of hip extensors concentrically acting to quickly swing the thigh back, while also acting as knee flexors to eccentrically decelerate the forward swing of the lower leg (Mann 1981). In the early contact phase, the hamstring muscles are activated concentrically as hip extensor and knee flexor against the ground reaction force, resulting in the generation of a forward propulsive force (Simonsen et al. 1985). The hamstring muscles release elastic energy during ground contact, which is stored during the swing phase when they function eccentrically (Simonsen et al. 1985). The generation of hip extension torque during the terminal swing through the stance phase is suggested to play an important role in achieving high running speeds (Hunter et al. 2004; Mann and Sprague 1980). Overall, the hamstring muscles primarily act as strong hip extensors in these phases.

25.2.2 *Activation Characteristics of the Lateral Versus. Medial Hamstring Muscles During Sprinting*

The hamstring muscles activate from mid-swing until the terminal stance of sprinting (Chumanov et al. 2011; Higashihara et al. 2010; Jonhagen et al. 1996; Kyrolainen et al. 1999, 2005; Simonsen et al. 1985; Yu et al. 2008), with peaks of activation during the terminal swing and early stance phases (Chumanov et al. 2011; Higashihara et al. 2010, 2013; Yu et al. 2008) (Fig. 25.1b). The hamstring muscles exhibit very different EMG activation characteristics within the sprinting gait cycle. During sprinting, the hamstring muscles play a role in overall function as strong hip extensors and knee flexors. However, the time series recruitment patterns of these muscles are not uniform. The mechanism underlying these activation characteristics may involve morphological differences between the hamstring muscles, which likely reflect each muscle's function during sprinting.

In a recent study we investigated the activation patterns of the lateral and medial hamstring muscles durning overground sprinting (Higashihara et al. 2013). The EMG activation of the ST muscle during the mid-swing phase was greater than that of the BFlh muscle (Fig. 25.2). In addition, the peak activation of the ST muscle occurred significantly earlier than that of the BFlh muscle during the swing phase (Fig. 25.1c). During the mid-swing phase, the biarticular hamstring muscles are stretched due to hip flexion and simultaneous knee extension. Studies using musculoskeletal simulation have shown that the biarticular hamstring muscles underwent lengthening contraction in this phase of sprinting and the ST muscle displayed the greatest musculotendon lengthening velocity among the hamstring muscles (Schache et al. 2012; Thelen et al. 2005; Yu et al. 2008). The ST muscle has long fibres that contain many sarcomeres in series, illustrating its potential to contract quickly over long distances (Heron and Richmond 1993). Therefore, the ST muscle is selectively recruited for controlling simultaneous knee extension and hip flexion because of its morphological ability to effectively handle strain during lengthening contractions. Furthermore, the activity of the ST muscle was significantly greater than that of the BFlh muscle during the late stance phase (Fig. 25.2), and the peak activation of the ST muscle occurrs significantly later than that of the BFlh muscle during the stance phase (Fig. 25.1c). An earlier study reported that an eccentric contraction of the hamstrings occurs during the late-stance phase of overground sprinting (Yu et al. 2008). Therefore, the ST muscle is selectively recruited for eccentric knee flexion during the late stance phase.

Activation of the BFlh muscle is greater during the early-stance and late-swing phases than during the late-stance and mid-swing phases, whereas the activity of the

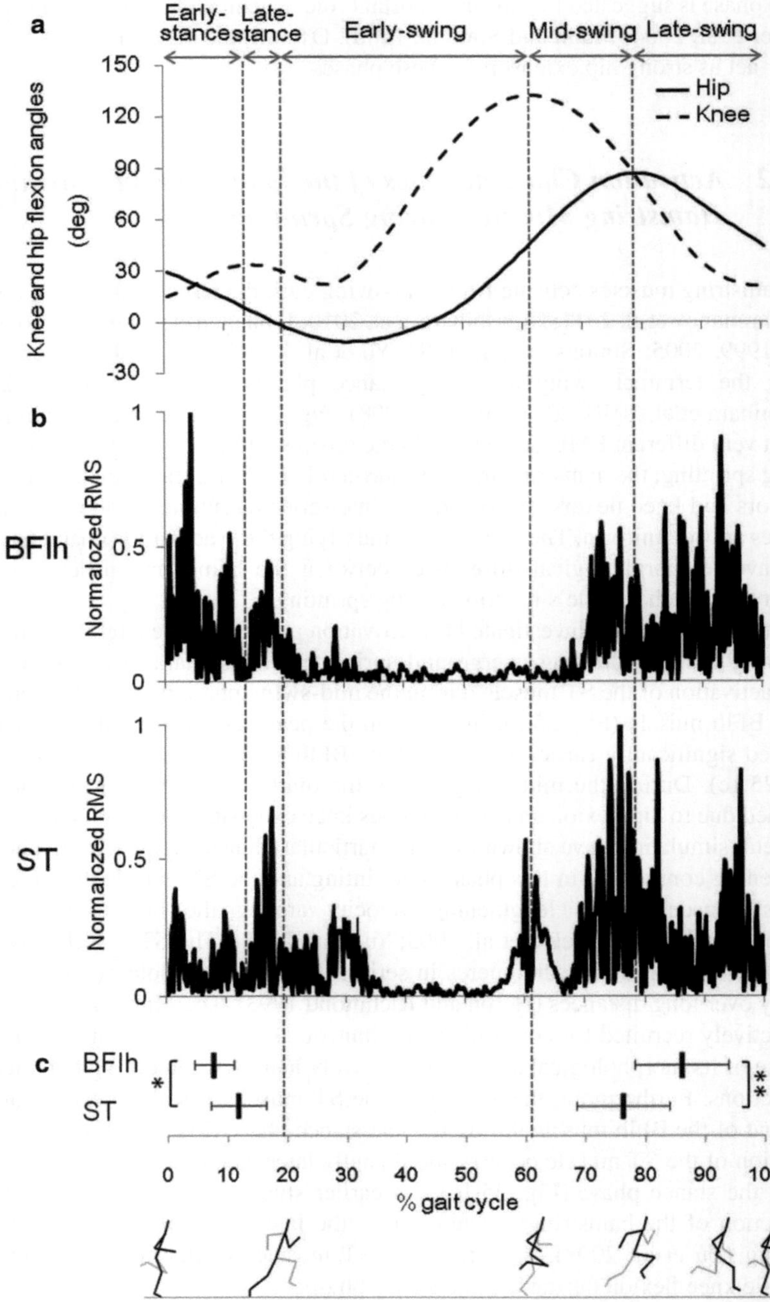

Fig. 25.1 (**a**) Mean hip flexion and knee flexion angles during sprinting. (**b**) Typical example of normalized EMGs for the biceps femoris long head (*BFlh*) and semitendinosus (*ST*) muscles during sprinting. The hamstring muscles were activated during the stance, mid-swing, and late-swing

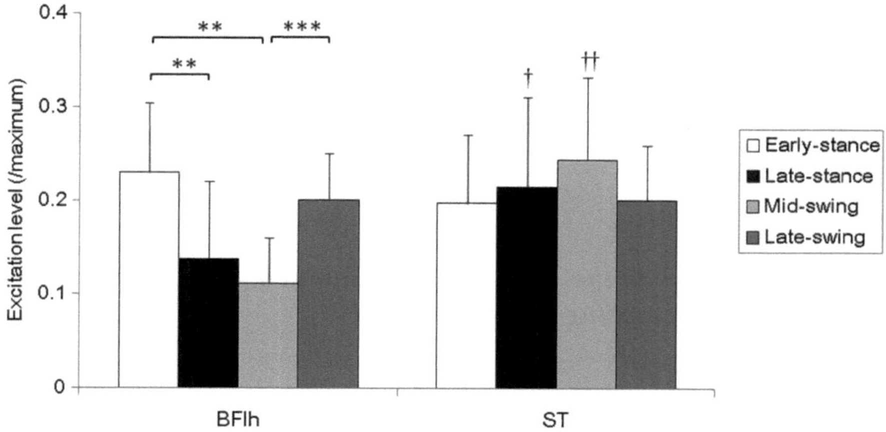

Fig. 25.2 Mean and standard deviation of the normalized EMG values during the early-stance, late-stance, mid-swing, and late-swing phases of sprinting. The two-way ANOVA indicated statistically significant interaction effects (muscle × phase; $p < 0.001$). The activation of the biceps femoris long head (*BFlh*) muscle during the early-stance and late-swing phases was significantly greater than during the mid-swing phase. The activation of the BFlh muscle during the early-stance phase was significantly greater than during the late-stance phase. The activity of the semitendinosus (*ST*) muscle was significantly greater than that of the BFlh muscle during the late-stance and mid-swing phases: **, $p < 0.01$; ***, $p < 0.001$; †, $p < 0.05$; ††, $p < 0.01$ vs. BFlh (Modified image from Conference Proceedings (Higashihara et al. 2013))

ST muscle exhibits no differences between each phase (Fig. 25.2) (Higashihara et al. 2013). In addition, peak activation of the BFlh muscle occurrs significantly closer to the point of foot contact compared to that of the ST muscle (Fig. 25.1c). Therefore, the activation demand of the BFlh muscle is high before and after foot contact. The hamstring muscles function as hip extensors that concentrically act to quickly swing the thigh back during the late-swing phase (Sugiura et al. 2008), and to provide the activation necessary for the generation of a forward propulsive force during the stance phase (Simonsen et al. 1985). In addition, the maximum torques for hip extension and knee flexion occur during foot contact in overground sprinting (Mann 1981). Therefore, during these phases of sprinting, the hamstring muscles primarily act as strong hip extensors (Novacheck 1998). One study that compared

Fig. 25.1 (continued) phases: the early-stance phase, beginning with the foot strike and ending with the maximum knee flexion during ground contact; late-stance phase, beginning with maximum knee flexion during the ground contact and ending with the toe-off; early-swing phase, beginning with toe off and ending with maximum knee flexion; mid-swing phase, beginning with maximum knee flexion and ending with maximal hip flexion; late-swing phase, beginning with maximum hip flexion and ending with foot strike. (**c**) Mean and standard deviation of the peak time, at which each RMS of the BFlh and ST reached their maximum value during the stance and swing phases. Statistically significant peak time differences between these two muscles were found: *, $p < 0.05$; **, $p < 0.01$

the recruitment patterns of the various hamstring muscles demonstrated that the BF muscle is selectively recruited during hip extension exercises because of the morphological potential for large torque production (Ono et al. 2011). In light of these previous investigations, the results of the present study indicate that the BF muscle is selectively recruited primarily as a strong hip extensor during the terminal swing and early stance phases.

25.2.3 Changes in the Activity of the Hamstring Muscles with Increasing Running Speed

Previous studies (Higashihara et al. 2010; Kyrolainen et al. 2005; Mero and Komi 1987; Schache et al. 2013) have examined activity of the lower extremity muscles during running at different speeds. Kyrolainen et al. (2005) found that the greatest change in the BF muscle activity pattern occurred as the speed increased from that of a slow jog to the maximum and indicated that increasing the running speed effectively requires increased EMG activity in the two-joint muscles during the entire running cycle.

We previously investigated activation patterns of the lateral and medial hamstring muscles during treadmill sprinting at different speeds and observed changes in activity patterns of the BFlh and ST muscles as the speed increased (Higashihara et al. 2010). There were no significant differences in the timing of peak activation of the BFlh and ST muscles at speeds below 95 % max. In contrast, at near maximum sprinting speed, peak activation of the BFlh muscle occurrs significantly closer to the point of foot contact as compared to that of the ST muscle. These findings indicate that a complex neuromuscular coordination pattern occurs during the running gait cycle at near maximum sprinting speeds. The increase to maximum speed is accomplished by utilizing each muscle according to its specific function.

When running at near maximum sprinting speed, the BFlh muscle is selectively recruited primarily as strong hip extensor because of the morphological potential for large torque production (see 25.2.2). The ground reaction forces at foot strike increase with increasing running speed (Kyrolainen et al. 2005; Mero and Komi 1987). Additionally, as running speed is increased, peak knee flexion torque during the terminal swing and peak hip extension torque during the terminal swing and initial stance are increased (Schache et al. 2011). Therefore, when running speed is increased toward the maximal level, the BFlh muscle strongly activates before and after foot strike to prepare for the intensive impact of the ground reaction force at foot strike, and also to play an important role in the generation of the forward propulsive force.

In contrast to the EMG activation of the BFlh muscle, peak activation of the ST muscle occurred significantly earlier than that of the BFlh muscle during the swing phase. During the mid-swing phase, the biarticular hamstring muscles are stretched, due to hip flexion along with simultaneous knee extension. Peak musculotendon lengthening velocities of the hamstrings occur at this phase and increase significantly with running speed (Schache et al. 2013; Thelen et al. 2005). In addition, lengthening velocity of the ST muscle is greater than that of the BFlh and SM muscles (Schache et al. 2012; Thelen et al. 2005; Yu et al. 2008). The ST muscle is therefore selectively recruited for simultaneously controlling knee extension and hip flexion because of its morphological property of effectively dealing with strain.

25.3 Insight into the Mechanism of Hamstring Strain Injuries During Sprinting

25.3.1 Mechanism of Hamstring Strain Injuries During Sprinting

It is well recognised that the hamstring muscles are susceptible to acute strain injury during high-speed sprinting (Askling et al. 2007). Several factors have been hypothesised to increase the risk for hamstring strains (Agre 1985; Hoskins and Pollard 2005). Commonly studied factors include hamstring muscle weakness (Sugiura et al. 2008), thigh muscle strength inbalance (Croisier et al. 2008; Sugiura et al. 2008), flexibility (Arnason et al. 2004; Heiderscheit et al. 2010), fatigue (Woods et al. 2004), and a history of previous injury (Croisier 2004). During sprinting, a complex neuromuscular coordination pattern in the hamstring muscles is accomplished according to each muscle's specific functional characteristics. We, therefore, hypothesise that the disruption of the functional balance of the hamstring muscles, which may occur because of interactions between several of the risk factors, may lead to hamstring muscle strain (Fig. 25.3).

Furthermore, a hamstring strain is likely to occur during overstriding, i.e. when the foot lands too far in front of the centre of mass when running at close to the maximum speed and trying to maintain or achieve even greater speed (Orchard 2002). The BFlh muscle is selectively recruited during the late-swing and early-stance phases during which the hamstring muscles contract concentrically to produce hip extension torque. Therefore, we hypothesize that the risk of acute hamstring strain injury that involve the BFlh increases when the athletes have more forward trunk lean or flexed hip during these phases. On the other hand, the ST muscle is selectively recruited when the hamstring muscles contract eccentrically to control knee extension. Therefore, the risk of the injuries that involve the ST will increase when athletes have more knee extension during swing phase. These external stresses on hamstring muscles which are due to rapid intensive changes in movement also contribute to an increased risk of injury (Fig. 25.3).

25.3.2 Relation to the Location of the Hamstring Strain Injury

Previous magnetic resonance imaging studies of hamstring strain injuries (De Smet and Best 2000; Gibbs et al. 2004; Slavotinek et al. 2002) suggest that the BFlh muscle is the most likely to be injured; these injuries occur more commonly

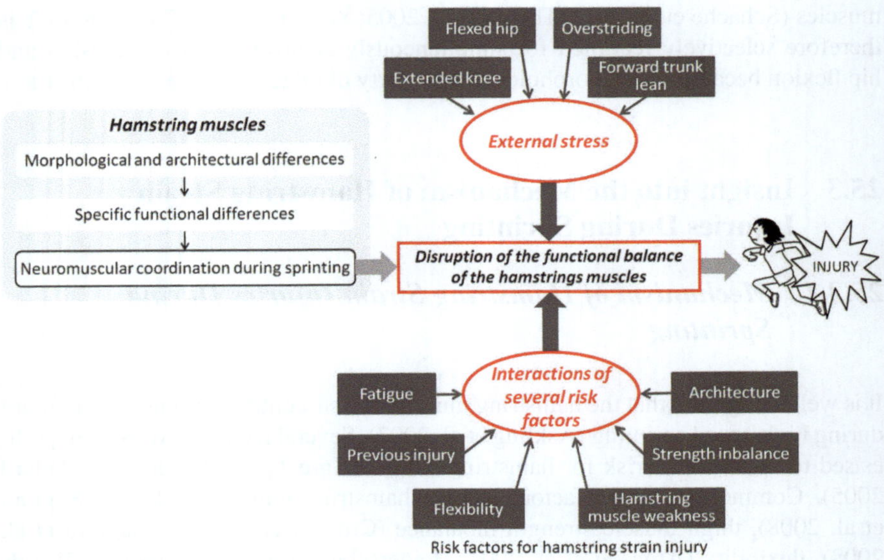

Fig. 25.3 One of the possible mechanisms of the hamstring strain injury which occurs during sprinting. During sprinting, a complex neuromuscular coordination pattern in the hamstring muscles would be accomplished by each muscle's specific functional differences due to their morphological and architectural differences. When the hamstring strain injury occurs, the disruption of the functional balance of the hamstring muscles, which may occur because of interactions between several risk factors and/or the external stress, may lead to hamstring muscle strain

at the proximal end of the BFlh muscle and in combination with the adjacent ST. These results can be explained anatomically by the fact that in the proximal hamstring, the BFlh and ST muscles have a combined tendon origin and proximal part of their muscle (Gibbs et al. 2004). Considering the relations between the anatomical features of the proximal hamstrings and injury location, differences in the activation patterns between the BFlh and ST muscles, which were found in our previous studies (Higashihara et al. 2010; Higashihara et al. 2013), might have significant effects on the mechanism of hamstring strain. During sprinting at near maximum speed, a complex neuromuscular coordination pattern is accomplished according to each hamstring muscle's specific functional differences. However, when the disruption of the functional balance of the lateral and medial hamstring muscles occurs due to the risk factors mentioned above, this may lead to an increase in load as well as force concentration. This combination could, result in increased risk for a hamstring strain injury. Future studies need to address interactions between the risk factors for hamstring strain injury (such as fatigue), as well as activation patterns of the specific involved muscles, to further explicate injury mechanisms.

25.4 Conclusion

During near maximal sprinting speed, the respective hamstring muscles exhibit very different characteristics for EMG activation within the sprinting gait cycle. The activation demand of the BFlh muscles is high before and after foot contact because of their function as hip extensors, while the ST muscle shows high activation primarily during the control of knee extension and hip flexion in the mid-swing phase. The mechanism underlying these activation characteristics may reflect the architectural differences in the hamstring muscles, which are in turn due to each muscle's function during sprinting. A complex neuromuscular coordination pattern of the hamstring muscles during sprinting is accomplished by a coordination of each muscle's activation pattern and its specific functional characteristics.

References

Agre JC (1985) Hamstring injuries. Proposed aetiological factors, prevention, and treatment. Sports Med 2(1), 21–33

Arnason A, Sigurdsson SB, Gudmundsson A, Holme I, Engebretsen L, Bahr R (2004) Risk factors for injuries in football. Am J Sports Med 32:5S–16S

Askling CM, Tengvar M, Saartok T, Thorstensson A (2007) Acute first-time hamstring strains during high-speed running: a longitudinal study including clinical and magnetic resonance imaging findings. Am J Sports Med 35:197–206. doi:10.1177/0363546506294679, 0363546506294679 [pii]

Bennell KL, Crossley K (1996) Musculoskeletal injuries in track and field: incidence, distribution and risk factors. Aust J Sci Med Sport 28:69–75

Brooks JH, Fuller CW, Kemp SP, Reddin DB (2005) Epidemiology of injuries in English professional rugby union: part 1 match injuries. Br J Sports Med 39:757–766

Brooks JH, Fuller CW, Kemp SP, Reddin DB (2006) Incidence, risk, and prevention of hamstring muscle injuries in professional rugby union. Am J Sports Med 34:1297–1306

Chumanov ES, Heiderscheit BC, Thelen DG (2011) Hamstring musculotendon dynamics during stance and swing phases of high-speed running. Med Sci Sports Exerc 43:525–532. doi:10.1249/MSS.0b013e3181f23fe8

Croisier JL (2004) Factors associated with recurrent hamstring injuries. Sports Med 34:681–695, 34105 [pii]

Croisier JL, Ganteaume S, Binet J, Genty M, Ferret JM (2008) Strength imbalances and prevention of hamstring injury in professional soccer players: a prospective study. Am J Sports Med 36:1469–1475

De Smet AA, Best TM (2000) Mr imaging of the distribution and location of acute hamstring injuries in athletes. Am J Roentgenol 174:393–399

Friederich JA, Brand RA (1990) Muscle fiber architecture in the human lower limb. J Biomech 23:91–95, 0021-9290(90)90373-B [pii]

Gibbs NJ, Cross TM, Cameron M, Houang MT (2004) The accuracy of mri in predicting recovery and recurrence of acute grade one hamstring muscle strains within the same season in Australian rules football players. J Sci Med Sport 7:248–258

Heiderscheit BC, Sherry MA, Silder A, Chumanov ES, Thelen DG (2010) Hamstring strain injuries: recommendations for diagnosis, rehabilitation, and injury prevention. J Orthop Sports Phys Ther 40:67–81. doi:10.2519/jospt.2010.3047, 2394 [pii]

Heron MI, Richmond FJ (1993) In-series fiber architecture in long human muscles. J Morphol 216:35–45. doi:10.1002/jmor.1052160106

Higashihara A, Ono T, Kubota J, Okuwaki T, Fukubayashi T (2010) Functional differences in the activity of the hamstring muscles with increasing running speed. J Sports Sci 28:1085–1092. doi:10.1080/02640414.2010.494308, 924847988 [pii]

Higashihara A, Ono T, Nagano Y, and Fukubayashi T (2013) Functional differences in the hamstring muscles during sprinting. In: 31th international conference on biomechanics in Sports 2013, Taipei

Hoskins W, Pollard H (2005) The management of hamstring injury–Part 1: Issues in diagnosis. Man Ther 10(2), 96–107

Hunter JP, Marshall RN, McNair P (2004) Reliability of biomechanical variables of sprint running. Med Sci Sports Exerc 36:850–861, 00005768-200405000-00018 [pii]

Jonhagen S, Ericson MO, Nemeth G, Eriksson E (1996) Amplitude and timing of electromyographic activity during sprinting. Scand J Med Sci Sports 6:15–21

Koulouris G, Connell D (2005) Hamstring muscle complex: an imaging review. Radiographics 25:571–586. doi:10.1148/rg.253045711, 25/3/571 [pii]

Kubota J, Ono T, Araki M, Torii S, Okuwaki T, Fukubayashi T (2007) Non-uniform changes in magnetic resonance measurements of the semitendinosus muscle following intensive eccentric exercise. Eur J Appl Physiol 101:713–720

Kyrolainen H, Komi PV, Belli A (1999) Changes in muscle activity patterns and kinetics with increasing running speed. J Strength Cond Res 13:400–406

Kyrolainen H, Avela J, Komi PV (2005) Changes in muscle activity with increasing running speed. J Sports Sci 23:1101–1109

Lieber RL, Friden J (2000) Functional and clinical significance of skeletal muscle architecture. Muscle Nerve 23:1647–1666. doi:10.1002/1097-4598(200011)23:11<1647::AID-MUS1>3.0.CO;2-M

Makihara Y, Nishino A, Fukubayashi T, Kanamori A (2006) Decrease of knee flexion torque in patients with acl reconstruction: combined analysis of the architecture and function of the knee flexor muscles. Knee Surg Sports Traumatol Arthrosc 14:310–317. doi:10.1007/s00167-005-0701-2

Mann RV (1981) A kinetic analysis of sprinting. Med Sci Sports Exerc 13:325–328

Mann R, Sprague P (1980) A kinetic analysis of the ground leg during sprint running. Res Q Exerc Sport 51:334–348

Mero A, Komi PV (1987) Electromyographic activity in sprinting at speeds ranging from submaximal to supra-maximal. Med Sci Sports Exerc 19:266–274

Novacheck TF (1998) The biomechanics of running. Gait Posture 7:77–95

Ono T, Okuwaki T, Fukubayashi T (2010) Differences in activation patterns of knee flexor muscles during concentric and eccentric exercises. Res Sports Med 18:188–198. doi:10.1080/15438627.2010.490185, 924057280 [pii]

Ono T, Higashihara A, Fukubayashi T (2011) Hamstring functions during hip-extension exercise assessed with electromyography and magnetic resonance imaging. Res Sports Med 19:42–52. doi:10.1080/15438627.2011.535769, 932527223 [pii]

Orchard J (2002) Biomechanics of muscle strain injury. N Z J Sports Med 30:92–98

Schache AG, Blanch PD, Dorn TW, Brown NA, Rosemond D, Pandy MG (2011) Effect of running speed on lower limb joint kinetics. Med Sci Sports Exerc 43:1260–1271. doi:10.1249/MSS.0b013e3182084929

Schache AG, Dorn TW, Blanch PD, Brown NA, Pandy MG (2012) Mechanics of the human hamstring muscles during sprinting. Med Sci Sports Exerc 44:647–658. doi:10.1249/MSS.0b013e318236a3d2

Schache AG, Dorn TW, Wrigley TV, Brown NA, Pandy MG (2013) Stretch and activation of the human biarticular hamstrings across a range of running speeds. Eur J Appl Physiol 113:2813–2828. doi:10.1007/s00421-013-2713-9

Simonsen EB, Thomsen L, Klausen K (1985) Activity of mono- and biarticular leg muscles during sprint running. Eur J Appl Physiol Occup Physiol 54:524–532

Slavotinek JP, Verrall GM, Fon GT (2002) Hamstring injury in athletes: using mr imaging measurements to compare extent of muscle injury with amount of time lost from competition. AJR Am J Roentgenol 179:1621–1628

Sugiura Y, Saito T, Sakuraba K, Sakuma K, Suzuki E (2008) Strength deficits identified with concentric action of the hip extensors and eccentric action of the hamstrings predispose to hamstring injury in elite sprinters. J Orthop Sports Phys Ther 38:457–464. doi:10.2519/jospt.2008.2575, 1409 [pii]

Thelen DG, Chumanov ES, Hoerth DM, Best TM, Swanson SC, Li L, Young M, Heiderscheit BC (2005) Hamstring muscle kinematics during treadmill sprinting. Med Sci Sports Exerc 37:108–114

Wickiewicz TL, Roy RR, Powell PL, Edgerton VR (1983) Muscle architecture of the human lower limb. Clin Orthop Relat Res 179:275–283

Woodley SJ, Mercer SR (2005) Hamstring muscles: architecture and innervation. Cells Tissues Organs 179:125–141. doi:10.1159/000085004, CTO2005179003125 [pii]

Woods C, Hawkins RD, Maltby S, Hulse M, Thomas A, Hodson A (2004) The football association medical research programme: an audit of injuries in professional football–analysis of hamstring injuries. Br J Sports Med 38:36–41

Yu B, Queen RM, Abbey AN, Liu Y, Moorman CT, Garrett WE (2008) Hamstring muscle kinematics and activation during overground sprinting. J Biomech 41:3121–3126. doi:10.1016/j.jbiomech.2008.09.005, S0021-9290(08)00455-7 [pii]

Slavotinek JP, Verrall GM, Fon GT (2002) Hamstring injury in athletes: using MR imaging measurements to compare extent of muscle injury with amount of time lost from competition. AJR Am J Roentgenol 179:1621–1628

Sugiura Y, Saito T, Sakuraba K, Sakuma K, Suzuki E (2008) Strength deficits identified with concentric action of the hip extensors and eccentric action of the hamstrings predispose to hamstring injury in elite sprinters. J Orthop Sports Phys Ther 38(8):457–464. doi:10.2519/jospt.2008.2575 [pii]

Thelen DG, Chumanov ES, Hoerth DM, Best TM, Swanson SC, Li L, Young M, Heiderscheit BC (2005) Hamstring muscle kinematics during treadmill sprinting. Med Sci Sports Exerc 37:108–114

Wickiewicz TL, Roy RR, Powell PL, Edgerton VR (1983) Muscle architecture of the human lower limb. Clin Orthop Relat Res 179:275–283

Woodley SJ, Mercer SR (2005) Hamstring muscles: architecture and innervation. Cells Tissues Organs 179(3):125–141. doi:10.1159/000085004. CTO2005179003125 [pii]

Woods C, Hawkins RD, Maltby S, Hulse M, Thomas A, Hodson A (2004) The Football Association medical research programme: an audit of injuries in professional football—analysis of hamstring injuries. Br J Sports Med 38:36–41

Yu B, Queen RM, Abbey AN, Liu Y, Moorman CT, Garrett WE (2008) Hamstring muscle kinematics and activation during overground sprinting. J Biomech 41(15):3121–3126. doi:10.1016/j.jbiomech.2008.09.005. S0021-9290(08)00455-7 [pii]

Chapter 26
The Relationship Between Eccentric Exercise and Muscle Damage in Hamstring Muscles

Takashi Ono

Abstract We have clarified the relationship between activation patterns during eccentric knee flexion/hip extension exercises and the muscle damage after exercise in hamstring muscles by utilizing electromyography (EMG) and magnetic resonance imaging (MRI). In the eccentric knee flexion exercise, subjects exercised with each unilateral limb at 120 % of 1 repetition of maximum (1RM) and 50 % of 1RM. EMG activity was recorded for the biceps femoris long head (BFlh), semitendinosus (ST), and semimembranosus (SM) muscles; MRI T2 values and cross-sectional areas (CSAs) of the same muscles were measured at rest, as well as immediately after, and 1, 2, 3 and 7 days after exercise. It was found that EMG of the ST was significantly higher than that of the SM during the exercises at 120 % 1RM. T2 value change in the ST was significantly higher than in the BFlh and SM in both exercises. In the eccentric hip extension exercise, other subjects performed 5 sets of 10 repetitions of hip extension. EMG activity during the exercise was recorded for the BFlh, ST, and SM muscles; MRI T2 values and CSAs of the same muscles were measured at rest, immediately after, and 2 and 7 days after the exercise. It was found that EMG of the BFlh and SM were significantly higher than that of the ST. Immediately after exercise, the T2 value and CSA changes in the SM showed a significant increase. Overall, it was concluded that the ST muscle has a sensitive to intensive eccentric knee flexion load and the BFlh and SM muscles to an eccentric hip extension load.

Keywords Muscle morphology • Recruitment pattern • Magnetic resonance imaging • Muscle strain

26.1 Introduction

Hamstring muscles, which include the biceps femoris long head (BFlh), semitendinosus (ST), and semimembranosus (SM) muscles, form a multiarticular muscle group that cross the knee and hip joints. These muscles work synergistically to

T. Ono (✉)
Faculty of Sport Sciences, Waseda University, Saitama, Japan
e-mail: ono-t@aoni.waseda.jp

K. Kanosue et al. (eds.), *Sports Injuries and Prevention*,
DOI 10.1007/978-4-431-55318-2_26

produce knee flexion and/or hip extension torque. Hamstring muscles have an important role in sports for sprinting or kicking and frequently are injured during such activities (Arnason et al. 2008; Askling et al. 2007; Brockett et al. 2004; Thelen et al. 2005; Woods et al. 2004). For the purpose of improving sports performance or preventing sports injuries, various strength training exercises are performed with hamstring muscles (Andersen et al. 2006; Arnason et al. 2008; Mesfar and Shirazi-Adl 2008). It has been demonstrated, however, that each hamstring muscle has inherent morphological features (Friederich and Brand 1990; Woodley and Mercer 2005), leading to different functional properties even in the case of a single joint movement.

Previous studies with electromyography (EMG) have revealed the activation patterns of individual hamstring muscles during knee flexion movement (Makihara et al. 2006; Mohamed et al. 2002; Onishi et al. 2002). These reports confirm that the EMG activity of each hamstring muscle during maximum knee flexion varies with the knee angle, and the differences might well be due to each muscle's distinct morphological features, such as muscle fascicle length, pennation angle, and moment arm. Various studies have investigated the morphological and functional properties of multiple muscles that participate in a single knee joint flexion; however, the applied contraction pattern in those studies was isometric or concentric, and little information about which muscle was involved during eccentric contraction is available.

Despite these detailed investigations of hamstring muscle function during knee flexion movement, there is little information about the function of these muscles during hip extension. Worrell et al. (2001) investigated the relationship between hip angle and EMG activity of the hamstring muscle during isometric hip extension in the prone position and at various hip angle positions and found that the EMG activity was constant at all hip angles. The hamstring muscles, however, were examined as a single functional group utilizing only a single pair of electrodes, and the individual muscle from which activity was being recorded was not precisely identified. In addition, the various functions of the hamstring muscles during hip extension while standing, that is, the movement involved in bending forward/backward from the hip joint, was not investigated. It is common knowledge that the hamstring muscles are frequently injured during various sports activities, such as football, rugby, and track and field events. Also documented is that most acute cases of hamstring strains involve the BFlh, whereas the ST and SM muscles are less often injured (Brooks et al. 2006; Woods et al. 2004). Most of the injuries occur during the late-swing phase and stance phase of sprinting (Montgomery et al. 1994; Verrall et al. 2001), and during these phases, the hamstring muscles contract eccentrically to decelerate the forward swing of the leg and concentrically to push off the ground. Thus, it seems important to investigate the activation patterns of the hamstring muscles during eccentric and concentric hip extension to clarify the mechanism underlying hamstring injury.

Recently, MRI has been used to provide a quantitative assessment of skeletal muscle activity and response during exercise (Adams et al. 1992; Kinugasa and Akima 2005); however, practical application of the neurophysiological mechanisms described, which include a unique muscle sensitivity during eccentric exercise, remain unclear. The MRI transverse relaxation time (T2), which reflects the detailed changes in the intramuscular water content, has been observed to increase proportionally in activated and damaged muscles with an increase in exercise intensity. Adams et al. (1992) showed that changes in the muscle functional MRI signal correlated with the integrated EMG activity in the case of both concentric and eccentric contractions in the biceps brachii. Further, using this technique, detailed information regarding the morphological changes in individual muscles can be obtained by calculating the physiological crosssectional areas (CSAs) following exercise. It has been shown that eccentric muscular contraction is more effective in the improvement of muscle activation level or hypertrophy than the other contraction patterns (Hather et al. 1991; Hortobágyi et al. 1996); therefore, clarifying the relationship between the activation patterns during eccentric knee flexion/hip extension exercises and the muscle damages after exercises in hamstring muscles would make it possible to increase the efficiency of strength training and thus improve sports performance or rehabilitation of sports injuries.

Based on the background information described above, we have utilized EMG recordings to investigate the recruitment patterns of each hamstring muscle during knee flexion/hip extension exercises as well as differences in the time-course changes in the MRI measurements, such as the T2 values and CSAs, after the exercise.

26.2 Activation Patterns of Hamstring Muscles During Eccentric Knee Flexion Exercises and the Differences in the Damage After Exercise

In 2010, we examined differences in activation patterns among hamstring muscles during two patterns of knee flexion exercises and clarified the degree of muscle damage after those exercises (Ono et al. 2010). The subjects were seven male volunteers who performed two different sessions of eccentric/concentric knee flexion exercises (the session for the right limb was followed by a session for the left). With the right leg, the subjects performed a session of eccentric knee flexion tasks consisting of 5 sets of 10 repetitions at 120 % of the 1-repetition maximum (RM) (120 % ECC). Subjects were instructed to lower the weight from a 90° flexed knee position to a 0° extended knee position in 3 s, while maintaining a constant lowering velocity that was synchronized with an electronic metronome beat. An examiner removed the weight at the bottom of the curl, and the subject returned the unweighted leg to the flexed position and then received the weight to begin the next repetition. The subjects were encouraged to generate maximal voluntary force at the starting position and to give maximal resistance to the knee-extending action throughout the range of motion. With the left leg, the subjects performed a session of concentric (50 % CON) and eccentric (50 % ECC) knee flexion tasks consisting of 5 sets of 10 repetitions at 50 % of the 1-RM (50 %1RM). These subjects were

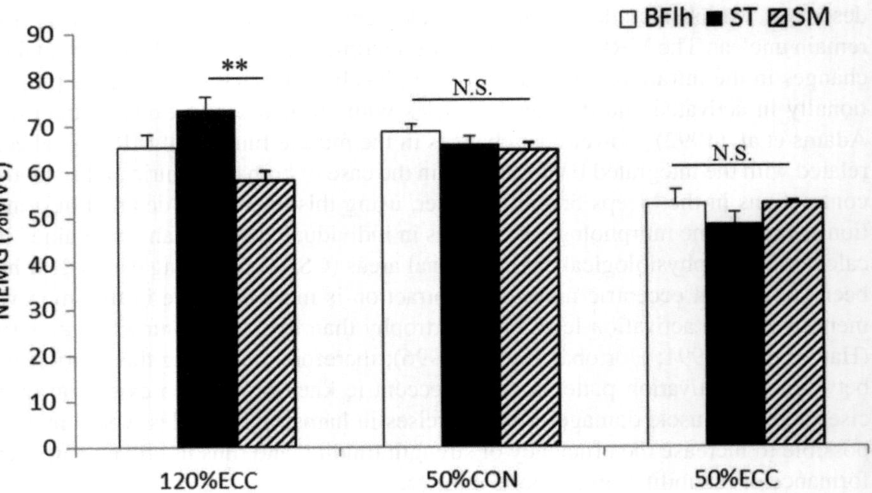

Fig. 26.1 The NiEMG values of hamstring muscles during 2 sessions of knee flexion exercises. Values refer to the mean ± the standard error (SE). **$P<0.01$ between the muscles, *N.S.* no significance. *BFlh* biceps femoris long head, *ST* semitendinosus, *SM* semimembranosus. *120 % ECC*, eccentric knee flexion at 120 % of the 1 RM; *50 % CON*, concentric knee flexion at 50 % of the 1 RM; *50 % ECC*, eccentric knee flexion at 50 % of the 1 RM

instructed to curl the weight from the fully extended knee position to a 90° flexed knee position and then to lower the weight from a 90° flexed knee position to fully extended knee position, while maintaining a constant curling and lowering velocity until the end of the session.

EMG activities during the exercise were recorded from the midbellies of the BFlh, ST, and SM muscles in the active limb at a sampling rate of 1 kHz. The digital EMG signals were full-wave rectified and integrated (iEMG) over each period of eccentric contraction, which was determined by the knee angle from a 90° flexed knee position to a full-extended knee position. The iEMG values were normalized to the values during maximum voluntary contraction (MVC) (NiEMG = iEMG/MVC [%]) for each subject. The NiEMG values during the sessions are shown in Fig. 26.1. During the 120 % ECC, the NiEMG values of the ST muscle were significantly greater than those of the SM ($P<0.01$). There were no significant differences among the hamstring muscle activities during the 50 % CON and 50 % ECC.

One week before, immediately (within 5 min) after, and 1, 2, 3, and 7 days after performing two exercise sessions, the subjects underwent a separate MRI of each limb (imaging of the right limb was followed by imaging of the left). The MRI T2 value was measured to evaluate the degree of muscle recruitment and damage. All MRI images of the participants' thighs were obtained by using a 1.5-T MR imaging system with a body coil. Transverse T2-weighted spin-echo images were subsequently obtained. The MRI data were evaluated for T2 relaxation time (T2 value) of the hamstring muscles. The signal intensity (SI) of each muscle (BFlh, ST, and SM)

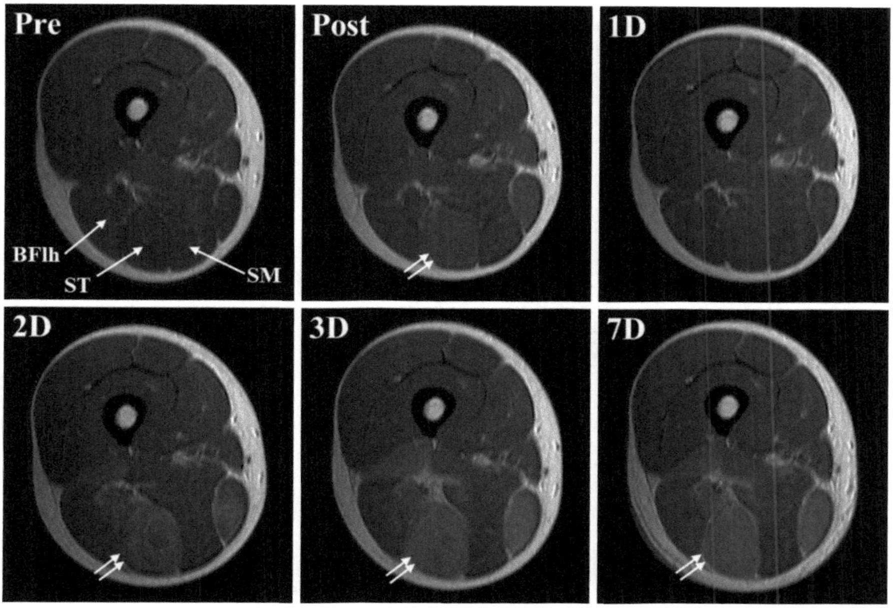

Fig. 26.2 Typical T2-weighted MR images (TR = 2,000 ms, TE = 30 ms) of the middle region (50 % of the thigh length) of the thigh at rest (*Pre*), immediately after (*Post*), and 1 (*1D*), 2 (*2D*), 3 (*3D*), and 7 (*7D*) days after the session of eccentric knee flexion at 120 % of the 1-RM. After the exercise, obvious SI increases were visible within the ST muscles (*double white arrow*). Images were obtained from 1 representative subject. *BFlh*, biceps femoris long head; *ST*, semitendinosus; SM, semimembranosus

was also measured. A T2 measurement sequence with four TEs (30, 45, 60, and 75 ms) was applied to measure the absolute T2 value. Images taken at different TEs were fit to a monoexponential time curve to extract the T2 values, based on the formula: $SI = M0 \times \exp(-TE/T2)$, where SI represents the SI at a given TE and M0 is the original MR SI. For comparison purposes, the averaged T2 mean value of each muscle was computed as a percentage change increase relative to the value measured before exercise. Typical T2-weighted MRI images of the thigh obtained from a subject at rest, immediately after, and 1, 2, 3, and 7 days after exercise are shown in Figs. 26.2 and 26.3; the absolute values for T2 at each exercise are shown in Tables 26.1 and 26.2. Figures 26.4 and 26.5 present changes in averaged T2 values of the individual muscles for each exercise. We found that the percentage changes in T2 immediately after the exercises were greater in all muscles at each exercise ($P < 0.05$). Two, 3, and 7 days after the exercise, a significant increase was identified only in the case of the ST muscle ($P < 0.05$) at 120 % ECC, and no significant change was identified in any muscle at 50 % 1RM. The degree of T2 value change immediately after the exercise were different between the muscles. The percentage change in the ST muscles were significantly greater than those in the BFlh and SM muscles for each exercise ($P < 0.01$). The absolute values of CSA for each exercise are shown

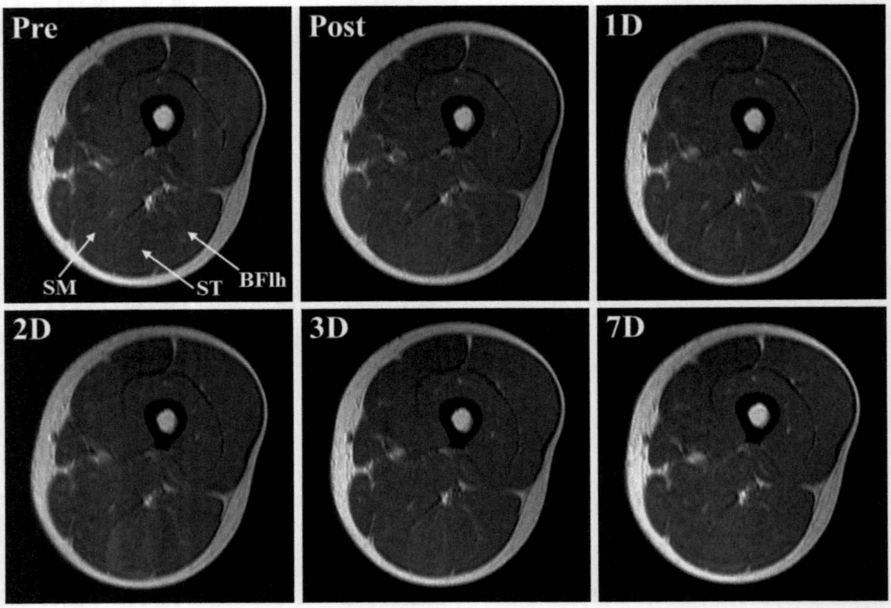

Fig. 26.3 Typical T2-weighted MR images (TR=2,000 ms, TE=30 ms) of the middle region (50 % of the thigh length) of the thigh at rest (*Pre*), immediately after (*Post*), and 1 (*1D*), 2 (*2D*), 3 (*3D*), and 7 (*7D*) days after the session of eccentric knee flexion at 50 % of the 1-RM. After the exercise, no obvious SI increases were visible within the hamstring muscles. Images were obtained from 1 representative subject. BFlh, biceps femoris long head; *ST* semitendinosus, *SM* semimembranosus

Table 26.1 The absolute values for T2 at rest (*Pre*), immediately after (*Post*), and 1 (*1D*), 2 (*2D*), 3 (*3D*), and 7 (*7D*) days after the session of eccentric knee flexion at 120 % of the 1-RM

	Pre	Post	1D	2D	3D	7D
T2(ms) BFlh	40.2±0.2	44.7±1.3	42.7±1.4	48.0±4.8	49.2±4.6	50.0±4.6
ST	37.4±0.6	53.3±2.5	49.0±6.8	62.9±9.8	74.0±7.3	80.9±9.9
SM	38.4±0.5	41.0±1.0	38.9±0.9	41.0±2.5	43.5±4.6	43.5±4.1

Table 26.2 The absolute values for T2 at rest (*Pre*), immediately after (*Post*), and 1 (*1D*), 2 (*2D*), 3 (*3D*), and 7 (*7D*) days after the session of eccentric knee flexion at 50 % of the 1-RM

	Pre	Post	1D	2D	3D	7D
T2(ms) BFlh	39.5±0.4	47.5±1.3	40.2±0.5	41.0±1.1	42.2±2.2	42.2±2.9
ST	37.0±0.8	55.7±1.8	39.3±0.6	41.0±1.7	40.7±1.4	40.3±1.3
SM	38.5±0.5	45.5±1.8	38.8±0.6	39.2±0.8	38.9±0.8	39.0±0.7

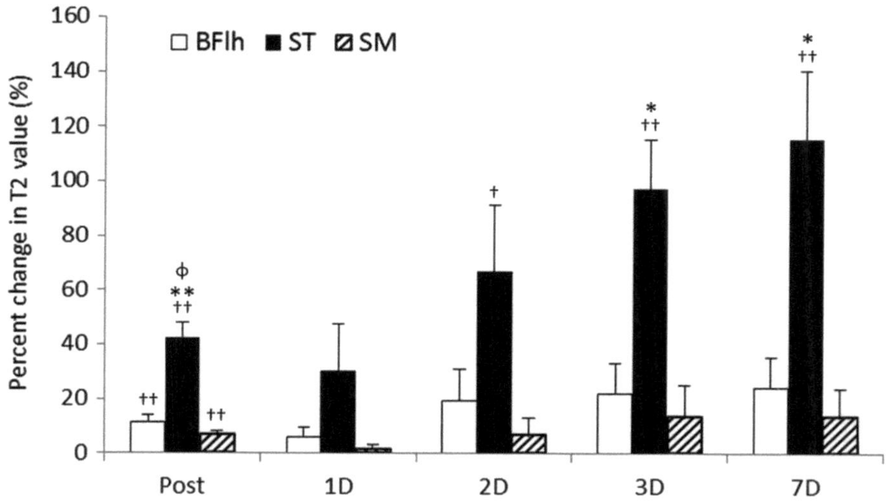

Fig. 26.4 Percentage change in MRI T2 value of each hamstring muscle immediately after (*Post*), and 1 (*1D*), 2 (*2D*), 3 (*3D*), and 7 (*7D*) days after the session of eccentric knee flexion at 120 % of the 1-RM. Values are the percentage change compared with the values at rest. $^{\dagger}P<0.05$, $^{\dagger\dagger}P<0.01$ vs the values at rest; $\varphi P<0.05$ vs BFlh; $^{*}P<0.05$, $^{**}P<0.01$ vs SM. *BFlh* biceps femoris long head, *ST* semitendinosus, *SM* semimembranosus

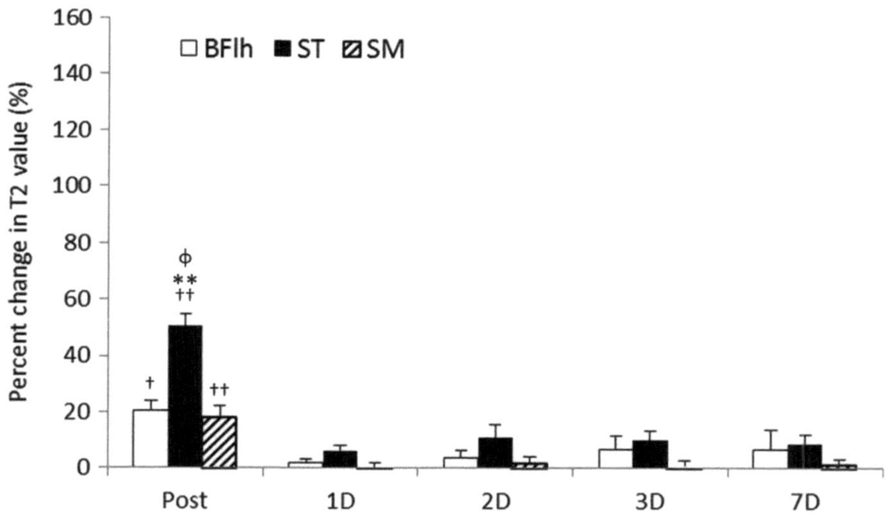

Fig. 26.5 Percentage change in MRI T2 value of each hamstring muscle immediately after (*Post*), and 1 (*1D*), 2 (*2D*), 3 (*3D*), and 7 (*7D*) days after the session of eccentric knee flexion at 50 % of the 1RM. Values are the percentage change compared with the values at rest. $^{\dagger}P<0.05$, $^{\dagger\dagger}P<0.01$ vs the values at rest; $\varphi P<0.05$ vs BFlh; $^{**}P<0.01$ vs SM. *BFlh*, biceps femoris long head; *ST* semitendinosus, *SM* semimembranosus

Table 26.3 The absolute values for CSA at rest (*Pre*), immediately after (*Post*), and 1 (*1D*), 2 (*2D*), 3 (*3D*), and 7 (*7D*) days after the session of eccentric knee flexion at 120 % of the 1-RM

	Pre	Post	1D	2D	3D	7D
CSA (mm^2) BFlh	1,313.0±67.4	1,327.4±67.6	1,345.1±65.0	1,416.1±73.6	1,404.3±81.4	1,354.9±80.1
ST	907.2±40.2	1,051.1±55.7	1,053.3±73.4	1,195.0±82.9	1,285.2±72.2	1,210±60.6
SM	910.6±36.8	907.4±34.7	959.1±41.0	975.8±35.6	953.5±50.5	975.6±36.7

Table 26.4 The absolute values for CSA at rest (*Pre*), immediately after (*Post*), and 1 (*1D*), 2 (*2D*), 3 (*3D*), and 7 (*7D*) days after the session of eccentric knee flexion at 50 % of the 1-RM

	Pre	Post	1D	2D	3D	7D
CSA (mm^2) BFlh	1,303.9±76.5	1,375.9±74.9	1,326.4±70.5	1,317.2±68.9	1,352.0±78.6	1,308.1±74.0
ST	784.7±43.7	955.1±51.5	809.5±40.3	854.6±50.1	836.8±48.5	832.9±39.2
SM	1,009.7±59.9	1,032.9±39.7	1,005.0±42.4	1,023.1±51.9	970.9±32.3	1,003.8±35.3

in Tables 26.3 and 26.4. Figures 26.6 and 26.7 present changes in averaged CSA values of the individual muscles for each exercise. The degrees of CSA value changes were different between the muscles and the percentage changes in the ST muscle was significantly greater than those in the BFlh and SM muscles 3 days after 120 % 1RM exercise and immediate after 50 % 1RM exercise ($P<0.01$).

Our results showed that the NiEMG values during 120 % ECC were greater in the ST muscle than in the SM muscle. In addition, immediately after exercise, the change in the T2 value in the ST muscle was greater than in the BFlh and SM muscles. This finding supports a selective recruitment of the ST muscle during intensive eccentric knee flexion exercise.

In general, since the maximum torque produced by a skeletal muscle depends on the physiological cross-sectional area, and the muscle fiber tension per an area is constant, pennate muscles are more suitable for torque production than fusiform muscles. Nevertheless, as shown by the results of this study, the ST muscle, which is a long, thin fusiform muscle, showed a higher level of activation during the intensive eccentric knee flexion exercise than the BFlh or SM muscles, which are bulky pennate muscles. That is, the ST might have a higher level of sensitivity to an intensive eccentric knee flexion load. The ST is characterized by a long fiber length containing many sarcomeres in series, and thus illustrates a notable potential for them to shorten quickly at long distances; therefore, it is considered that such morphological properties allow it to selectively deal with this type of exercise more efficiently. Baczkowski et al. (2006) reported that the ST muscle showed a great increase in MR SI after 100 Australian Rules football kicks, in which the hamstring muscles of both legs, during kicking, act eccentrically, whereas the BFlh showed minimal change. This finding also implies a minor functional characteristic of the ST muscle that contributes to eccentric knee flexion. In the present study, however, the T2 shifts in SI that reflected changes in fluid distribution within the muscle were

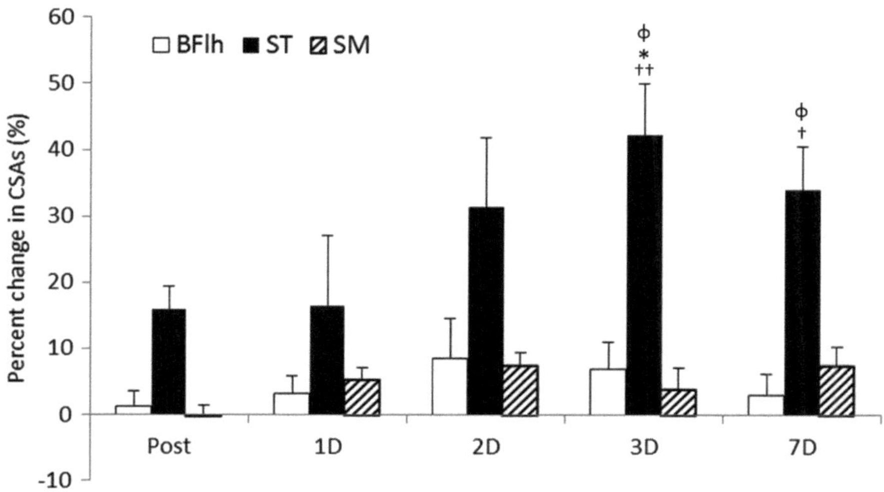

Fig. 26.6 Percentage change in the CSAs of each muscle with time immediately after (*Post*), and 1 (*1D*), 2 (*2D*), 3 (*3D*), and 7 (*7D*) days after the session of eccentric knee flexion at 120 % of the 1RM. Values are the percentage change compared with the value at rest. $^{\dagger}P<0.05$, $^{\dagger\dagger}P<0.01$ vs the values at rest; $\varphi P<0.05$ vs BFlh; $^{*}P<0.05$ vs SM. *BFlh* biceps femoris long head, *ST* semitendinosus, *SM* semimembranosus

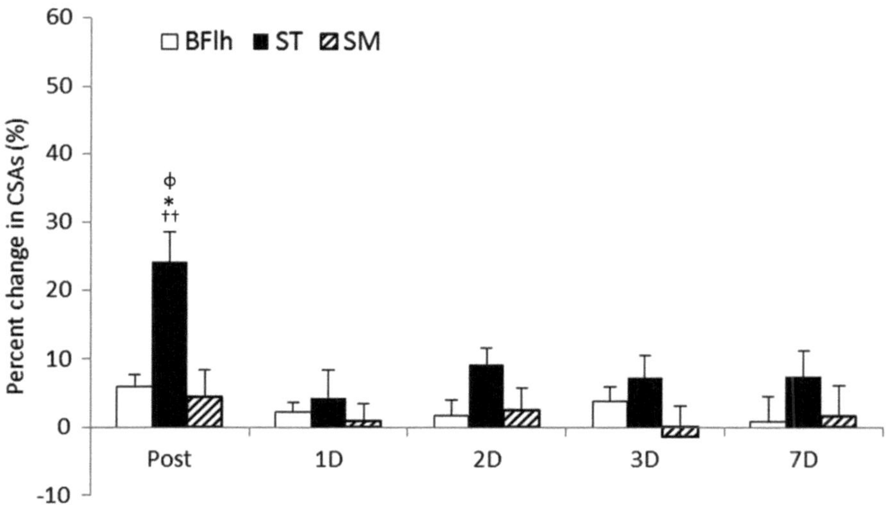

Fig. 26.7 Percentage change in the CSAs of each muscle with time immediately after (*Post*), and 1 (*1D*), 2 (*2D*), 3 (*3D*), and 7 (*7D*) days after the session of eccentric knee flexion at 50 % of the 1RM. Values are the percentage change compared with the value at rest. $^{\dagger\dagger}P<0.01$ vs the values at rest; $\varphi P<0.05$ vs BFlh; $^{*}P<0.05$ vs SM. *BFlh* biceps femoris long head, *ST* semitendinosus, *SM* semimembranosus

not consistent with the EMG activity during the 50 % 1RM exercise. Kinugasa and Akima (2005) suggested that mfMRI signals and iEMG activity were correlated with workload in individual triceps surae muscles during a repetitive plantar-flexion exercise, and mfMRI signals were associated with neuromuscular activity reflected in iEMG. From the results of this study, we determined that the activation patterns among the hamstring muscles during eccentric/concentric knee flexion exercises were nonuniform, and that the effect on each muscle's morphology and function caused by eccentric exercise also would be different. In sports scenes, various strength training exercises are performed that involve the hamstring muscles, such as the "leg curl" or "Nordic hamstring lowers," for the purpose of performance improvement as well as injury prevention and rehabilitation (Andersen et al. 2006; Arnason et al. 2008; Mesfar and Shirazi-Adl 2008). Clinically, the most commonly reported strain of the hamstring muscles has been in the BFlh (Askling et al. 2007; Brockett et al. 2004; Woods et al. 2004). The results of the present study, however, showed that the degree of activation was greater in the ST muscle than in the BFlh and SM muscles during eccentric/concentric knee flexion exercises. Consequently, we have suggested that knee flexion exercise might be insufficient for prevention and rehabilitation of strain injury in the BFlh muscle. It may be possible to improve muscle function more efficiently by selecting the task-considered load intensity, contraction pattern, and morphological property of target muscle recruited during the exercise.

26.3 Activation Patterns of Hamstring Muscles During Eccentric Hip Extension Exercises and Differences in the Damage After Exercise

In 2011, we examined the recruitment patterns of each hamstring muscle during a hip extension exercise and evaluated differences in the time-course changes of the MRI measurements after the exercise (Ono et al. 2011). The subjects were six male volunteers who underwent a session of hip extension exercises: These exercises are generally called "stiff-leg deadlift." A session consisted of 5 sets of 10 repetitions of the exercise. Initially, the subjects were instructed to stand upright with their feet shoulder-width apart, torso erect (chest out, scapulae in an adducted position), and to grasp a bar loaded at 60 % of their body weight. With the weight hanging at arm's length, the subjects were instructed to bend forward from the hip over a 2 s count to smoothly lower the weight while maintaining the torso rigid throughout the hip flexion. Hip flexion continued until the trunk was approximately parallel to the floor. From this position, the subjects were told to smoothly return to the initial position over a 2 s count while maintaining a rigid torso. This exercise was properly performed by maintaining a flat trunk and bending from the hip to a position where the trunk was parallel to the floor and the knees fully extended. There was a 1-min rest between each set of 10 repetitions during which the subjects rested in a standing position while not holding the weight.

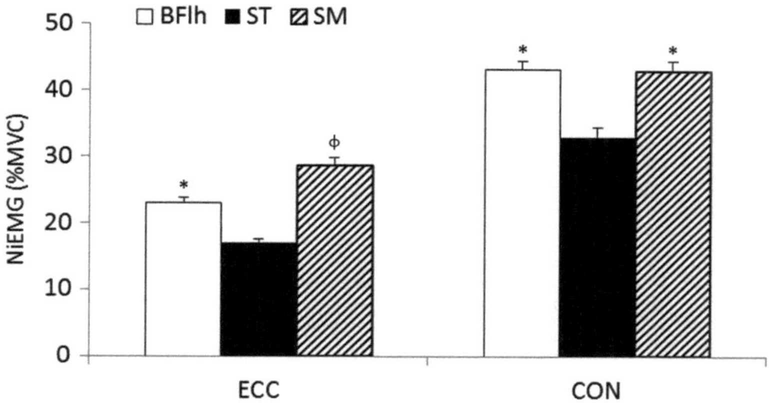

Fig. 26.8 The NiEMG values in eccentric (*ECC*) and concentric (*CON*) phases of hip extension exercise. Values refer to the mean ± the standard error (*SE*). $^{*}P<0.01$ vs ST, $\varphi P<0.01$ vs the other muscles. *BFlh* biceps femoris long head, *ST* semitendinosus, *SM* semimembranosus

EMG activities during the exercise were recorded from the midbellies of the BFlh, ST, and SM muscles in the right limb at a sampling rate of 1 kHz. The digital EMG signals were full-wave rectified and integrated (iEMG) over each period of eccentric and concentric contraction (10 repetitions × 5 sets = 50 for each muscle in total), which was determined by the digital signal markers recorded with the EMG signals. The iEMG values were normalized to the values during maximum voluntary contraction (MVC) (NiEMG = iEMG/MVC [%]) for each subject. NiEMG values during the exercise are shown in Fig. 26.8. During the eccentric phases, the NiEMG values of the BFlh and SM muscles were significantly higher than that of the ST muscle ($P<0.01$),. Activity of the SM muscle was also higher than that of the BFlh muscle ($P<0.01$). During the concentric phases, the activity of the BFlh and SM muscles were still significantly higher than that of the ST muscle ($P<0.01$), but no significant difference was observed between BFlh and SM muscle activity.

Immediately (within 5 min) after performing the exercise session, the subjects underwent MRI of the right limb. The MRI T2 value was measured to evaluate the degree of muscle recruitment and damage. All MRI images of the participants' thighs were obtained by using a 1.5-T MR body coil imaging system. Transverse T2-weighted spin-echo images were subsequently obtained. The MRI data were evaluated for theT2 relaxation time (T2 value) of the hamstring muscles and the signal intensity (SI) of each muscle (BFlh, ST, and SM) was measured. A T2 measurement sequence with four TEs (25, 50, 75, and 100 ms) was applied to measure the absolute T2 value. Images taken at different TEs were fit to a monoexponential time curve to extract the T2 values based on the formula: $SI = M0 \times \exp(-TE/T2)$, where SI represents the SI at a given TE and M0 is the original MR SI. For comparison purposes, the averaged T2 mean value of each muscle was computed as a percentage change increase relative to the value measured before exercise. Typical T2-weighted MR images of the right thigh before and after the exercise are shown in Fig. 26.9; the absolute values for T2

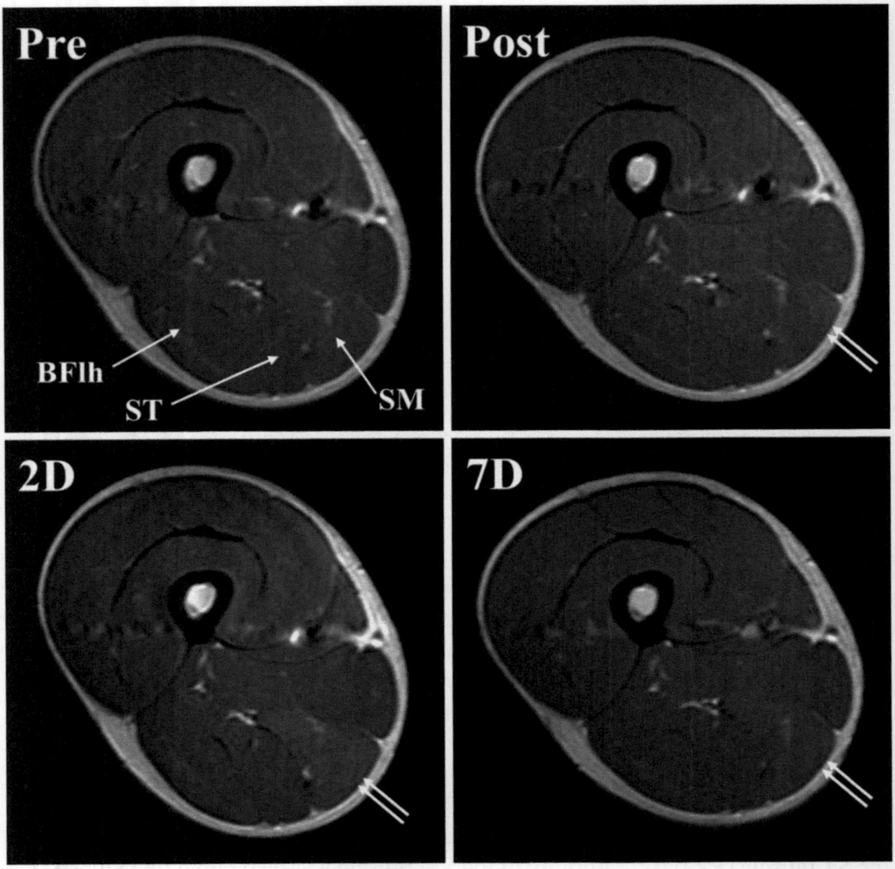

Fig. 26.9 Typical T2-weighted MR images (TR = 3,000 ms, TE = 25 ms) of the middle region (50 % of the thigh length) of the thigh at rest (*Pre*), immediately after (*Post*), and 2 (*2D*), and 7 (*7D*) days after the session of hip extension exercise. After the exercise, obvious SI increases were visible within the SM muscles (*double white arrow*). Images were obtained from 1 representative subject. *BFlh* biceps femoris long head, *ST* semitendinosus, *SM* semimembranosus

at the exercise are shown in Table 26.5. Figure 26.10 depicts changes in average T2 values of each muscle as a function of time. The averaged T2 mean value immediately after evaluation increased only in the case of the SM muscle ($P<0.05$) and no significant change was identified in the other muscles. There was no significant change in any muscle at 2 and 7 days after the exercise. The absolute values of CSA for the exercise are shown in Table 26.6. Figure 26.11 displays the time course of the changes in the CSAs of each muscle. Immediately after the exercise no significant change was detected in any muscle, but 2 and 7 days after exercise, a significant increase was identified in the case of the SM muscle ($P<0.05$). No significant changes was identified in any of the other muscles.

Table 26.5 The absolute values for T2 at rest (*Pre*), immediately after (*Post*), and 2 (*2D*), and 7 (*7D*) days after the session of hip extension exercise

	Pre	Post	2D	7D
T2 (ms) BFlh	31.2±0.9	33.7±1.0	31.2±1.0	31.6±0.9
ST	29.9±0.4	31.5±0.7	30.1±0.6	30.1±0.7
SM	30.8±0.7	34.7±1.3	31.6±0.8	32.3±0.9

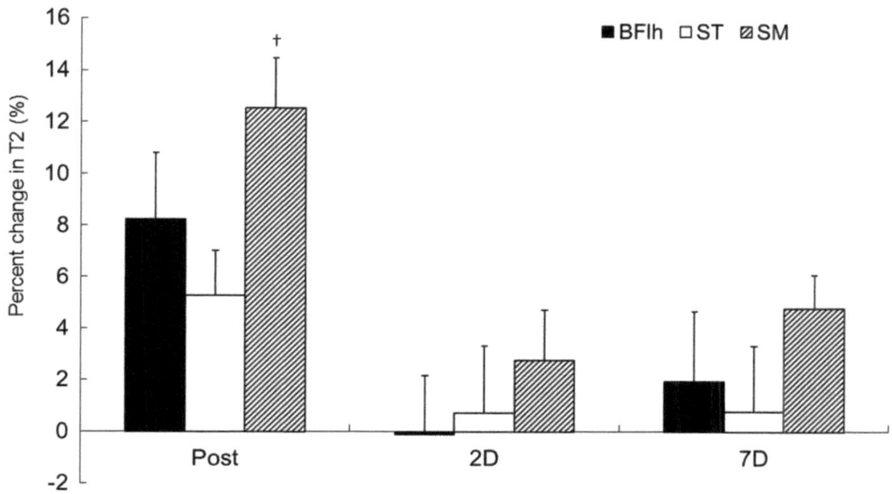

Fig. 26.10 Percentage change in the T2 value of each muscle with time immediately after (*Post*), and 2 (*2D*), and 7 (*7D*) days after the session of hip extension exercise. Values are the percentage change compared with the value at rest (*Pre*); $^{\dagger}P<0.05$ vs *Pre*

Table 26.6 The absolute values for CSA at rest (*Pre*), immediately after (*Post*), and 2 (*2D*), and 7 (*7D*) days after the session of hip extension exercise

	Pre	Post	2D	7D
CSA (mm^2) BFlh	1,255.6±106.2	1,255.2±76.3	1,277.1±51.5	1,275.4±87.5
ST	1,004.4±24.4	1,016.6±21.3	1,004.2±28.3	1,010.1±21.9
SM	856.0±88.1	1,012.3±89.0	1,060.6±94.3	1,085.7±79.4

These results suggest that the recruitment patterns of each hamstring muscle during hip extension are not uniform among the hamstring muscles, and that the BFlh and SM muscles are selectively recruited to a greater extent. Immediately after the exercise, the T2 mean value increased, especially in the case of the SM muscle. As Kinugasa and Akima (2005) report, the MRI signal correlated with EMG activity during repetitive exercise, and the relationships were associated with both neuromuscular and metabolic factors during the exercise. Hence, the T2 increase in the present study is in accordance with the EMG data, and it supports this data from the

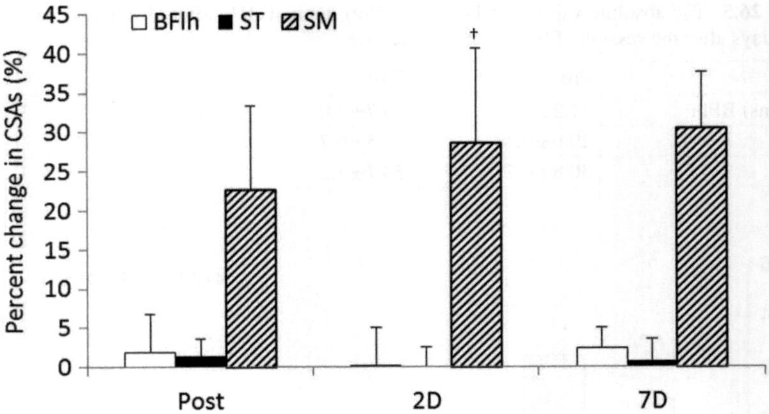

Fig. 26.11 Percentage change in the CSA value of each muscle with time immediately after (*Post*), and 2 (*2D*), and 7 (*7D*) days after the session of hip extension exercise. Values are the percentage change compared with the value at rest (*Pre*); $^{\dagger}P<0.05$ vs Pre

metabolic viewpoint in vivo. In our previous study (Kubota et al. 2007), we showed that the ST muscle responded sensitively to eccentric knee flexion load. The rationale behind the selective recruitment of this muscle was that the morphological properties of this muscle are such that it can effectively deal with the strain that occurs during this type of exercise. The ST muscle possesses long fibers containing many sarcomeres in series, illustrating its notable potential to contract quickly over large distances (Heron and Richmond 1993). In contrast, during the hip extension exercise, the degree of activation was higher for the BFlh and SM muscles than for the ST muscle in both the eccentric and concentric phases. The BFlh and SM muscles are pennate muscles, which have large CSAs and are more suitable for torque production than are the fusiform muscles (Koulouris and Connell 2005; Lieber and Friden 2000; Woodley and Mercer 2005). Architecturally, the BFlh and SM muscles originate from the ischial tuberosity and control the movement of the pelvis and torso. On the other hand,, the architecture of the SM muscle is unique. The SM muscle possesses the longest proximal tendon of all the hamstring muscles. It originates beneath the common tendon of the BFlh and ST muscles. The belly of the SM muscle overlaps that of the ST muscle, and the SM muscle then inserts into the medial tibial condyle as a distal tendon. Thus, the SM muscle has a relatively small moment arm at the hip joint and a large moment arm at the knee joint. On the basis of these morphological and architectural properties, our findings suggest that the BFlh and SM muscles are selectively recruited in order to deal with hip joint movements during standing, bending forward, and extending backward from the hip, as these movements demand a high muscle torque. The CSAs values reveal that the SM muscle exhibits a conspicuous response following hip extension exercise, and we thus assumed that this muscle is recruited and stimulated during repetitive eccentric and concentric hip extension exercise. There were no significant changes, however, in the T2 values for any muscle 2 and 7 days after exercise, which indicates that there was no muscle damage. To explain this result, i.e., the unchanged T2 values, we assume that the activities of the hamstring muscles during the exercise were not high relative to those during MVC, especially during the eccentric phase; so even the highest values in the case of the SM muscle were approximately 30 % of the MVC value. From the results of this study and our previous study (Kubota et al. 2007; Ono et al. 2010), we argue that the function of each hamstring muscle is different during hip extension and knee flexion movement and that the BFlh and SM muscles are selectively recruited in order to deal with hip joint movements during standing, bending forward, and extending backward from the hip. In contrast, it is generally known that most acute hamstring strains involve the BFlh, whereas the ST and SM muscles are less often injured (Brooks et al. 2006; Woods et al. 2004) and that most of the injuries occur during the late-swing phase and stance phase of sprinting (Montgomery et al. 1994; Verrall et al. 2001) during which the hamstring muscles contract eccentrically to decelerate the forward swing of the leg and concentrically to push off the ground. In light of these considerations, we suggest that most of the acute cases of hamstring strains that involve the BFlh happen during the stance phase of sprinting during which the hamstring muscles contract concentrically to push off the ground, producing hip extension torque. Further studies are needed to clarify the activation pattern of each hamstring muscle and kinematics of sprinting on the ground. This clarification will allow us to understand the mechanism of hamstring muscle strain injuries and thus to develop more practical training methods to prevent them.

References

Adams GR, Duvoisin MR, Dudley GA (1992) Magnetic resonance imaging and electromyography as indexes of muscle function. J Appl Physiol 73:1578–1583

Andersen LL, Magnusson SP, Nielsen M, Haleem J, Poulsen K, Aagaard P (2006) Neuromuscular activation in conventional therapeutic exercises and heavy resistance exercises: Implications for rehabilitation. Phys Ther 86:683–697

Arnason A, Andersen TE, Holme I, Engebretsen L, Bahr R (2008) Prevention of hamstring strains in elite soccer: an intervention study. Scand J Med Sci Sports 18:40–48

Askling CM, Tengvar M, Saartok T, Thorstensson A (2007) Acute first-time hamstring strains during high-speed running. Am J Sports Med 35:197–206

Baczkowski K, Marks P, Silberstein M, Schneider-Kolsky ME (2006) A new look into kicking a football: an investigation of muscle activity using MRI. Australas Radiol 50:324–329

Brockett CL, Morgan DL, Proske U (2004) Predicting hamstring strain injury in elite athletes. Med Sci Sports Exerc 36:379–387

Brooks JH, Fuller CW, Kemp SP, Reddin DB (2006) Incidence, risk, and prevention of hamstring muscle injuries in professional rugby union. Am J Sports Med 34:1297–1306

Friederich JA, Brand RA (1990) Muscle fiber architecture in the human lower limb. J Biomech 23:91–95

Hather BM, Tesch PA, Buchanan P, Dudley GA (1991) Influence of eccentric actions on skeletal muscle adaptations to resistance training. Acta Physiol Scand 143:177–185

Heron MI, Richmond FJR (1993) In-series fiber architecture in long human muscles. J Morphol 216:35–45

Hortobágyi T, Barrier J, Beard D, Braspenningx J, Koens P, Devita P, Dempsey L, Lambert J (1996) Greater initial adaptations to submaximal muscle lengthening than maximal shortening. J Appl Physiol 81:1677–1682

Kinugasa R, Akima H (2005) Neuromuscular activation of triceps surae using muscle functional MRI and EMG. Med Sci Sports Exerc 37:593–598

Koulouris G, Connell D (2005) Hamstring muscle complex: an imaging review. Radiographics 25:571–586

Kubota J, Ono T, Araki M, Tawara N, Torii S, Okuwaki T, Fukubayashi T (2007) Nonuniform changes in magnetic resonance measurements of the semitendinosus muscle following intensive eccentric exercise. Eur J Appl Physiol 101:713–720

Lieber RL, Friden J (2000) Functional and clinical significance of skeletal muscle architecture. Muscle Nerve 23:1647–1666

Makihara Y, Nishino A, Fukubayashi T, Kanamori A (2006) Decrease of knee flexion torque in patients with ACL reconstruction: combined analysis of the architecture and function of the knee flexor muscles. Knee Surg Sports Traumatol Arthrosc 14:310–317

Mesfar W, Shirazi-Adl A (2008) Knee joint biomechanics in open-kinetic-chain flexion exercises. Clin Biomech 23:477–482

Mohamed O, Perry J, Hislop H (2002) Relationship between wire EMG activity, muscle length, and torque of the hamstrings. Clin Biomech 17:569–579

Montgomery WH 3rd, Pink M, Perry J (1994) Electromyographic analysis of hip and knee musculature during running. Am J Sports Med 22:272–278

Onishi H, Yagi R, Oyama M, Akasaka K, Ihashi K, Handa Y (2002) EMG-angle relationship of the hamstring muscles during maximum knee flexion. J Electromyogr Kinesiol 12:399–406

Ono T, Okuwaki T, Fukubayashi T (2010) Differences in activation patterns of knee flexor muscles during eccentric exercises. Res Sports Med Int J 18:188–198

Ono T, Higashihara A, Fukubayashi T (2011) Hamstring functions during hip-extension exercise assessed with electromyography and magnetic resonance imaging. Res Sports Med Int J 19:42–52

Thelen DG, Chumanov ES, Hoerth DM, Best TM, Swanson SC, Li L, Young M, Heiderscheit BC (2005) Hamstring muscle kinematics during treadmill sprinting. Med Sci Sports Exerc 37:108–114

Verrall GM, Slavotinek JP, Barnes PG, Fon GT, Spriggins AJ (2001) Clinical risk factors for hamstring muscle strain injury: a prospective study with correlation of injury by magnetic resonance imaging. Br J Sports Med 35:435–439; discussion 440

Woodley SJ, Mercer SR (2005) Hamstring muscles: architecture and innervation. Cells Tissues Organs 179:125–141

Woods C, Hawkins RD, Maltby S, Hulse M, Thomas A, Hodson A (2004) The football association medical research programme: an audit of injuries in professional football: analysis of hamstring injuries. Br J Sports Med 38:36–41

Worrell TW, Karst G, Adamczyk D, Moore R (2001) Influence of joint position on electromyographic and torque generation during maximal voluntary isometric contractions of the hamstrings and gluteus maximus muscles. J Orthop Sports Phys Ther 31:730–740

Chapter 27
Risk Factors and Prevention of Hamstring Strain

Chihiro Fukutome and Toru Fukubayashi

Abstract Muscle strain is common in many sports and occurs most often in the hamstring muscle. Because hamstring strain occurrence and recurrence rates are high, many studies of risk factors and prevention programs have been reported to date. In this chapter we review the current knowledge about hamstring strains and reveal some new findings. A number of studies have suggested that increasing age and previous muscle strain increase the muscle strain recurrence rate. However, at present there is little evidence-based research relating to muscle strength, flexibility or fatigue. Due to this lack, effective hamstring strain prevention programs have not been established. Future studies are needed to enable the establishment of effective strain prevention protocols.

Keywords Hamstring strain • Risk factor • Prevention

27.1 Introduction

Muscle strain is a common injury in sports. Epidemiological data obtained from a professional soccer team in Europe over the past 11 years suggests that muscle strain is the most common injury (Ekstrand et al. 2013). Hamstring injuries comprise the most common type of muscle strain (Hawkins et al. 2001; Alonso et al. 2009). Many hamstring strains occur in Australian football (Orchard and Seward 2002; Gabbe et al. 2006a), rugby (Devlin 2000; Brooks et al. 2006), American football (Carling et al. 2011; Elliott et al. 2011), soccer (Kucera et al. 2011; Ekstrand et al. 2011), and track and field (Alonso et al. 2009; Opar et al. 2014).

An unfortunate characteristic of muscle strain is its high rate of recurrence. Orchard and Seward (2002) reported that 34 % of athletes who experience hamstring

C. Fukutome (✉)
Graduate School of Sport Sciences, Waseda University, Saitama, Japan
e-mail: cfukutome.46@gmail.com

T. Fukubayashi
Faculty of Sport Sciences, Waseda University, Saitama, Japan
e-mail: fukubayashi@waseda.jp

K. Kanosue et al. (eds.), *Sports Injuries and Prevention*,
DOI 10.1007/978-4-431-55318-2_27

strain are injured again in the same season. Although it is clear that muscle strain, and strain of the hamstring in particular, has high injury and re-injury rates, the rate of muscle strain injuries has neither increased nor decreased (Opar et al. 2012; Ekstrand et al. 2013).

Many researchers have investigated the risk factors of hamstring strain, and there have been a number of proposed prevention programs. In this chapter, we discuss both of these topics in detail.

27.2 Risk Factors

27.2.1 Anatomy

It is well known that the incidence of biceps femoris injury, and that of the long head in particular, is much greater than that of the other hamstring muscles (Gibbs et al. 2004; Cohen et al. 2011). The long head of the biceps femoris has a unique architecture consisting of a short fascicle, muscle tendon junction, and aponeurosis (Woodley and Mercer 2005). Longer fascicles enable greater muscle lengthening and reduce the risk of over-lengthening during eccentric contractions (Brockett et al. 2004; Butterfield 2010). However, the long head of the biceps femoris has short fascicles.

Magnetic resonance imaging has shown that muscle strain often occurs at the muscle tendon junction (De Smet and Best 2000). The biceps femoris muscle has a larger muscle tendon junction (Woodley and Mercer 2005). A three-dimensional model showed that the area near the proximal muscle tendon junction was the most stretched during muscle lengthening (Rehorn and Blemker 2010). In this chapter we report that the proximal aponeurosis of the biceps femoris is narrower than the distal aponeurosis. These unique anatomical features of the long head of the biceps femoris most likely contribute to the high incidence of strain injury.

27.2.2 Hamstring Muscle Strength

Hamstring muscle strength is thought to be a risk factor for strain (Agre 1985). Many investigators have investigated the correlation between muscle strength and muscle strain. To date, they have suggested possible causes of hamstring strain that include imbalance of the bilateral leg (Croisier et al. 2008) as well as a lower strength compared to that of the quadriceps (H:Q ratio) (Yeung et al. 2009).

Some prospective studies have shown that in the presence of muscle imbalances between the right and left legs, hamstring strain occurred in the weaker leg. These findings suggested that an 8–15 % difference in hamstring strength between the two sides was a cause of increased hamstring injury rates in a number of sports (Orchard et al. 1997; Croisier et al. 2008). One causal factor could be that hamstring strength asymmetry alters running biomechanics and thus affects hamstring loading during the terminal swing phase (Opar et al. 2012).

Investigators commonly state that a hamstrings:quadriceps (H:Q) ratio ≥0.6 reduces the hamstring strain rate (Yeung et al. 2009). H:Q ratio has been often measured in concentric contraction. However, conventional H:Q ratios do not reflect the role of the hamstring during the terminal swing phase of running, the point at which hamstring strain occurs frequently (Verrall et al. 2001). As such, some investigators have recently measured the functional H:Q ratio (Fousekis et al. 2011; Sugiura et al. 2008). The functional H:Q ratio compares the function of the eccentric hamstrings to the concentric quadriceps. A study of professional soccer players found that players with a strength imbalance (functional H:Q ratio ≤0.89) had a significantly higher hamstring strain rate than players without strength imbalances (Croisier et al. 2008). However, the correlation between the isokinetic H:Q ratio and hamstring strain is weak if the measured speed and contraction type are not considered. Orchard et al. (1997) report that although H:Q ratios measured at 180 and 300°/s did not differ between the injured and non-injured groups, that measured at 60°/s was significantly different between groups . Thus, both contraction type and measuring speed should be considered in future studies.

27.2.3 Fatigue

Some studies have reported that hamstring strain occurs at a higher rate in the latter half of matches (Woods et al. 2004; Brooks et al. 2006). Due to such reports, fatigue has been proposed as a risk factor for hamstring strain. Repeated over-lengthening of the muscles causes microscopic damage which may lead to muscle strain (Morgan 1990). Another study showed that in muscles fatigued by repeated eccentric contraction, the capacity of absorbing energy to resist overstretching was decreased compared with that of non-fatigued muscles (Mair et al. 1996). These studies used animal muscles.

In humans, fatigued hamstring muscles lead to an increased magnitude of knee extension angle during the terminal swing phase of running (Pinniger et al. 2000). An increasing knee extension angle may lead to overstretching during the running phase.

Problems involving muscle fatigue may emanate from other organ systems such as the nervous control network (Opar et al. 2012). Fatigued athletes display reduced motor control, which may lead to mistimed contractions of the biceps femoris (Agre 1985). Fatigue also decreases an athlete's ability to concentrate, which increases their risk of injury in many regards (Devlin 2000). However, further studies examining the correlation between fatigue and hamstring strain are needed.

27.2.4 Flexibility

Poor hamstring flexibility may increase the risk of strain injury due to a reduced capacity of the muscle–tendon unit to absorb lengthening forces (Clark 2008). Some studies have reported that poor hamstring flexibility is related to muscle injury (Jonhagen et al. 1994; Witvrouw et al. 2003), while other studies found no correlation between poor hamstring flexibility and strain risk (Arnason et al. 2004; Gabbe et al. 2006a). Accordingly, it is unclear whether flexibility is a risk factor of hamstring strain. Unfortunately, earlier reports used different methods to assess flexibility, there is as of yet no "gold-standard" method for precisely measuring either hamstring flexibility or length (Foreman et al. 2006). Tests such as the sit-and-reach or the passive/active knee extension are influenced by hip or pelvic motion, so they cannot precisely measure hamstring flexibility. Thus, a more quantitative, or at a minimum, standardized method of determining flexibility is required if its potential role as a risk factor is to be determined.

27.2.5 Age

Increasing age is considered a risk factor for hamstring strain by many investigators (Woods et al. 2004; Gabbe et al. 2006a). Unlike other risk factors, many studies have demonstrated that increasing age is related to hamstring strain prevalence. In fact, using multivariate analysis, a number of studies have clearly indicated that increasing age is a definite risk factor (Verrall et al. 2001; Orchard 2001).

Freckleton et al. (2013) performed a meta-analysis to evaluate the correlation between increasing age and hamstring strain. They reported that increasing age increases the risk of hamstring injury (odds ratio, 2.46; 95 % confidence interval, 0.98–6.14; $p=0.06$)***Assuming a relatively linear time course, an rough estimate of the rate of increase in hamstring strain probability over time, say over a 10 or 20 year increase in age, would be of interest to most readers. Ie: "Their data indicate that a 10 year increase in age increases the likelihood of hamstring strain by a factor of x.x."***

The reason for the increase in injury rate with increasing age is unclear. One study suggested that the lumbar nerve roots of L5/S1 tend to be more affected by age-related spinal degeneration than other lumbar nerve roots (Orchard et al. 2004). The lumbar nerve roots of L5/S1 control the hamstring muscles, so nerve degeneration caused by aging is thought to be a reason for the increasing hamstring injury rate. However, there is no scientific evidence of this suggestion. As such, further studies investigating why older age increases the occurrence rate of hamstring strain are required to elucidate the correlation between age and hamstring strain rate.

27.2.6 *Previous Injury*

A number of studies have suggested that a previous muscle strain increases the muscle strain recurrence rate (Orchard 2001; Croisier et al. 2008). In fact, meta-analyses indicate that athletes with a previous hamstring injury are 4.06 times more likely to experience repeat injury (Freckleton and Pizzari 2013). Previous injuries induce a change in knee angle produced peak torque (Brockett et al. 2004), and decreased tendon tissue compliance (Silder et al. 2010). Neuromuscular control is also altered by previous injury. Opar et al. (2013) suggest that the affected limb of athletes with previous unilateral hamstring strain has a lower surface electromyogram level than that of the contralateral limb. From this result, they state that previous muscle strain alters neuromuscular function. Thus, adequate rehabilitation is needed to prevent future injury.

27.3 Prevention

To date, many researchers have attempted to establish viable prevention programs for hamstring strain (Engebretsen et al. 2008; Arnason et al. 2008). Many of the proposed protocols have focused on muscle strength or flexibility.

Compared with concentric training (i.e., leg curl exercise), eccentric training increases the eccentric torque (Mjolsnes et al. 2004). Therefore, some studies that focused on eccentric muscle training showed that it decreases muscle strain rates (Askling et al. 2003). In eccentric training, Nordic hamstring training, a technique that uses one's own body weight, has gained popularity. An athlete assumes a kneeling position, a partner fixes their ankles, and the athlete slowly moves their upper body toward the ground. A randomized controlled trial examined the effect of the Nordic hamstring training on hamstring strain rates (Petersen et al. 2011). Fifty Danish male soccer teams were divided into training or control groups randomly. Ten-weeks of eccentric exercise with the Nordic hamstring training was performed during the players' midseason break. The rate of acute hamstring injury per 100 player-seasons in the training versus control group was 3.8 versus 13.1. On the other hand, Gabbe et al. (2006b), who studied the effect of Nordic hamstring training on community-level Australian football players, found that the hamstring strain rate did not different significantly between the training and control groups. Although the protocol and the participants' sports differed between these two previous reports, the effect of Nordic hamstring training on muscle strain injury rate is obviously not clear and should be clarified by further research.

Another approach to preventing hamstring strain is by increasing flexibility. An earlier study reported that a program composed of stretching and Nordic Hamstring training reduced the rate of hamstring strain (Arnason et al. 2008). However the stretching program itself affected the hamstring strain rate. As stated above, evidence that lower flexibility increases the rate of hamstring strain has not been established. Thus, the effect of lack of flexibility on the hamstring strain rate should be investigated to determine whether improving flexibility can prevent hamstring strain.

27.4 Conclusion

Many researchers have investigated risk factors and proposed potential prevention programs for hamstring strain. However, their opinions have been divided on the correlations between the various risk factors and hamstring strain. Despite the fact that the overall hamstring strain risk is generally felt to involve a number of separate risk factors, many earlier reports have focused on the effect of a single risk factor. In turn, many of the prevention programs have also focused on a specific risk factor. This could be one of the reasons why no clearly effective prevention program has yet been established.

In the future, more detailed studies are needed to reveal the relative effect of the various risk factors on the occurrence rate of hamstring strain. Then, a prevention program could be created which took into account the relative importance of the involved risk factors which lead to hamstring strain.

References

Agre JC (1985) Hamstring injuries. Proposed aetiological factors, prevention, and treatment. Sports Med 2(1):21–33

Alonso JM, Junge A, Renström P, Engebretsen L, Mountjoy M, Dvorak J (2009) Sports injuries surveillance during the 2007 IAAF World Athletics Championships. Clin J Sport Med 19(1):26

Arnason A, Sigurdsson SB, Gudmundsson A, Holme I, Engebretsen L, Bahr R (2004) Risk factors for injuries in football. Am J Sports Med 32(1 Suppl):5S–16S

Arnason A, Andersen TE, Holme I, Engebretsen L, Bahr R (2008) Prevention of hamstring strains in elite soccer: an intervention study. Scand J Med Sci Sports 18(1):40–48. doi:10.1111/j.1600-0838.2006.00634.x, SMS634 [pii]

Askling C, Karlsson J, Thorstensson A (2003) Hamstring injury occurrence in elite soccer players after preseason strength training with eccentric overload. Scand J Med Sci Sports 13(4):244–250, doi:0312 [pii]

Brockett CL, Morgan DL, Proske U (2004) Predicting hamstring strain injury in elite athletes. Med Sci Sports Exerc 36(3):379–387, doi:00005768-200403000-00005 [pii]

Brooks JHM, Fuller CW, Kemp SPT, Reddin DB (2006) Incidence, risk, and prevention of hamstring muscle injuries in professional rugby union. Am J Sports Med 34(8):1297–1306

Butterfield TA (2010) Eccentric exercise in vivo: strain-induced muscle damage and adaptation in a stable system. Exerc Sport Sci Rev 38(2):51–60. doi:10.1097/JES.0b013e3181d496eb

Carling C, Le Gall F, Orhant E (2011) A four-season prospective study of muscle strain reoccurrences in a professional football club. Res Sports Med 19(2):92–102. doi:10.1080/15438627.2011.556494, 936097287 [pii]

Clark RA (2008) Hamstring injuries: risk assessment and injury prevention. Ann Acad Med Singapore 37(4):341–346

Cohen SB, Towers JD, Zoga A, Irrgang JJ, Makda J, Deluca PF, Bradley JP (2011) Hamstring injuries in professional football players: magnetic resonance imaging correlation with return to play. Sports Health Multidiscip Approach 3(5):423–430

Croisier J-L, Ganteaume S, Binet J, Genty M, Ferret J-M (2008) Strength imbalances and prevention of hamstring injury in professional soccer players: a prospective study. Am J Sports Med 36(8):1469–1475

De Smet AA, Best TM (2000) MR imaging of the distribution and location of acute hamstring injuries in athletes. AJR Am J Roentgenol 174(2):393–399

Devlin L (2000) Recurrent posterior thigh symptoms detrimental to performance in rugby union: predisposing factors. Sports Med 29(4):273–287

Ekstrand J, Hagglund M, Walden M (2011) Epidemiology of muscle injuries in professional football (soccer). Am J Sports Med 39(6):1226–1232. doi:10.1177/0363546510395879, 0363546510395879 [pii]

Ekstrand J, Hägglund M, Kristenson K, Magnusson H, Waldén M (2013) Fewer ligament injuries but no preventive effect on muscle injuries and severe injuries: an 11-year follow-up of the UEFA Champions League injury study. Br J Sports Med 47(12):732–737

Elliott MC, Zarins B, Powell JW, Kenyon CD (2011) Hamstring muscle strains in professional football players: a 10-year review. Am J Sports Med 39(4):843–850. doi:10.1177/0363546510394647

Engebretsen AH, Myklebust G, Holme I, Engebretsen L, Bahr R (2008) Prevention of injuries among male soccer players: a prospective, randomized intervention study targeting players with previous injuries or reduced function. Am J Sports Med 36(6):1052–1060

Foreman T, Addy T, Baker S, Burns J, Hill N, Madden T (2006) Prospective studies into the causation of hamstring injuries in sport: a systematic review. Phys Ther Sport 7(2):101–109

Fousekis K, Tsepis E, Poulmedis P, Athanasopoulos S, Vagenas G (2011) Intrinsic risk factors of non-contact quadriceps and hamstring strains in soccer: a prospective study of 100 professional players. Br J Sports Med 45(9):709–714. doi:10.1136/bjsm.2010.077560, bjsm.2010.077560 [pii]

Freckleton G, Pizzari T (2013) Risk factors for hamstring muscle strain injury in sport: a systematic review and meta-analysis. Br J Sports Med 47(6):1–9

Gabbe B, Bennell K, Finch C, Wajswelner H, Orchard J (2006a) Predictors of hamstring injury at the elite level of Australian football. Scand J Med Sci Sports 16(1):7–13

Gabbe BJ, Branson R, Bennell KL (2006b) A pilot randomised controlled trial of eccentric exercise to prevent hamstring injuries in community-level Australian Football. J Sci Med Sport 9(1–2):103–109. http://dx.doi.org/10.1016/j.jsams.2006.02.001

Gibbs NJ, Cross TM, Cameron M, Houang MT (2004) The accuracy of MRI in predicting recovery and recurrence of acute grade one hamstring muscle strains within the same season in Australian Rules football players. J Sci Med Sport 7(2):248–258. http://dx.doi.org/10.1016/S1440-2440(04)80016-1

Hawkins RD, Hulse MA, Wilkinson C, Hodson A, Gibson M (2001) The association football medical research programme: an audit of injuries in professional football. Br J Sports Med 35(1):43–47

Jonhagen S, Nemeth G, Eriksson E (1994) Hamstring injuries in sprinters. The role of concentric and eccentric hamstring muscle strength and flexibility. Am J Sports Med 22(2):262–266

Kucera KL, Marshall SW, Bell DR, Distefano MJ, Goerger CP, Oyama S (2011) Validity of soccer injury data from the National Collegiate Athletic Association's Injury Surveillance System. J Athl Train 46(5):489–499

Mair SD, Seaber AV, Glisson RR, Garrett WE (1996) The role of fatigue in susceptibility to acute muscle strain injury. Am J Sports Med 24(2):137–143. doi:10.1177/036354659602400203

Mjolsnes R, Arnason A, Osthagen T, Raastad T, Bahr R (2004) A 10-week randomized trial comparing eccentric vs. concentric hamstring strength training in well-trained soccer players. Scand J Med Sci Sports 14(5):311–317. doi:10.1046/j.1600-0838.2003.367.xSMS367 [pii]

Morgan DL (1990) New insights into the behavior of muscle during active lengthening. Biophys J 57(2):209–221. http://dx.doi.org/10.1016/S0006-3495(90)82524-8

Opar DA, Williams MD, Shield AI (2012) Hamstring strain injuries: factors that lead to injury and re-injury. Sports Med 42(3):209–226

Opar DA, Williams MD, Timmins RG, Dear NM, Shield AJ (2013) Knee flexor strength and bicep femoris electromyographical activity is lower in previously strained hamstrings. J Electromyogr Kinesiol 23(3):696–703. http://dx.doi.org/10.1016/j.jelekin.2012.11.004

Opar DA, Drezner J, Shield A, Williams M, Webner D, Sennett B, Kapur R, Cohen M, Ulager J, Cafengiu A, Cronholm PF (2014) Acute hamstring strain injury in track-and-field athletes: A 3-year observational study at the Penn Relay Carnival. Scand J Med Sci 24(4):254–259. Sports:n/a-n/a. doi:10.1111/sms.12159

Orchard JW (2001) Intrinsic and extrinsic risk factors for muscle strains in Australian Football Neither the author nor the related institution has received any financial benefit from research in this study. Am J Sports Med 29(3):300–303

Orchard J, Seward H (2002) Epidemiology of injuries in the Australian Football League, seasons 1997–2000. Br J Sports Med 36(1):39–44

Orchard J, Marsden J, Lord S, Garlick D (1997) Preseason hamstring muscle weakness associated with hamstring muscle injury in Australian footballers. Am J Sports Med 25(1):81–85

Orchard JW, Farhart P, Leopold C (2004) Lumbar spine region pathology and hamstring and calf injuries in athletes: is there a connection? Br J Sports Med 38(4):502–504

Petersen J, Thorborg K, Nielsen MB, Budtz-Jørgensen E, Hölmich P (2011) Preventive effect of eccentric training on acute hamstring injuries in men's soccer: a cluster-randomized controlled trial. Am J Sports Med 39(11):2296–2303

Pinniger GJ, Steele JR, Groeller H (2000) Does fatigue induced by repeated dynamic efforts affect hamstring muscle function? Med Sci Sports Exerc 32(3):647–653

Rehorn MR, Blemker SS (2010) The effects of aponeurosis geometry on strain injury susceptibility explored with a 3D muscle model. J Biomech 43(13):2574–2581. http://dx.doi.org/10.1016/j.jbiomech.2010.05.011

Silder A, Reeder SB, Thelen DG (2010) The influence of prior hamstring injury on lengthening muscle tissue mechanics. J Biomech 43(12):2254–2260. http://dx.doi.org/10.1016/j.jbiomech.2010.02.038

Sugiura Y, Saito T, Sakuraba K, Sakuma K, Suzuki E (2008) Strength deficits identified with concentric action of the hip extensors and eccentric action of the hamstrings predispose to hamstring injury in elite sprinters. J Orthop Sports Phys Ther 38(8):457–464. doi:10.2519/jospt.2008.2575, 1409 [pii]

Verrall GM, Slavotinek JP, Barnes PG, Fon GT, Spriggins AJ (2001) Clinical risk factors for hamstring muscle strain injury: a prospective study with correlation of injury by magnetic resonance imaging. Br J Sports Med 35(6):435–439; discussion 440

Witvrouw E, Danneels L, Asselman P, D'Have T, Cambier D (2003) Muscle flexibility as a risk factor for developing muscle injuries in male professional soccer players. Am J Sports Med 31(1):41–46

Woodley SJ, Mercer SR (2005) Hamstring muscles: architecture and innervation. Cells Tissues Organs 179(3):125–141. doi:10.1159/000085004, 85004 [pii]

Woods C, Hawkins RD, Maltby S, Hulse M, Thomas A, Hodson A (2004) The Football Association Medical Research Programme: an audit of injuries in professional football–analysis of hamstring injuries. Br J Sports Med 38(1):36–41

Yeung SS, Suen AMY, Yeung EW (2009) A prospective cohort study of hamstring injuries in competitive sprinters: preseason muscle imbalance as a possible risk factor. Br J Sports Med 43(8):589–594

Part VI
Function in Arches of the Feet and Occurrence Mechanisms of Foot Disorders, Characteristics, and Their Prevention

Chapter 28
Kinematics of the Foot and Ankle

Mako Fukano

Abstract This chapter is designed to provide a better understanding of foot and ankle kinematics. In the first half of this Chapter I describe foot arch kinematics and related gender differences, and in the last half I provide a new approach for evaluating talocrural and subtalar joint kinematics.

Keywords Foot longitudinal arch • Talocrural joint • Subtalar joint

28.1 Introduction

Civilized society was built upon humans developing their brain and hand function. Concurrently, humans evolved the ability to stand upright and perform bipedal locomotion. The human foot has important roles for performing the daily activities required in the modern world, and foot problems may lead to a decline in the quality of life such as life-space-constriction and/or a decreasing in the optimal activity level. Therefore, foot health is considered important in maintaining a high quality of daily and cultural life.

In recent years, the objective of medicine has shifted from a stress on "treatment" to an emphasis on "prevention." A similar trend has also been observed in sports medicine; that is, the focus has shifted from how to treat injuries to how to prevent injuries. In 1992, van Mechelen proposed a four-step concept for sports-related injury prevention was proposed as follows: (i) reveal the severity and frequency of injury, (ii) clarify the injury mechanism, (iii) conduct an intervention, (iv) verify the effects of the intervention, and then return to the first step. Efforts to prevent injuries, especially anterior cruciate ligament injury, have been made following these four steps and have achieved some positive results.

While research in the field of injury prevention has progressed, advances in the area of foot and ankle research have been insufficient because of methodological difficulties. Because the terms for describing foot and ankle movements have not

M. Fukano (✉)
Faculty of Sport Sciences, Waseda University, Saitama, Japan
e-mail: mako.fukano@aoni.waseda.jp

K. Kanosue et al. (eds.), *Sports Injuries and Prevention*,
DOI 10.1007/978-4-431-55318-2_28

been standardized, both in Japan and worldwide, developing a basic understanding foot and ankle kinematics has been difficult. Global efforts toward the unification of terms began with the first conference on International Foot and Ankle Biomechanics which was held in Bologna, Italy, in 2008. Thus, the study of the foot and ankle in the field of sports medicine is currently developing, and basic data to clarify injury mechanics are required for the prevention of foot and ankle injuries.

28.2 Foot Kinematics

The human foot consists of 33 joints and 26 bones, which form three biomechanically functional arch structures: the medial longitudinal arch (MLA), lateral longitudinal arch (LLA), and transversal arch. These arch structures have roles as impact attenuators, through their functional deformation upon the application of load and on impact during dynamic activities.

Thus far, most studies have focused on the MLA. The earliest description stating that the inside of the forefoot has a slight movement in the vertical direction was made by Duchenn in the 1880s. Early knowledge about the foot arch was largely limited to the MLA and very little information is available about the LLA.

The most convenient and traditional measurement method in foot research is manual measurement. The navicular drop test (NDT) is the most widespread method for foot-arch evaluation among the manual tests, and some reports have demonstrated its relevance for lower-extremity injuries (Bandholm et al. 2008; Bennett et al. 2001; Loudon et al. 1996). However, NTD has a low reliability because of within- and between-examiner errors. Moreover, examinations based on kinematics data during dynamic activities are necessary in sports-related injuries.

Measurement techniques for biomechanical analysis have advanced remarkably in recent years. Motion analysis using reflective markers attached to landmarks on the human body is becoming popular and allows the investigator to obtain dynamic kinematics data during activities such as running and/or landing. This method is minimally invasive and allows the calculation of joint movements from the position coordinate values. However, obtaining foot kinematic data from reflective skin markers is difficult because of errors due to skin and fat pad deformation.

Radiographic image analysis with fluoroscopy or cineradiography has recently attracted attention, as it demonstrates relative bony motions during dynamic motion. These devices were originally developed for use in operative treatment and/or examination of the cardiovascular system. Studies on joint kinematics using time-series radiographic images were initiated in the 1980s and became much more widespread during the 1990s. The advantage of this method is that it allows the observation of precise bone motion during dynamic movements (Smith et al. 1986) and avoids measurement errors due to skin and soft tissue deformation artifacts. Fluoroscopic studies in the 1990s mainly focused on the lumbar and the cervical regions (Kanayama et al. 1996; Cholewicki and McGill 1992; Croft et al. 1994; Roozmon et al. 1993) of the spine, and few studies were focused on foot and ankle kinematics.

Recently, fluoroscopic image analysis has been applied to foot and ankle studies. Perlman et al. confirmed that the inclination angle of the calcaneus was decreased in the sagittal plane during the stance phase, by analyzing three points—early, middle, and final—of the stance phase for 27 gait cycles in 15 subjects in 1996. In 1999, Wearing et al. reported that the calcaneal inclination angle was positively correlated with arch height and was a useful indicator for classifying feet on the basis of biomechanical aspects (Wearing et al. 1999). Gefen et al. conducted biomechanical analysis using fluoroscopic images of foot bones and soft tissue taken in mimic-walking conditions (Gefen 2003; Gefen et al. 2000). However, a thorough kinematic examination of the foot during dynamic motion has not been accomplished. Therefore, we conducted foot-arch analysis during drop landing (Fukano and Fukubayashi 2009) and examined differences due to gender.

In our study, to illustrate the sagittal plane motion of the MLA and LLA, we analyzed the first and fifth metatarsal bone positions with respect to the calcaneus (Fig. 28.1) using fluoroscopy during a 10-cm landing trial. Foot images from the

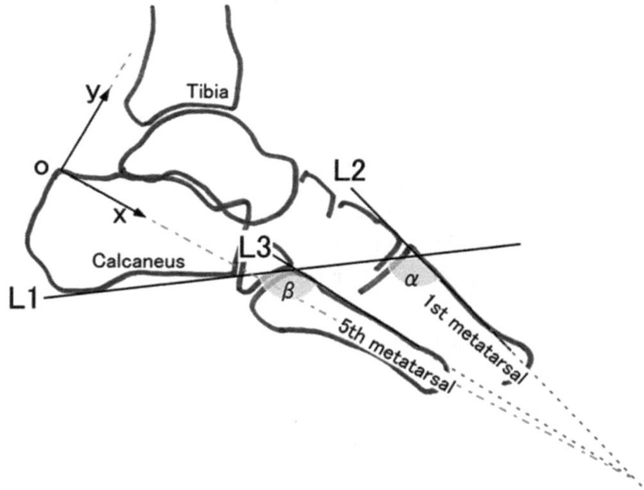

Fig. 28.1 The definition of the foot coordination system. The origin is located on the calcaneus and is at the upper edge of the posterior surface. The calcaneus and the first and fifth metatarsal extension lines define the X-axis, and the positive direction of the X-axis is anterior. The Y-axis is orthogonal to the X-axis and its positive direction is superior. *L1* represents the straight line that links the calcaneal tubercle and the anteroinferior aspect of the calcaneus. *L2* represents the dorsal aspect of the first metatarsal shaft. *L3* represents the dorsal aspect of the fifth metatarsal shaft. The medial arch angle is defined as the obtuse angle formed by L1 and L2. The lateral arch angle is defined as the obtuse angle formed by L1 and L3 (Fukano and Fukubayashi 2012). Notice: This is the author's version of a work accepted for publication by Elsevier. Changes resulting from the publishing process, including peer review, editing, corrections, structural formatting, and other quality control mechanisms, may not be reflected in this document. Changes may have been made to this work since it was submitted for publication. The definitive version has been published in *The Foot*, 22(1):2011. doi:10.1016/j.foot.2011.08.002

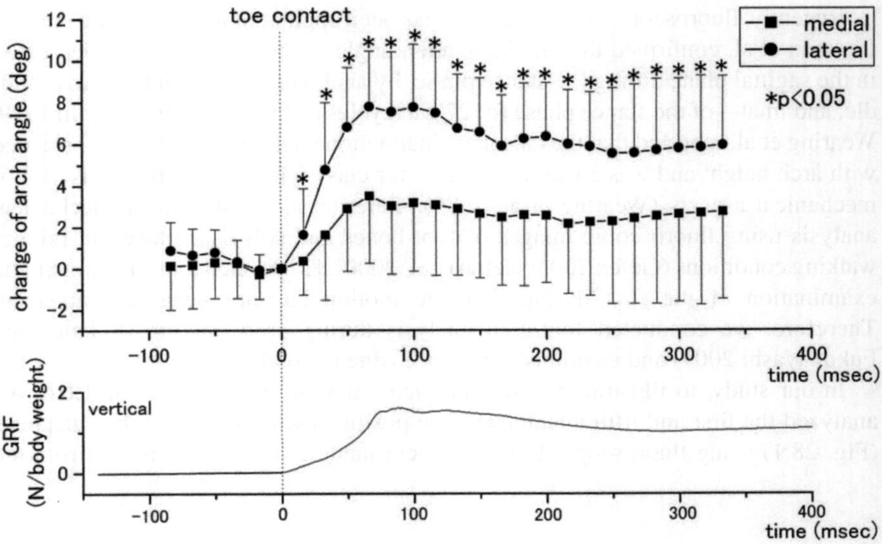

Fig. 28.2 The mean results of the quantitation of the angular change of the medial and lateral longitudinal arches. The *square plot* shows the medial longitudinal arch. The *circle plot* shows the lateral longitudinal arch (Fukano and Fukubayashi 2009)

sagittal plane were obtained at a rate of 60 Hz while 10 healthy male subjects performed a single-leg landing. The MLA and LLA angles increased with time for 80–100 ms in all subjects. The magnitude of the angular change of the MML after landing was significantly larger than that of the LLA (Fig. 28.2). With regard to the anteroposterior displacement, the first metatarsal was displaced anteriorly. In contrast, the fifth metatarsal had little displacement. The first metatarsal demonstrated an inferior displacement. In contrast, the fifth metatarsal was slightly displaced superiorly (Fig. 28.3). These results suggest that each longitudinal arch has a different deformation pattern upon landing (Fig. 28.4) (Fukano and Fukubayashi 2009).

To determine the gender differences in arch deformation, we compared the first and fifth metatarsal bone positions with respect to the calcaneus (Fig. 28.1) between male and female subjects using fluoroscopic images taken during landing. Foot images in the static condition (no-load and loaded [single-leg standing] position) and during a single-leg landing from a 10-cm height were obtained using lateral fluoroscopy at 60 Hz in 11 healthy male and 8 healthy female subjects. In the static condition, the MLA and LLA of the female subjects were significantly greater than those of the male subjects in the loaded position; however, each arch was comparable between the two groups in the no-loaded condition (Table 28.1). The angular change of both arches was significantly greater in the female than in the male subjects (Fig. 28.5). These results suggest that females have a greater range of arch motion than males under static and dynamic loaded conditions (Fukano and Fukubayashi 2012).

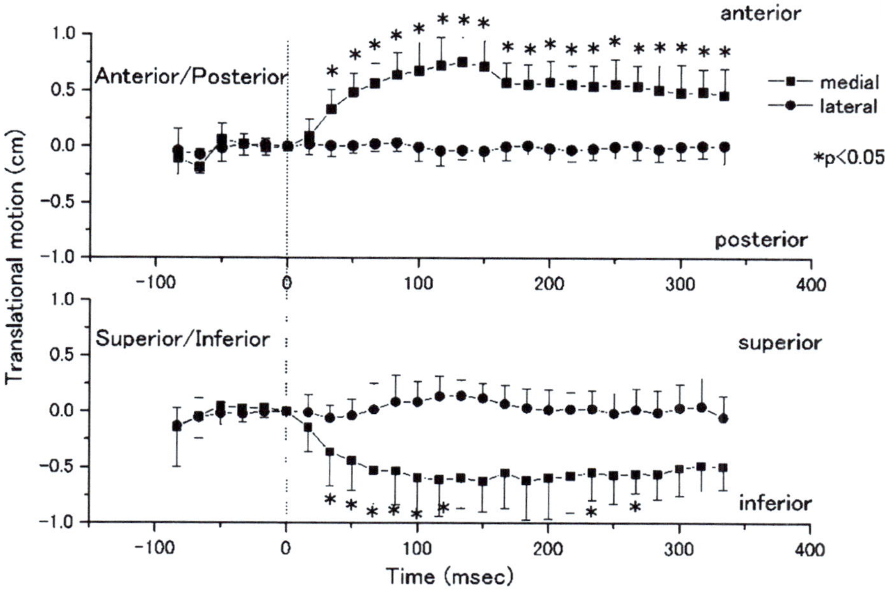

Fig. 28.3 The mean results of the translational motion of the medial and lateral longitudinal arches. The *square plot* shows the displacement of the distance of the calcaneus to the first metatarsal. The *circle plot* shows the displacement of the distance of the calcaneus to the fifth metatarsal (Fukano and Fukubayashi 2009)

Fig. 28.4 Motion difference of the first and fifth metatarsals with respect to the calcaneus (Fukano and Fukubayashi 2009)

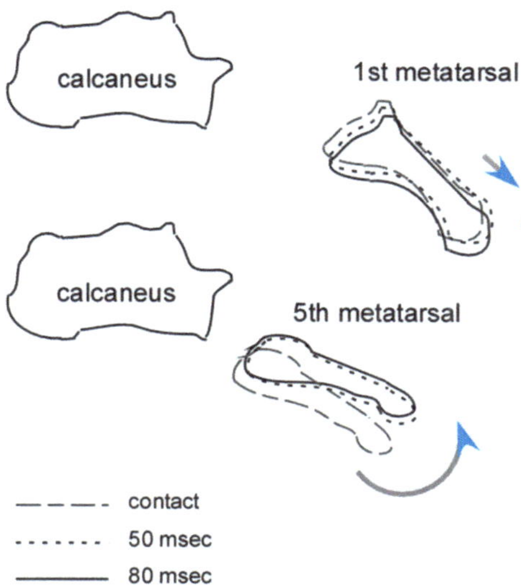

Table 28.1 Average values and standard deviations of the arch angles of male and female subjects in the static condition

	Arch angle (in degrees)	Mean (SD) males	Mean (SD) females	Significance	(p-Value)
No load	Medial	129.5 (5.8)	129.8 (2.6)	n. s.	(p=0.99)
	Lateral	138.2 (6.2)	141.2 (1.3)	n. s.	(p=0.46)
Loaded	Medial	130.3 (5.9)	136.2 (2.3)	*	(p=0.01)
	Lateral	144.9 (6.1)	148.9 (1.3)	*	(p=0.03)

Notice: This is the author's version of a work accepted for publication by Elsevier. Changes resulting from the publishing process, including peer review, editing, corrections, structural formatting, and other quality control mechanisms, may not be reflected in this document. Changes may have been made to this work since it was submitted for publication. The definitive version has been published in *The Foot*, 22(1), 2011. doi:10.1016/j.foot.2011.08.002
*p<0.05

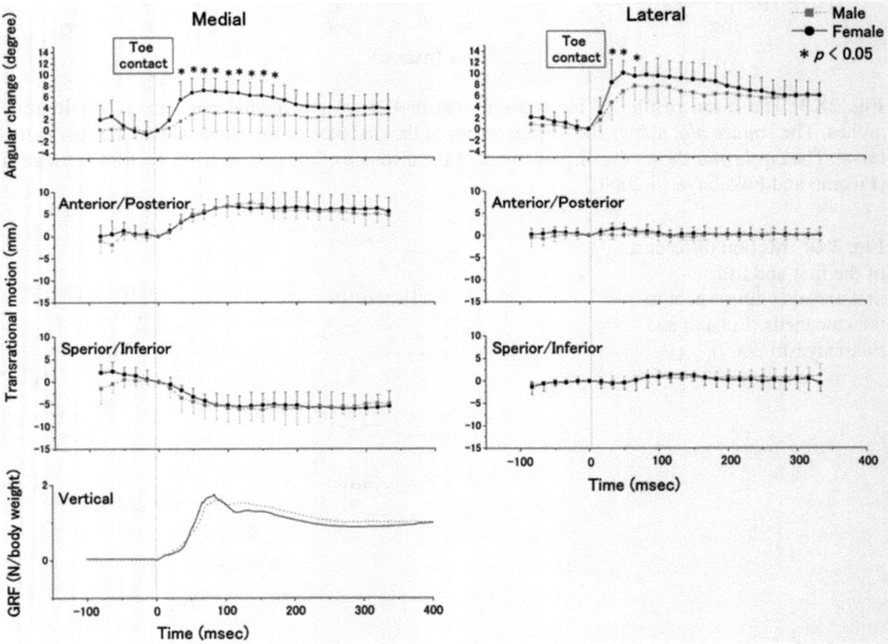

Fig. 28.5 Comparison of the degree of angular change and the translational motion of the medial and lateral longitudinal arches between males and females. The *gray square* and *black circle* plots are the values for the males and females, respectively. *GRF* ground reaction force (Fukano and Fukubayashi 2012). Notice: This is the author's version of a work accepted for publication by Elsevier. Changes resulting from the publishing process, including peer review, editing, corrections, structural formatting, and other quality control mechanisms, may not be reflected in this document. Changes may have been made to this work since it was submitted for publication. The definitive version has been published in *The Foot*, 22(1), 2011. doi:10.1016/j.foot.2011.08.002

28.3 Ankle Kinematics

In recent years, quantitative evaluations of the ankle joint, which consists of the talocrural and subtalar joints, have been inadequate. However, this subject is important in the field of orthopedic and sports medicine because: (i) ankle joints function as a shock attenuator together with the foot; (ii) subtalar joint kinematics may be associated with subjective ankle instability in patients with chronic ankle instability; and (iii) subjective ankle instability (which may be associated with subtalar joint kinematics) could be an impediment to sports performance. Unfortunately, there is no international standard for evaluating talocrural and subtalar joint kinematics. A highly accurate method for measuring the movement of the talus has not been established, since errors due to deformation of the skin and the plantar soft tissue are large with respect to the bony movement of the foot. Thus no reference data from healthy subjects are available. Therefore, discussions based on quantitative data are difficult, and a functional knowledge about injury mechanics or prevention is based only on qualitative observations and textbook descriptions.

Studies with radiographic images acquired by fluoroscopy or cineradiography to analyze the detailed kinematics of joints during dynamic movement are increasingly available. The three-dimensional (3D)-2D model-based registration technique (MBRT) is a method that provides for a detailed analysis of 3D *in vivo* joint kinematics, by matching bone models created using 3D computed tomography and time-series radiographic images. MBRT was mainly used for the knee joint after 2000, but has been more recently applied to ankle joints in the development of total ankle replacement surgery. Some groups have performed a highly accurate analysis using orthogonal fluoroscopes and have estimated the length of the anterior talofibular ligament (Caputo et al. 2009).

However, the MBRT has several technical limitations: (i) the image acquisition frequency is often limited to 7.5–30 Hz because these devices are generally used to diagnose cardiovascular problems; (ii) the imaging area is normally limited to approximately 25×25 cm; (iii) MBRT analysis is time and effort intensive. Table 28.2 presents studies on ankle joint kinematics with MBRT from 2006 to 2013. As shown in Table 28.2, simulated movements, which were performed at a speed far lower than that in daily movement or sports activities or multiple trials in static conditions with changing load and/or joint angles, were analyzed, as in-vivo dynamic motion analysis. Precise joint kinematics during natural movement has not been demonstrated to date. To fill this gap, we attempt to analyze *in vivo* ankle kinematics using 60-Hz cineangiographic images and hope that our results will contribute to a better understanding of ankle joint kinematics during sports activities.

Table 28.2 Studies on ankle kinematics with the model-based registration technique

Reference	Number of planes	Trial	Sampling frequency	Subjects
Cenni et al. (2013)	Single	Stair climbing and descending	10	Implanted
Cenni et al. (2012)	Single	Stair climbing and descending, and flexion/extension against gravity	?	Total ankle replacement
List et al. (2012)	?	Level gait, walking downhill and walking uphill	25	Total ankle arthroplasty
List et al. (2012)	?	Gait, uphill, downhill, and cross-slope walking	25	Total ankle arthroplasty
Yamaguchi et al. (2012)	Single	Walking	7.5	Total ankle arthroplasty
Caputo et al. (2009)	Bi-plane	Static loading (25–100 % body weight)	–	ATFL deficient vs. intact ankles
Yamaguchi et al. (2009)	Single	Dorsi-plantar flexion	7.5	Healthy
Leszko et al. (2008)	Single	Normal gait, step-up	?	Salto total ankle arthroplasty
De Asla et al. (2006)	Single	Mimic walking (early-mid-last stance phase)	?	Healthy vs. total ankle arthroplasty
Conti et al. (2006)	Single	Walking (heel-strike, 33 %, and 66 % of stance phase)	?	Nonimplanted and implanted
Komistek et al. (2000)	Bi-plane	Dorsi-plantar flexion	–	Healthy

References

Bandholm T, Boysen L, Haugaard S, Zebis MK, Bencke J (2008) Foot medial longitudinal-arch deformation during quiet standing and gait in subjects with medial tibial stress syndrome. J Foot Ankle Surg 47(2):89–95. doi:10.1053/j.jfas.2007.10.015, S1067-2516(07)00430-9 [pii]

Bennett JE, Reinking MF, Pluemer B, Pentel A, Seaton M, Killian C (2001) Factors contributing to the development of medial tibial stress syndrome in high school runners. J Orthop Sports Phys Ther 31(9):504–510

Caputo AM, Lee JY, Spritzer CE, Easley ME, DeOrio JK, Nunley JA 2nd, DeFrate LE (2009) In vivo kinematics of the tibiotalar joint after lateral ankle instability. Am J Sports Med 37(11):2241–2248. doi:10.1177/0363546509337578

Cenni F, Leardini A, Belvedere C, Bugane F, Cremonini K, Miscione MT, Giannini S (2012) Kinematics of the three components of a total ankle replacement: in vivo fluoroscopic analysis. Foot Ankle Int 33(4):290–300. doi:10.3113/FAI.2012.0290

Cenni F, Leardini A, Pieri M, Berti L, Belvedere C, Romagnoli M, Giannini S (2013) Functional performance of a total ankle replacement: thorough assessment by combining gait and fluoroscopic analyses. Clin Biomech (Bristol, Avon) 28(1):79–87. doi:10.1016/j.clinbiomech.2012.10.008

Cholewicki J, McGill SM (1992) Lumbar posterior ligament involvement during extremely heavy lifts estimated from fluoroscopic measurements. J Biomech 25(1):17–28

Conti S, Lalonde KA, Martin R (2006) Kinematic analysis of the agility total ankle during gait. Foot Ankle Int 27(11):980–984

Croft AC, Krage JS, Pate D, Young DN (1994) Videofluoroscopy in cervical spine trauma: an interinterpreter reliability study. J Manipulative Physiol Ther 17(1):20–24

de Asla RJ, Wan L, Rubash HE, Li G (2006) Six DOF in vivo kinematics of the ankle joint complex: Application of a combined dual-orthogonal fluoroscopic and magnetic resonance imaging technique. J Orthop Res 24(5):1019–1027. doi:10.1002/jor.20142

Fukano M, Fukubayashi T (2009) Motion characteristics of the medial and lateral longitudinal arch during landing. Eur J Appl Physiol 105(3):387–392. doi:10.1007/s00421-008-0915-3

Fukano M, Fukubayashi T (2012) Gender-based differences in the functional deformation of the foot longitudinal arch. Foot (Edinb) 22(1):6–9. doi:10.1016/j.foot.2011.08.002, S0958-2592(11)00086-1 [pii]

Gefen A (2003) Plantar soft tissue loading under the medial metatarsals in the standing diabetic foot. Med Eng Phys 25(6):491–499, doi:S1350453303000298 [pii]

Gefen A, Megido-Ravid M, Itzchak Y, Arcan M (2000) Biomechanical analysis of the three-dimensional foot structure during gait: a basic tool for clinical applications. J Biomech Eng 122(6):630–639

Kanayama M, Abumi K, Kaneda K, Tadano S, Ukai T (1996) Phase lag of the intersegmental motion in flexion-extension of the lumbar and lumbosacral spine. An in vivo study. Spine (Phila Pa 1976) 21(12):1416–1422

Komistek RD, Stiehl JB, Buechel FF, Northcut EJ, Hajner ME (2000) A determination of ankle kinematics using fluoroscopy. Foot Ankle Int 21(4):343–350

Leszko F, Komistek RD, Mahfouz MR, Ratron YA, Judet T, Bonnin M, Colombier JA, Lin SS (2008) In vivo kinematics of the salto total ankle prosthesis. Foot Ankle Int 29(11):1117–1125. doi:10.3113/FAI.2008.1117

List R, Gerber H, Foresti M, Rippstein P, Goldhahn J (2012) A functional outcome study comparing total ankle arthroplasty (TAA) subjects with pain to subjects with absent level of pain by means of videofluoroscopy. Foot Ankle Surg 18(4):270–276. doi:10.1016/j.fas.2012.04.001

Loudon JK, Jenkins W, Loudon KL (1996) The relationship between static posture and ACL injury in female athletes. J Orthop Sports Phys Ther 24(2):91–97

Roozmon P, Gracovetsky SA, Gouw GJ, Newman N (1993) Examining motion in the cervical spine. I: Imaging systems and measurement techniques. J Biomed Eng 15(1):5–12

Smith LS, Clarke TE, Hamill CL, Santopietro F (1986) The effects of soft and semi-rigid orthoses upon rearfoot movement in running. J Am Podiatr Med Assoc 76(4):227–233

van Mechelen W, Hlobil H, Kemper HC (1992) Incidence, severity, aetiology and prevention of sports injuries. A review of concepts. Sports Med 14(2):82–99

Wearing SC, Urry S, Perlman PR, Dubois P, Smeathers JE (1999) Serial measurement of calcaneal pitch during midstance. J Am Podiatr Med Assoc 89(4):188–193

Yamaguchi S, Sasho T, Kato H, Kuroyanagi Y, Banks SA (2009) Ankle and subtalar kinematics during dorsiflexion-plantarflexion activities. Foot Ankle Int 30(4):361–366. doi:10.3113/FAI.2009.0361, 50022909 [pii]

Yamaguchi S, Tanaka Y, Banks S, Kosugi S, Sasho T, Takahashi K, Takakura Y (2012) In vivo kinematics and articular surface congruency of total ankle arthroplasty during gait. J Biomech 45(12):2103–2108. doi:10.1016/j.jbiomech.2012.05.043

Chapter 29
Biomechanical Analysis of the Effects of Footwear

Mako Fukano

Abstract In this chapter I utilize a biomechanical approach in order to better understand the effects of footwear. In the first half of the chapter I describe the changes in lower extremity kinematics that occur during walking and running. In the second half I provide a new approach for the examination of foot and ankle kinematics and focus on the effects of footwear.

Keywords Footwear • Foot • Ankle • Kinematics

29.1 Introduction

Footwear is one of the key pieces of equipment required for sports and exercise. It serves the role of an interface between foot and ground. The correct footwear can not only protect the foot from injury, it can also enhance performance. A constant growth of research interest in this area over the last 30 years has led to the development and sale of numerous types of athletic footwear.

29.2 Changes in Lower Extremity Kinematics During Running

Tibial rotation that accompanies hindfoot pronation during locomotion serves a shock absorbing function, but has also been proposed as a key factor in the medial tibial stress syndrome (Moen et al. 2009; Willems et al. 2006) and/or patellofemoral pain (Cheung et al. 2006) caused by repetitive mechanical stress. However, there is little data on the effects of footwear on tibial rotation since the typical markers are set only for selected anatomical landmarks and do not provide accurate information

M. Fukano (✉)
Faculty of Sport Sciences, Waseda University, Saitama, Japan
e-mail: mako.fukano@aoni.waseda.jp

K. Kanosue et al. (eds.), *Sports Injuries and Prevention*,
DOI 10.1007/978-4-431-55318-2_29

on the in vivo tibiofemoral kinematics, especially that of rotation, because of skin error artifacts. We employed the point cluster technique (PCT) (Andriacchi et al. 1998) in order to obtain detailed data on knee joint kinematics in 15 healthy subjects because a cluster system eliminates the noise resulting from skin movement. The results we obtained for running subjects are shown in Fig. 29.1 (Fukano et al. 2009).

Figure 29.2 shows that subjects experienced internal tibial rotation during the period ranging from foot strike to approximately the first half of the stance phase and external rotation during the second half of the stance phase. The data demonstrate that the degree of tibial rotation was reduced with athletic footwear (Table 29.1). This study suggests that the degree of tibial rotation that accompanies foot pronation could be reduced by wearing athletic footwear during running. To improve our understanding of the interaction between the human body and footwear, additional analyses of muscle activation, plantar pressure distribution, and different types and designs of footwear are required.

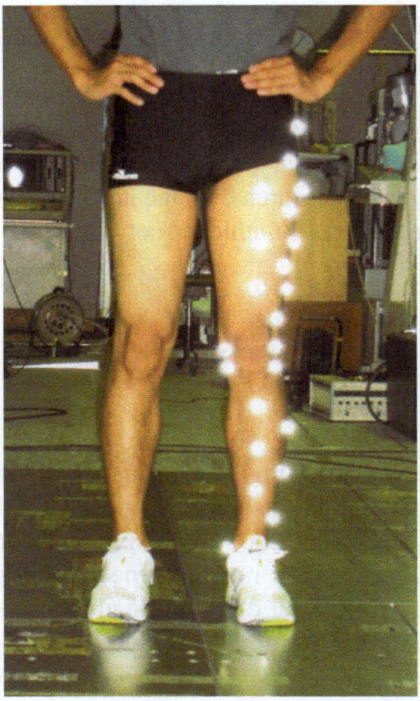

Fig. 29.1 Arrangement of markers for the point cluster technique. Twenty-five reflective markers were secured to the left lower extremity of each subject (*This is the Author's Accepted Manuscript of an article published* as "Change in tibial rotation of barefoot versus shod running." *as published in Footwear Science Vol.1, No. 1 March 2009, 19–23 copyright Tylor & Francis, available online at* http://www.tandfonline.com/, doi:10.1080/19424280902950456)

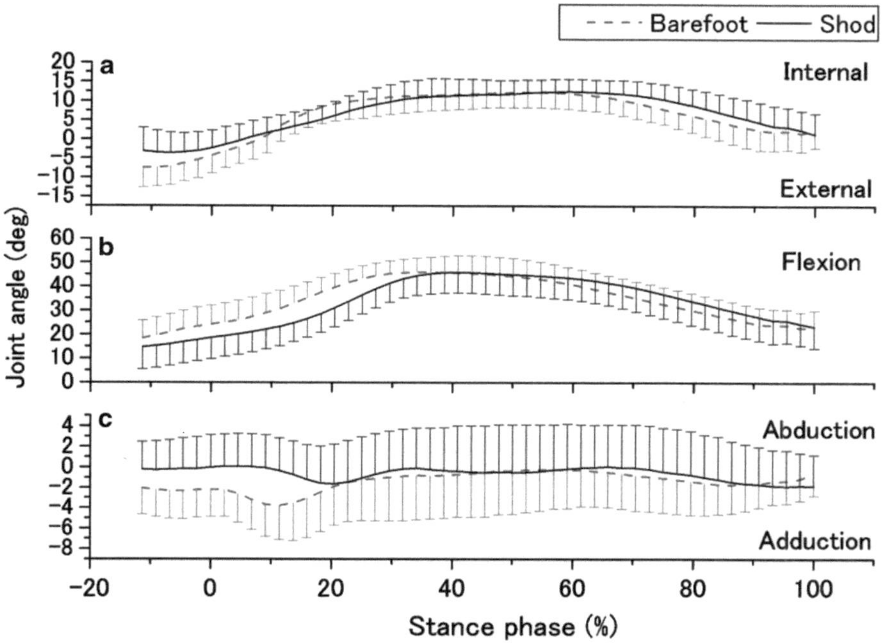

Fig. 29.2 Average kinematic data of the knee joint during running. Data are presented for knee internal/external rotation (**a**), knee flexion (**b**) and adduction/abduction (**c**) (*This is the Author's Accepted Manuscript of an article published* as "Change in tibial rotation of barefoot versus shod running." *as published in Footwear Science Vol.1, No. 1 March 2009, 19–23 copyright Tylor & Francis, available online at* http://www.tandfonline.com/ doi:10.1080/19424280902950456)

Table 29.1 Angular change in the knee joint during the 100-ms period after foot strike

Condition	Barefoot		Shod		Difference	
Variable	Mean	(SD)	Mean	(SD)		p value
Internal rotation (deg)	16.0	(4.1)	13.7	(5.3)	*	*0.02*
Flexion (deg)	20.8	(4.9)	26.5	(4.2)	*	*0.00*
Adduction (deg)	1.1	(4.1)	1.2	(0.4)	–	*0.60*

This is an Author's Accepted Manuscript of an article published in "Change in tibial rotation of barefoot versus shod running." *as published in the Footwear Science Vol.1, No. 1 March 2009, 19–23 copyright Tylor & Francis, available online at* http://www.tandfonline.com/ doi:10.1080/19424280902950456

Note: *SD* Standard deviation

29.2.1 A New Approach for Analyzing Foot and Ankle Kinematics with Footwear

Researchers face two major problems when analyzing foot and ankle movement. As discussed in the previous chapter, one of these is the skin movement artifact, which is a commonly encountered issue in the analysis of foot and ankle motion. The second problem is that the researchers are unable to observe the foot when the subjects are using footwear. Consequently, researchers must employ strategies such as use of sandals with straps as footwear or cutting "windows" in the footwear to attach reflective markers on anatomical landmarks; however, these tactics may change the mechanical properties of the footwear.

29.2.2 Foot Arch

The foot arch plays the important role of an impact attenuator during activity. However, there is little information regarding the effects of footwear on foot arch kinematics during dynamic conditions because of the related methodological difficulties. To solve this problem, we used cineradiographic image sequences to determine the effects of footwear on the medial and lateral longitudinal arch kinematics during single leg landing (Fukano and Fukubayashi 2012a).

Six healthy male volunteers participated in the study (age, 21.7 ± 1.0 years; height, 172.5 ± 6.8 cm; weight, 61.9 ± 6.0 kg). All subjects performed single leg landings from a 10 cm height with their knee extended under three conditions: barefoot, with running footwear (Adidas Response Cushion, Adidas, Herzogenaurach, Germany), and with a soccer boot (Pathiqe 05 TRX HGJP, Adidas, Herzogenaurach, Germany). Lateral image sequences were obtained using cineradiography (INTEGRIS BH5000R.1, Philips) at 60 Hz and 2/1,000 shutter speed. The footwear used in the study was marketed for running and football, respectively. The images were analyzed using graphic software (CANVAS™ X, ACD system). The arch definition is shown in Fig. 29.3. The sagittal plane movement of the medial and lateral arches was defined as the angular change observed after landing. Repeated ANOVA

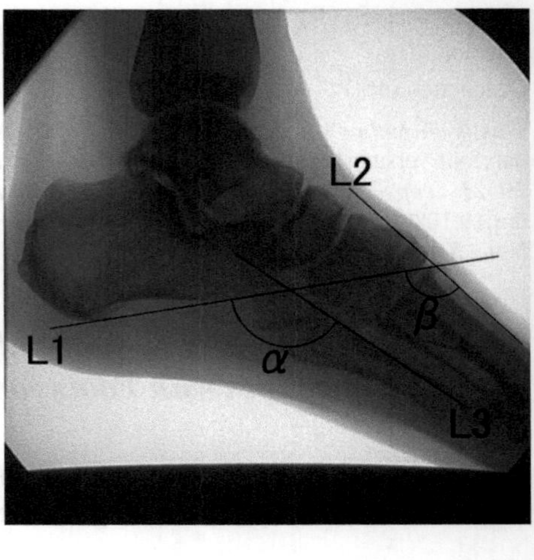

Fig. 29.3 Definition of arch angles. The medial arch angle is represented by the obtuse angle formed by the *L1* and *L2* lines. The lateral arch angle is represented by the obtuse angle formed by the *L1* and *L3* lines (Image from Published Paper (Fukano and Fukubayashi 2012a) entitled "The effects of shoe type on sagittal arch kinematics during landing." 靴の医学 vol. 26 No. 2 26–30)

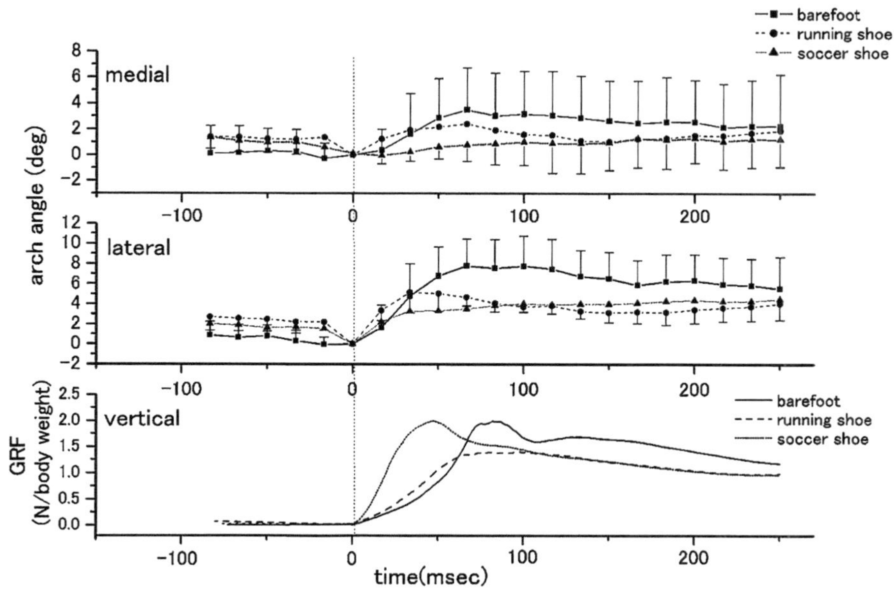

Fig. 29.4 Time-course kinematics of the medial and lateral longitudinal arches as well as the vertical ground reaction force data during landing (Image from Published Paper (Fukano and Fukubayashi 2012a) entitled "The effects of shoe type on sagittal arch kinematics during landing." 靴の医学 vol. 26 No. 2 26–30)

and post-hoc tests were performed to detect statistical differences among the three conditions. Significance was set at $p < 0.05$.

Figure 29.4 illustrates the time-course kinematics of the medial and lateral longitudinal arches during landing from a height of 10 cm. The arch angles at toe contact, angular change of the two arches, maximum vertical ground reaction force, and the implus during the 100-ms period after toe contact are shown in Table 29.2. The medial and lateral arch angles at toe contact were comparable among the three conditions. The angular change of the medial arch in the barefoot condition was greater than that in the soccer boot condition, and the angular change in the lateral arch in the barefoot condition was greater than that in the running footwear and soccer boot conditions. The maximum ground reaction force in the running footwear condition was smaller than that in the other two conditions, and the implus for 100-ms after toe contact in the barefoot and running footwear conditions was smaller than that in the soccer boot condition.

The foot arch functions as an impact attenuator by functional deformation that occurs owing to the changing spatial relationship between the bones at each joint in the foot. In this study, the shock absorption effect of foot arch deformation was reduced by the footwear, and we speculated that the footwear cushion played the role of an impact attenuator instead.

The maximum vertical ground reaction force in the barefoot condition was significantly smaller than that in the two other conditions, and the time of the maximum

Table 29.2 Arch angles at toe contact, angular change, maximum vertical ground reaction force (max Fz), and implus for 100-ms after toe contact

		Barefoot	Running footwear	Soccer boot	
Arch angles at the toe contact (deg)	Medial	126.6±4.5	127.9±4.3	126.7±5.3	n.s. (*p* = *0.24*)
	Lateral	137.9±4.6	140.2±4.3	139.0±4.8	n.s. (*p* = *0.15*)
Angular change (deg)	Medial	3.5±3.3	2.4±1.6	1.3±2.0	* (*p* = *0.04*)
	Lateral	7.9±3.2	5.1±2.0	4.5±1.6	* (*p* = *0.01*)
Max Fz (N/body weight)		2.0±0.3	1.4±0.3	2.0±0.3	* (*p* = *0.01*)
Implus (at toe contact ~100 ms) (N/body weight·s)		83.8±27.9	91.2±37.1	145.2±22.7	* (*p* = *0.01*)

Fukano and Fukubayashi (2012a)
n.s. not significant
*$p < 0.05$

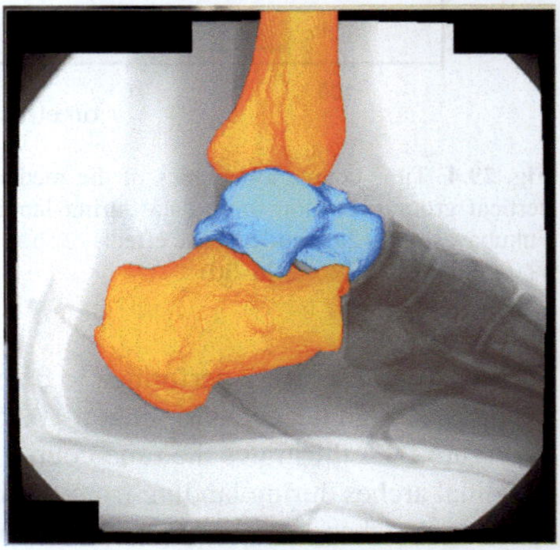

Fig. 29.5 The 3D-2D model-image registration for analysis of talocrural and subtalar joint kinematics while using footwear

vertical ground reaction force was premature in the soccer boot condition. This is possibly due to the footwear characteristics such as hardness of the sole material used, which implies that the kinetic energy of landing is attenuated by footwear sole deformation. Therefore, we conclude that the foot arch kinematics were altered by the type of footwear.

29.2.3 Talocrural and Subtalar Joints

In recent years, the radiographic shape-matching technique (i.e., 3 dimensional-2 dimensional (3D-2D) model-to-image registration) has been applied to foot and ankle motion analysis (Fukano and Fukubayashi 2009, 2012b; Yamaguchi et al. 2009). This technique has also been utilized to examine the effects of footwear on talocrural and subtalar joint movement, as shown in Fig. 29.5. This method is able to provide precise kinematic data on the talocrural and subtalar joints for subjects using footwear. Although an article on the use of 3D-2D model-image registration technique for examination of talocrural and subtalar kinematic changes with footwear has not been published yet, this method is expected to be proven effective in the near future.

References

Andriacchi TP, Alexander EJ, Toney MK, Dyrby C, Sum J (1998) A point cluster method for in vivo motion analysis: applied to a study of knee kinematics. J Biomech Eng 120(6):743–749

Cheung RT, Ng GY, Chen BF (2006) Association of footwear with patellofemoral pain syndrome in runners. Sports Med 36(3):199–205, doi:3632 [pii]

Fukano M, Fukubayashi T (2009) Motion characteristics of the medial and lateral longitudinal arch during landing. Eur J Appl Physiol 105(3):387–392. doi:10.1007/s00421-008-0915-3

Fukano M, Fukubayashi T (2012a) The effects of shoe type on sagittal arch kinematics during landing. 靴の医学 26(2):26–30

Fukano M, Fukubayashi T (2012b) Gender-based differences in the functional deformation of the foot longitudinal arch. Foot (Edinb) 22(1):6–9. doi:10.1016/j.foot.2011.08.002, S0958-2592(11)00086-1 [pii]

Fukano M, Nagano Y, Ida H, Fukubayashi T (2009) Change in tibial rotation of barefoot versus shod running. Footwear Sci 1(1):19–23

Moen MH, Tol JL, Weir A, Steunebrink M, De Winter TC (2009) Medial tibial stress syndrome: a critical review. Sports Med 39(7):523–546, doi:2 [pii]

Willems TM, De Clercq D, Delbaere K, Vanderstraeten G, De Cock A, Witvrouw E (2006) A prospective study of gait related risk factors for exercise-related lower leg pain. Gait Posture 23(1):91–98. doi:10.1016/j.gaitpost.2004.12.004, S0966-6362(04)00274-7 [pii]

Yamaguchi S, Sasho T, Kato H, Kuroyanagi Y, Banks SA (2009) Ankle and subtalar kinematics during dorsiflexion-plantarflexion activities. Foot Ankle Int 30(4):361–366. doi:10.3113/FAI.2009.0361, 50022909 [pii]

Chapter 30
Risk Factors and Mechanisms of Fifth Metatarsal Stress Fracture

Sho Matsuda and Toru Fukubayashi

Abstract Fractures of the proximal fifth metatarsal are troublesome to treat and often carry the risk of nonunion, delayed union, and recurrent fracture. For this reason, researchers have classified fractures of the proximal fifth metatarsal on the basis of which of the three fracture zones is involved and have verified that the healing process, mechanism of injury, and recommended treatment differ among the fracture types. Two studies have identified varus hindfoot as a risk factor for fifth metatarsal stress fracture. The pathogenetic mechanism of this injury is repetitive loading stress on the fifth metatarsal.

Keywords Fifth metatarsal • Jones fracture • Stress fracture

30.1 Introduction

Sir Robert Jones, in 1902, was the first to report on fractures of the proximal fifth metatarsal (Jones 1902). There has since been much controversy and confusion regarding fractures at this site because of disagreements about the difficulty of treatment (Carp 1927; Fetzer and Wright 2006; Lawrence and Botte 1993; Porter et al. 2005; Reese et al. 2004; Rettig et al. 1992; Smith et al. 1992; Torg et al. 1984; Wright et al. 2000; Zelko et al. 1979). The situation can be clarified, to some degree, by considering the location of fractures of the fifth metatarsal and utilizing anatomical criteria to categorize the mechanisms of and risk factors for such injuries.

S. Matsuda (✉)
Graduate School of Sport Sciences, Waseda University, Saitama, Japan
e-mail: showmatsuda@akane.waseda.jp

T. Fukubayashi
Faculty of Sport Sciences, Waseda University, Saitama, Japan

K. Kanosue et al. (eds.), *Sports Injuries and Prevention*,
DOI 10.1007/978-4-431-55318-2_30

30.2 Anatomy

The fifth metatarsal bone consists of the base, tuberosity (styloid process), shaft (diaphysis), neck, and head (Dameron 1995; Fetzer and Wright 2006; Quill 1995; Rosenberg and Sferra 2000; Strayer et al. 1999). The lateral portion of the Lisfranc joint includes three articulations: (1) the cuboid-fourth metatarsal, (2) cuboid-fifth metatarsal, and (3) fourth and fifth intermetatarsal articulations (Lawrence and Botte 1993; Nunley 2001; Rosenberg and Sferra 2000). The peroneus tertius tendon inserts on the dorsal surface of the metatarsal shaft, distal to the tuberosity. The peroneus brevis tendon inserts on the dorsolateral aspect of the tuberosity. The lateral band of the plantar aponeurosis inserts on the tip of the tuberosity.

30.3 Blood Supply

Smith described the intraosseous blood supply to the fifth metatarsal based on information from a cadaver model (Smith et al. 1992). The metaphyseal arteries, periosteal arteries, and nutrient artery are all potential sources of blood supply to the fifth metatarsal. The nutrient artery gives rise to longitudinal intramedullary branches and supplies the proximal diaphysis, whereas metaphyseal arteries arising from the surrounding soft tissue supply the head and neck region. This distribution of arterial supply may explain the predisposition of proximal diaphyseal fractures to delayed union or nonunion (Fetzer and Wright 2006; Smith et al. 1992).

30.4 Classification, Mechanism, and Treatment

Fractures of the proximal fifth metatarsal have been classified according to which of the three fracture zones is involved as tuberosity avulsion fractures (Zone I), Jones fractures (Zone II), and diaphyseal stress fractures (Zone III) (Lawrence and Botte 1993) (Fig. 30.1).

30.4.1 Tuberosity Avulsion Fracture (Zone I)

Tuberosity avulsion fractures are the most common acute fracture of the proximal fifth metatarsal (Lawrence and Botte 1993).

This fracture was once thought to occur via avulsion of the tuberosity by the contracting peroneus brevis muscle during inversion of the hindfoot. However, a cadaveric study indicated that the firm attachment of the lateral band of the plantar aponeurosis is more likely than the peroneus brevis to cause tuberosity avulsion

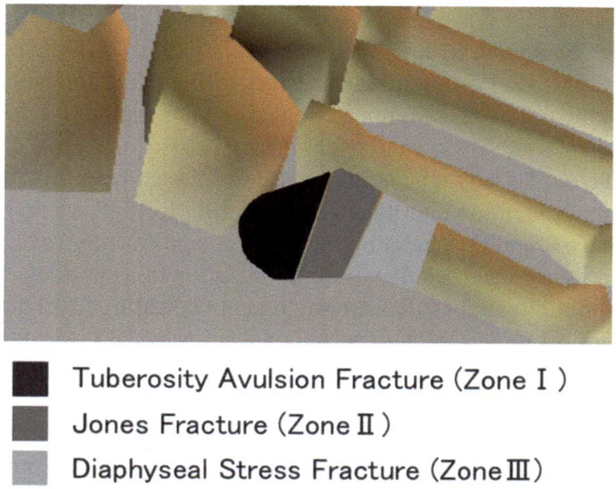

Fig. 30.1 Schematic representation of the fracture zones in the proximal fifth metatarsal bone

fractures (Richli and Rosenthal 1984). As yet, there is no consensus regarding the mechanism of this fracture.

Lawrence reported that fractures in Zone I have an excellent healing potential and should be treated symptomatically with a hard-soled shoe, a cast, or, in some patients, a protective wrap or brace (Lawrence and Botte 1993; Nunley 2001). Most fractures heal or become asymptomatic within 3 weeks, and radiographic union can be confirmed after 7–8 weeks (Nunley 2001; Rosenberg and Sferra 2000). Symptomatic nonunion occurs only rarely. In a prospective randomized study in which 60 patients were treated with either a short leg cast or a soft Jones dressing, all showed radiographic evidence of fracture healing by 65 days (mean, 44 days) post-treatment and returned to full weight bearing and full physical activity within 96 days (Wiener et al. 1997). However, Rettig et al. (1992) reported that eight athletes developed symptomatic nonunion of the tuberosity of the proximal fifth metatarsal. These occasional cases of symptomatic nonunion of intra-articular fracture of the tuberosity with significant articular step-off should be treated surgically (Fetzer and Wright 2006; Rammelt et al. 2004; Rosenberg and Sferra 2000; Zwitser and Breederveld 2010).

30.4.2 *Jones Fracture (Zone II)*

Sir Robert Jones is credited with the first description of this fracture, in 1902 (Jones 1902). The Jones fracture was defined as a transverse fracture at the diaphyseal-metaphyseal junction without extension distal to the fourth-fifth intermetatarsal articulation (Stewart 1960). Fractures in this zone take longer to heal than do more proximal fractures (Dameron 1995).

Jones fractures occur when a large adduction force is applied to the forefoot with the ankle in plantar flexion (Dameron 1995; Stewart 1960). A large bending motion produced by a high load on the plantar aspect of the fifth metatarsal head causes fracture at the junction of the proximal diaphysis and metaphysis (Fetzer and Wright 2006).

In a nonathlete who is amenable to conservative treatment, this fracture can be treated with immobilization and non-weight bearing for 6–8 weeks. A prospective randomized study found that functional treatment significantly reduced the time to return to the preinjury level relative to immobilization in a short leg cast, whereas a combination of both treatments led to complete bone healing in all patients (Wiener et al. 1997). Clapper et al. (1995) reported on a series of 235 true acute Jones fractures and found that 8 weeks of non-weight bearing followed by weight bearing as tolerated while wearing a cast was successful in 72 % of patients (mean time until union, 21.2 weeks), whereas 28 % showed clinical and radiographic evidence of nonunion 25 weeks after injury. In the seven patients in whom conservative treatment failed, intramedullary screw fixation achieved 100 % union in all patients after a mean of 12.1 weeks. True Jones fractures require a prolonged healing period when treated with 8 weeks of non-weight bearing followed by weight bearing as tolerated while wearing a cast. Dameron (1995) reported that the patient's lifestyle and desired activity level should determine the choice of treatment of Jones fracture, which can include a functional metatarsal brace, a stiff-soled shoe, a short-leg cast, or even internal fixation with a screw.

30.4.3 Proximal Diaphyseal Stress Fracture (Zone III)

Proximal diaphyseal stress fracture involves the proximal 1.5 cm of the diaphysis. DeLee et al. (1983) defined the criteria for a stress fracture of the proximal fifth metatarsal to include (1) a history of prodromal symptoms affecting the lateral aspect of the foot, (2) radiographic evidence of a stress lesion in the bone, and (3) no history of treatment for a fracture of the fifth metatarsal.

The mechanism of this fracture usually involves stress or fatigue produced by repeated normal loading of the plantar surface of the fifth metatarsal head over a relatively short period.

Torg et al. (1984) suggested a classification system for predicting the healing potential of proximal diaphyseal fifth metatarsal fractures, which includes (1) acute fractures, (2) delayed union, and (3) nonunion. In this system, the characteristic features of (1) – acute fractures – are no history of fracture, absence of intramedullary sclerosis, and radiographic evidence of sharp fracture margins, minimal or no periosteal reaction, and minimal cortical hypertrophy; the distinguishing features of (2) – delayed union – are a previous injury or fracture and radiographic evidence of some periosteal reaction, widened fracture margins, and some intramedullary sclerosis; and the features of (3) – non-union – are clinical history of repetitive trauma or recurrent symptoms and radiographic evidence of sclerosis obliterating the medullary canal and blunted fracture edges. Mologne et al. (2005) conducted a randomized, controlled clinical trial comparing surgery and casting as treatments for acute fractures located in Zone III. Treatment failure occurred in only 1 of the 19 patients in the surgery group. In contrast, 8 of 18 (44 %) cases in the cast group were considered treatment failures: five nonunions, one delayed union, and two recurrent fractures. The median times to union and return to sports were 7.5 and 8.0 weeks, respectively, in the surgery group but 14.5 and 15.0 weeks, respectively, in the cast group. Therefore, early screw fixation of acute fractures located in Zone III results in faster bony union and return to sports than does casting. The other types of fracture in this region, delayed union and non-union, usually require surgical treatment consisting of intramedullary screw fixation, tricortical-inlay bone grafting, or a combination of these two techniques (Quill 1995).

30.5 Risk Factors

Raikin et al. (2008) reported clinical and radiographic data from 20 patients (21 ft) with a history of Jones fracture. Their clinical hindfoot assessment included the Achilles-calcaneal axis angle and peek-a-boo heel sign (ability to visualize the medial heel pad when the patient stands with the feet aligned straight ahead) . Their radiographic assessment included the calcaneal pitch angle (the angle between a line drawn from the plantar-most surface of the calcaneus to the inferior border of the distal articular surface and the transverse plane) and talar-first metatarsal angle (the angle between the long axis of the talus and the first metatarsal) measured from a weight-bearing lateral view (Fig. 30.2).

Clinical evidence of a varus hindfoot was apparent in 16 ft (76.2 %), whereas radiographic measurements revealed varus alignment in 18 ft (85.7 %). The authors proposed that varus hindfoot alignment is a risk factor for the development of Jones fracture.

Lee et al. (2011) reported radiographic data from 50 consecutive athletes with a diagnosis of fifth metatarsal stress fracture (case group). Relative to the control group, the case group exhibited a significantly larger talocalcaneal angle and calcaneal pitch on lateral radiographs (Fig. 30.3).

On the anteroposterior radiographic view, the case group exhibited significantly greater lateral deviation of the fifth metatarsal (Fig. 30.4a) and a significantly

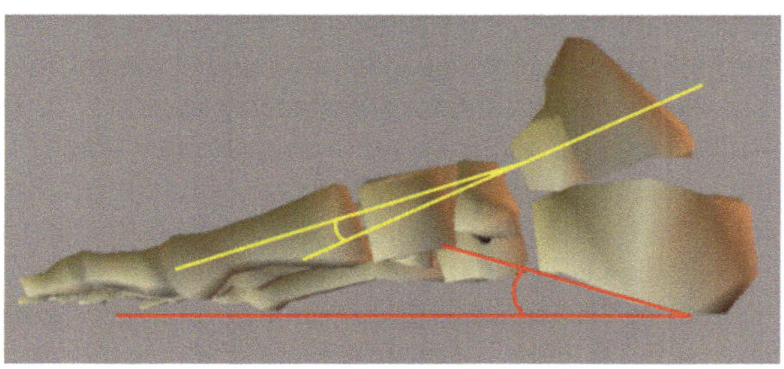

Fig. 30.2 Calcaneal pitch angle (*red line*) and talar-first metatarsal angle (*yellow line*)

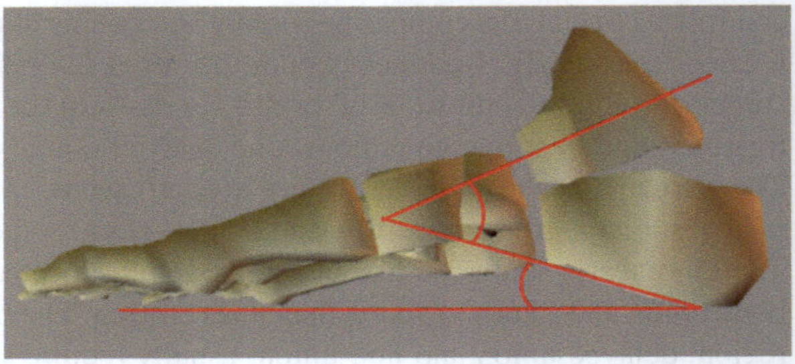

Fig. 30.3 Calcaneal pitch angle and talocalcaneal angle

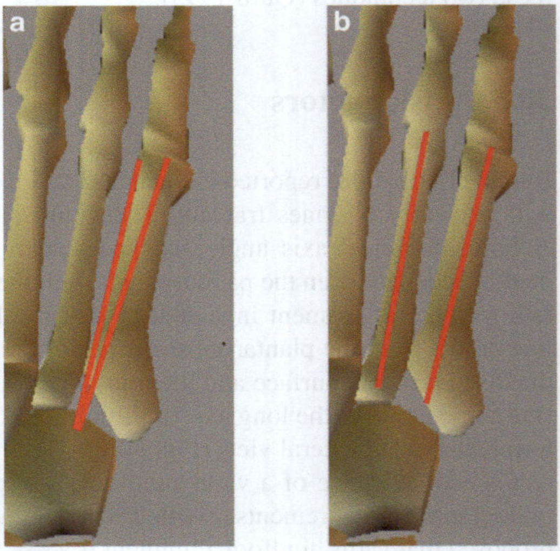

Fig. 30.4 (**a**) Fifth metatarsal lateral deviation angle (*left*) and (**b**) fourth-fifth intermetatarsal angle (*right*)

smaller fourth-fifth intermetatarsal angle (Fig. 30.4b). The authors concluded that a large talocalcaneal angle and calcaneal pitch angle, representing a cavus foot, and increased curvature of the fifth metatarsal were associated with fifth metatarsal stress fracture.

Yoho et al. (2012) reported a relationship between a transverse plane forefoot and Jones fracture. The forefoot adduction angle in the transverse plane was assessed from anteroposterior radiographs. The Jones fracture group exhibited a significantly larger forefoot adductus angle than the control group. Another study also measured the radiographic forefoot adductus angle in patients with stress fractures of the fifth and fourth metatarsal bones (Theodorou et al. 1999). The author noted that patients with metatarsus adductus are overrepresented among those with stress fractures of the lateral metatarsal bone. Therefore, forefoot adductus may contribute to the risk for Jones fracture and stress fracture of the fifth metatarsal.

Williams et al. (2001) investigated the differences in injury pattern between 20 high-arched runners and 20 low-arched runners. The high-arched runners sustained three stress fractures of the fifth metatarsal. On the other hand, Hetsroni et al. (2010)

reported that static measurements of the foot and arch structure revealed no differences among feet with fifth metatarsal stress fracture, the same patients' contralateral uninjured feet, and control feet. Therefore, scientific evidence does not consistently support a relationship between clinical assessment of foot structure and proximal fifth metatarsal fracture (including Jones fracture and stress fracture).

30.6 The Stress on the Lateral Part of the Foot

Lawrence et al. (1993) reported that Jones fracture is caused by adduction of the forefoot and avulsion fracture by inversion of the hindfoot. Sir Robert Jones described his own injury (Jones fracture) as having occurred when he trod on the outer side of his foot, at a moment in which his heel was off the ground (Jones 1902). Kavenaugh et al. (1978) performed force platform analysis and reported that concentration of vertical and mediolateral forces over the fifth metatarsal causes fracture of the proximal part of the diaphysis of the fifth metatarsal, thus confirming that this injury is not caused by inversion of the foot.

The strain on the fifth metatarsal has been measured using a strain gauge during a simulation of normal walking (Donahue et al. 2000). Peak fifth metatarsal strain was affected by the peak Achilles tendon tension and peak vertical ground reaction force at a point approximately 80 % into the stance phase of walking. Gu et al. (2010) investigated stresses on the metatarsals during landing using a three-dimensional (3D) finite element foot model. They revealed that inversion of the foot by 20° increased the stress on the fifth metatarsal relative to a normal landing.

Orendurff et al. (2009) recorded the plantar pressure on the fifth metatarsal of the right foot during running in a straight line, a jump take-off, a jump landing, as well as while cutting right, cutting left, and accelerating. These authors reported that the highest peak pressure on the fifth metatarsal head occurred during acceleration, and that the peak pressure was significantly lower during cutting left than under all other conditions. The peak pressure at the base of the fifth metatarsal was significantly greater during cutting right than during all other maneuvers. The greatest bending moment of the fifth metatarsal was observed during acceleration. Queen et al. (2007) recorded plantar pressure during side cutting, acceleration, and crossover-cutting tasks and reported that the crossover-cutting task produced the greatest pressure on the lateral forefoot.

References

Carp L (1927) Fracture of the fifth metatarsal bone: with special reference to delayed union. Ann Surg 86(2):308–320

Clapper MF, O'Brien TJ, Lyons PM (1995) Fractures of the fifth metatarsal. Analysis of a fracture registry. Clin Orthop Relat Res 315:238–241

Dameron TB Jr (1995) Fractures of the proximal fifth metatarsal: selecting the best treatment option. J Am Acad Orthop Surg 3(2):110–114
DeLee JC, Evans JP, Julian J (1983) Stress fracture of the fifth metatarsal. Am J Sports Med 11(5):349–353
Donahue SW, Sharkey NA, Modanlou KA, Sequeira LN, Martin RB (2000) Bone strain and microcracks at stress fracture sites in human metatarsals. Bone 27(6):827–833
Fetzer GB, Wright RW (2006) Metatarsal shaft fractures and fractures of the proximal fifth metatarsal. Clin Sports Med 25(1):139–150, x
Gu YD, Ren XJ, Li JS, Lake MJ, Zhang QY, Zeng YJ (2010) Computer simulation of stress distribution in the metatarsals at different inversion landing angles using the finite element method. Int Orthop 34(5):669–676
Hetsroni I, Nyska M, Ben-Sira D, Mann G, Segal O, Maoz G, Ayalon M (2010) Analysis of foot structure in athletes sustaining proximal fifth metatarsal stress fracture. Foot Ankle Int 31(3):203–211
Jones R (1902) Fracture of the base of the fifth metatarsal bone by indirect violence. Ann Surg 35:697–700
Kavanaugh JH, Brower TD, Mann RV (1978) The Jones fracture revisited. J Bone Joint Surg Am 60(6):776–782
Lawrence SJ, Botte MJ (1993) Jones' fractures and related fractures of the proximal fifth metatarsal. Foot Ankle 14(6):358–365
Lee KT, Kim KC, Park YU, Kim TW, Lee YK (2011) Radiographic evaluation of foot structure following fifth metatarsal stress fracture. Foot Ankle Int 32(8):796–801
Mologne TS, Lundeen JM, Clapper MF, O'Brien TJ (2005) Early screw fixation versus casting in the treatment of acute Jones fractures. Am J Sports Med 33(7):970–975
Nunley JA (2001) Fractures of the base of the fifth metatarsal: the Jones fracture. Orthop Clin North Am 32(1):171–180
Orendurff MS, Rohr ES, Segal AD, Medley JW, Green JR 3rd, Kadel NJ (2009) Biomechanical analysis of stresses to the fifth metatarsal bone during sports maneuvers: implications for fifth metatarsal fractures. Phys Sportsmed 37(2):87–92
Porter DA, Duncan M, Meyer SJ (2005) Fifth metatarsal Jones fracture fixation with a 4.5-mm cannulated stainless steel screw in the competitive and recreational athlete: a clinical and radiographic evaluation. Am J Sports Med 33(5):726–733
Queen RM, Haynes BB, Hardaker WM, Garrett WE Jr (2007) Forefoot loading during 3 athletic tasks. Am J Sports Med 35(4):630–636
Quill GE Jr (1995) Fractures of the proximal fifth metatarsal. Orthop Clin North Am 26(2):353–361
Raikin SM, Slenker N, Ratigan B (2008) The association of a varus hindfoot and fracture of the fifth metatarsal metaphyseal-diaphyseal junction: the Jones fracture. Am J Sports Med 36(7):1367–1372
Rammelt S, Heineck J, Zwipp H (2004) Metatarsal fractures. Injury 35(Suppl 2):SB77–SB86
Reese K, Litsky A, Kaeding C, Pedroza A, Shah N (2004) Cannulated screw fixation of Jones fractures: a clinical and biomechanical study. Am J Sports Med 32(7):1736–1742
Rettig AC, Shelbourne KD, Wilckens J (1992) The surgical treatment of symptomatic nonunions of the proximal (metaphyseal) fifth metatarsal in athletes. Am J Sports Med 20(1):50–54
Richli WR, Rosenthal DI (1984) Avulsion fracture of the fifth metatarsal: experimental study of pathomechanics. AJR Am J Roentgenol 143(4):889–891
Rosenberg GA, Sferra JJ (2000) Treatment strategies for acute fractures and nonunions of the proximal fifth metatarsal. J Am Acad Orthop Surg 8(5):332–338
Smith JW, Arnoczky SP, Hersh A (1992) The intraosseous blood supply of the fifth metatarsal: implications for proximal fracture healing. Foot Ankle 13(3):143–152
Stewart IM (1960) Jones's fracture: fracture of base of fifth metatarsal. Clin Orthop 16:190–198
Strayer SM, Reece SG, Petrizzi MJ (1999) Fractures of the proximal fifth metatarsal. Am Fam Physician 59(9):2516–2522

Theodorou DJ, Theodorou SJ, Boutin RD, Chung C, Fliszar E, Kakitsubata Y, Resnick D (1999) Stress fractures of the lateral metatarsal bones in metatarsus adductus foot deformity: a previously unrecognized association. Skeletal Radiol 28(12):679–684
Torg JS, Balduini FC, Zelko RR, Pavlov H, Peff TC, Das M (1984) Fractures of the base of the fifth metatarsal distal to the tuberosity. Classification and guidelines for non-surgical and surgical management. J Bone Joint Surg Am 66(2):209–214
Wiener BD, Linder JF, Giattini JF (1997) Treatment of fractures of the fifth metatarsal: a prospective study. Foot Ankle Int 18(5):267–269
Williams DS 3rd, McClay IS, Hamill J (2001) Arch structure and injury patterns in runners. Clin Biomech (Bristol, Avon) 16(4):341–347
Wright RW, Fischer DA, Shively RA, Heidt RS Jr, Nuber GW (2000) Refracture of proximal fifth metatarsal (Jones) fracture after intramedullary screw fixation in athletes. Am J Sports Med 28(5):732–736
Yoho RM, Carrington S, Dix B, Vardaxis V (2012) The association of metatarsus adductus to the proximal fifth metatarsal jones fracture. J Foot Ankle Surg 51(6):739–742
Zelko RR, Torg JS, Rachun A (1979) Proximal diaphyseal fractures of the fifth metatarsal–treatment of the fractures and their complications in athletes. Am J Sports Med 7(2):95–101
Zwitser EW, Breederveld RS (2010) Fractures of the fifth metatarsal; diagnosis and treatment. Injury 41(6):555–562

Part VII
Lumbar Disorder and an Approach for Its Prevention

Chapter 31
Low Back Disorders Among Athletes and Its Prevention

Koji Kaneoka

Abstract There are several pathomechanisms of lower back disorders among athletes such as lumbar intervertebral disc herniations, lumbar spondylolysis, facet joint disorders, sacroiliac joint disorders, muscle, fascia and its insertion disorders. In many cases except intervertebral disc herniations, there are less radiological findings so they should be diagnosed using functional classification. In order to decrease the mechanical stress to the pathological portion, it is thought to be important to maintain the proper spinal alignment (Neutral position) and the spinal segmental stability. The stabilization exercises which facilitates the deep trunk muscles (transversus abdominis, lumbar multifidus) function and the stretching for the muscles attached to the pelvis are recommended to achieve these tasks.

Keywords Low back pain • Athlete • Sports • Stabilization exercise • Deep trunk muscle

31.1 Introduction

The various physical activities engaged in by sports athletes physically stress the lumbar vertebrae, and carry a risk of developing lower back pain (LBP). We surveyed the relationship between sports competition history and incidence of LBP in 4,667 freshmen belonging to a physical education department (Hangai et al. 2010). The results showed a 50 % LBP incidence in the non-exercise group of subjects with no history of sports competition. In contrast, subjects in the moderate stress group, who had played competitive sports in either primary, junior high, or high school, showed an incidence of 62 %, whereas the high stress group, who had played competitive sports throughout primary, junior high, and high school, had an incidence of 72 %. Thus, more years in competitive sports led to higher LBP rates, indicating that physical stress from competitive sports is a risk factor for LBP. Furthermore, the survey results on LBP experience rate by specific

K. Kaneoka (✉)
Faculty of Sport Sciences, Waseda University, Saitama, Japan
e-mail: kaneoka@waseda.jp

K. Kanosue et al. (eds.), *Sports Injuries and Prevention*,
DOI 10.1007/978-4-431-55318-2_31

competitive sports indicated that volleyball players were the most likely to experience LBP (odds ratio: 3.8), followed by baseball, track and field, basketball, swimming, kendo fencing, tennis, and soccer. These results indicate that the physical movements and practice styles specific to each sport lead to different levels of LBP frequency.

31.2 The Pathomechanisms of LBP (Fig. 31.1)

The pathomechanisms of LBP vary according to age. Sports-related stress on immature spinal vertebrae during the growth phase can cause end-plate disorders, and stress fractures of the vertebral arch (spondylolysis). Intervertebral disc degeneration begins in youth and adulthood, and minute damage to intervertebral discs can allow nerve tissue to growth into the annulus fibrosus, causing intervertebral discogenic pain. Discogenic LBP can also be brought about by mechanical weakness that causes the nucleus pulposus to puncture the annulus fibrosus. Further, a simultaneous reduction in intervertebral disc height can decrease disc load functions, which increases stress on the facet joints and can easily cause facet joint pain. Degenerative changes progress with age, leading to osteoarthritis of the spine and stenosis of the spinal canal.

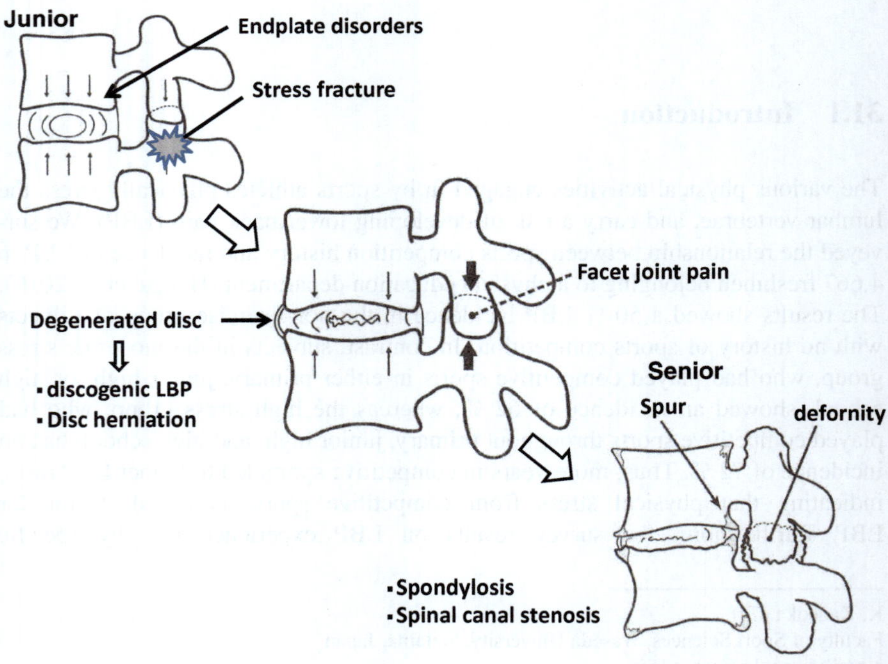

Fig. 31.1 Generational characteristics of LBP

31.2.1 Lumbar Intervertebral Disc Disorders/Disc Herniation

Reduced levels of proteoglycans, which generally occurs during the aging process, decrease water content in the nucleus pulposus, causing further degeneration of the lumbar intervertebral discs. Several studies are currently investigating the risk factors associated with intervertebral disc degeneration that accompanies spinal degeneration. The results of an epidemiological survey we conducted on middle-aged and elderly subjects showed that a history of sports activity is a risk factor, in addition to advanced age, high BMI, high levels of low-density lipoprotein (LDL) cholesterol, and heavy labor (Hangai et al. 2008). We also conducted a cross-sectional survey comparing lumbar intervertebral disc degeneration rates among members of sports clubs at a physical education university and subjects with no sports history to clarify the relationship between sports activity and intervertebral disc degeneration. The results showed significantly higher rates of disc degeneration in baseball players (60 %) and competitive swimmers (58 %) compared to the control group (31 %) (Hangai et al. 2009). Comparing LBP severity with the presence of intervertebral disc degeneration showed higher rates of disc degeneration as LBP severity increased. Several previous surveys have failed to clarify the relationship between LBP and lumbar intervertebral disc degeneration. However, such a relationship has been suggested in subjects with ages close to or around 20 years.

We are carrying out a continuous cross-sectional survey on intervertebral disc degeneration based on the type of sport. The results have shown high rates of disc degeneration in volleyball players (69 %), weightlifters (62 %), and rowers (60 %). These rate differences in disc degeneration among sports led us to believe that the specific physical actions of each sport apply stress to the intervertebral discs, which contribute to degeneration.

Intervertebral disc herniation is broadly divided into bulging type, subligamentous protrusion type, transligamentous protrusion type (protrusion into the spinal canal through the posterior longitudinal ligament), and sequestration type, with each type having a different natural process, prognosis, and treatment. Further, in cases of congenital spinal canal stenosis, even small herniated discs can compress the nerve root between the herniation and the vertebral arch. In these cases, even if the severe early symptoms of nerve root stimulation improve, later stimulation such as through exercise can cause continuous LBP and leg pain. Such compression-type nerve root disorders are difficult to treat and many require surgery. The Kemp test often generates positive results in these cases.

31.2.2 Extension Type Low Back Disorders

Lumbar vertebral arch stress fracture (lumbar spondylolysis), facet joint disorders, and repeated extension or rotation of the lower back and trunk focuses stress on the area between the articular process of the vertebral arch (Fig. 31.2), which can cause

Fig. 31.2 Repeated extension or rotation of the lower back and trunk focuses stress on the area between the articular process of the vertebral arch

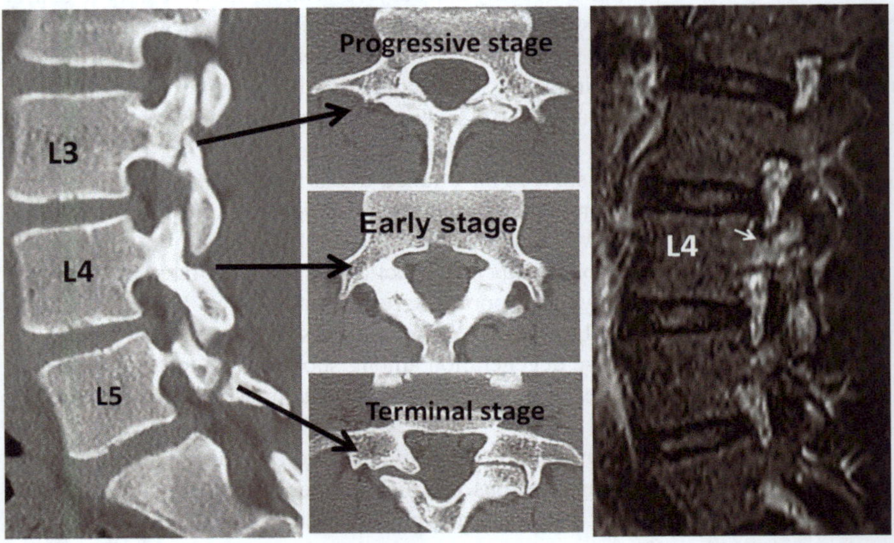

Fig. 31.3 Lumbar spondylolysis is classified into three stages: early stage, progressive stage, and terminal stage. MRI indicates high intensity change at L4 pedicle (early stage)

stress fractures in this area during the growth stage (Terai et al. 2010) After the bones mature, this kind of stress on the facet joint can lead to disorders involving this joint that present with pain when extension of the lumbar vertebrae is restricted and at the end of extension, as well as pressure pain on the spinous processes of the affected vertebrae. LBP can also be induced by restricting extension and rotation of the lumbar vertebrae (Kemp maneuver) (Note: When there is nerve root compression due to spinal canal stenosis in the lower back, this maneuver will induce pain in the legs and the result of the Kemp test will be positive). Because the nerve root runs ventrally through the area where the vertebral arch separates, growth of fibrous cartilage in this area can put pressure on the nerve root, inflammation of the facet joint, or area of separation can spread to the nerve root, and any of the above can cause leg pain or numbness. Lumbar spondylolysis is classified into three stages: early stage, progressive stage, and terminal stage (Fig. 31.3). Because it is difficult to render the area of separation using simple radiographic oblique views in the early and progressive stages, CT imaging is needed for early diagnosis. Synostosis can be achieved through conservative treatments in the early and progressive stages, in which high intensity changes which indicates bone edema are visible in the interarticular portion of the vertebral arch in magnetic resonance imaging (MRI). Orthotic treatments that restrict extension and rotation of the trunk are used in such cases. However, conservative treatments are not likely to achieve synostosis in terminal stage cases that exhibit clear signs of pseudoarthrosis in simple X-ray images. Thus, treatment of terminal stage lumbar spondylolysis focuses on LBP countermeasures such as stretching the muscle groups of the trunk and legs, and training the muscles of the trunk. Even in cases where separation cannot be confirmed in image findings, cases with similar spinal findings should be considered as a preliminary stage (early stage spondylolysis) .

31.2.3 Disorders of Muscle, Fascia, and Muscle Insertion

Stress on the muscles and fascia of the lumbar area from sports activities can cause muscle- or fascia-type LBP. Disorders frequently occur in the area in which the erector spinae muscles attach to the iliac crest.

31.2.4 Sacroiliac Joint Disorders

Putting weight on one leg puts a large amount of stress on the sacroiliac joint, which can cause disorders of this joint. Most of these cases present with a chief complaint of LBP, but when these patients are asked to point to the area of pain with the index finger, it is often limited to the posterior superior iliac spine (one finger test) (Murakami. 2007). Spinal findings vary among cases, with anteflexion causing pain in some cases, whereas extension induces the same effect in others. Severe cases present with pain when standing on one leg or running. Various tests that stress the sacroiliac joint are used, such as the revised Newton test, which puts pressure on the ilium in the prone position; the Patrick test, which forces abduction in flexion of the hip joint in the supine position; and the Gaenslen test, which forces extension of the hip joint. Some cases present with related leg pain, and because stressing the sacroiliac joint using the SLR test can induce pain, such cases must be differentiated from intervertebral disc herniation. Diagnosis is made through physical findings because no characteristic image findings have been established for this condition.

31.3 Treatment and Prevention

Stabilization of the trunk is considered as an important factor in increasing sports performance, and many athletes train the deep trunk muscles, which includes the transversus abdominis muscle attached to the transverse process of the lumbar vertebrae through the lumbar fascia surrounding the trunk and the lumbar multifidus muscle connecting the spinous processes of the lumbar vertebrae with the transverse processes of the neighboring vertebrae. The transversus abdominis muscles are mainly activated by drawing-in the abdomen, and contracting these muscles increase the rigidity of the lumbar spine by 44 % in cadaver experiments (Barker et al. 2006). Figure 31.4 shows a cross-sectional MR image of the trunk of a well-trained athlete. The lateral abdominal muscle groups of this athlete are well developed, and the white arrow shows an increased thickness of the transversus abdominis muscle upon drawing-in of the abdomen. Training these deep muscle groups in the trunk increases stability of the trunk during exercise, and is thought to prevent low back disorders.

Further, the symptoms of intervertebral discogenic LBP and intervertebral disc herniation are increased when pressure is applied to the intervertebral discs by anteflexion of the lumbar vertebrae, and the symptoms of facet joint-type LBP and lumbar spondylolysis are increased by extension. Thus, increasing lumbar lordosis is desirable in cases of anteflexion-type LBP, and decreasing lordosis is sought in extension-type LBP. The angle of inclination of the pelvis is a factor in regulating lumbar vertebrae alignment, and the contractile strength of the muscle groups that attach to the pelvis is thought to influence pelvic inclination. We analyzed the activity of the muscles surrounding the pelvis during active anterior tilting and posterior tilting of the pelvis by attaching wire electrodes to the deep trunk muscles (transversus abdominis muscle, multifidus muscle) and surface electrodes to the other muscle groups surrounding the pelvis (Takaki et al. 2010). This showed strong activity

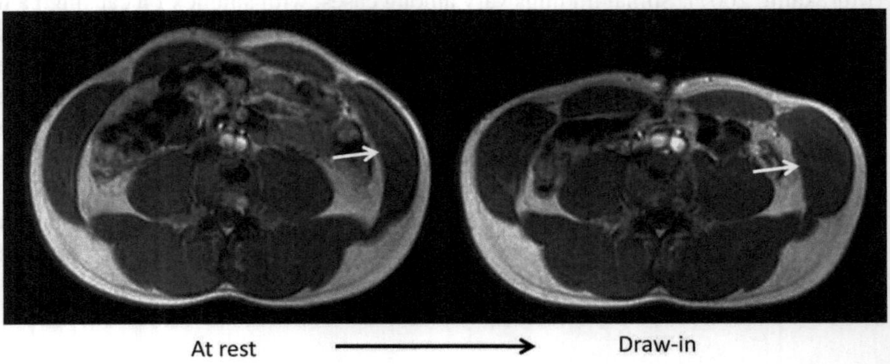

Fig. 31.4 Cross-section MRI image of the trunk of a well-trained athlete

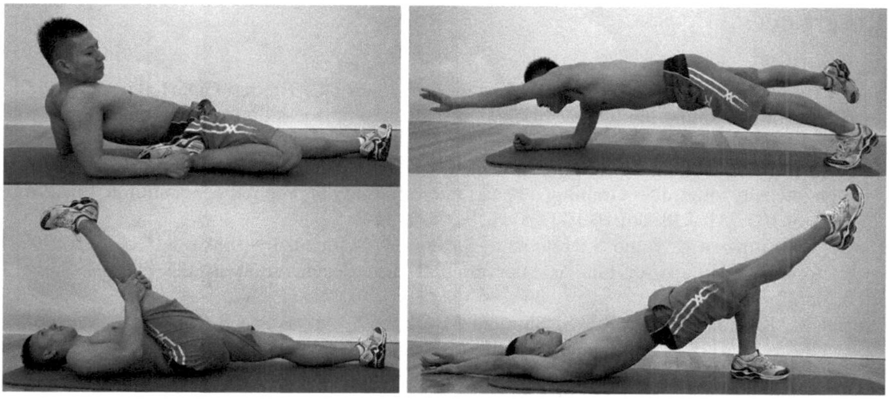

Fig. 31.5 Representative leg stretching and trunk muscle training methods. *Upper left*:streching for quadriceps femoris. *Lower left*: streching for hamstrings. *Upper right*: elbow-toe bridge: activates transversus abdominis. *Lower right*: back bridge: activates multifidus

of the bilateral multifidus muscle and erector spinae muscle during active anterior tilting of the pelvis, indicating that these muscle groups lift the pelvis posteriorly during anterior tilting. Further, significant muscle activity was observed in the transversus abdominis muscle during posterior tilting, showing that contraction of these muscles, which broadly surround the trunk and attach to the wings of the ilium, lift the pelvis anteriorly during posterior tilting.

Theoretically, this shows that to maintain lumbar lordosis in anteflexion-type LBP, it is important to strengthen the multifidus muscle and stretch the hamstrings to induce anteversion of the pelvis. Similarly, to reduce lordosis in extension-type LBP, it is important to strengthen the transverse abdominis muscle and to stretch the quadriceps to induce retroversion of the pelvis.

Figure 31.5 shows representative leg stretching and trunk muscle training methods. Generally, the deep trunk muscles trained by maintaining unstable postures, as shown in Fig. 31.5 on the right, and a variety of postures are used in upper sports and rehabilitation. We analyzed muscle activity in subjects exercising the deep trunk muscles by attaching wire electrodes to the transversus abdominis and multifidus muscles. The results showed maximum activity in the transverse abdominis muscle in the elbow-toe position (Fig. 31.5, upper right) and maximum multifidus muscle activity in the back bridge one-leg raised position (Fig. 31.5, lower right) (Okubo et al. 2010). Training methods based on these results could not only more effectively prevent LBP, but also improve sports performance.

Exercise therapies described in this report are designed not only for athletes, but are also indicated in regular LBP patients, and we expect them to become more commonly used. However, because the training methods shown in Fig. 31.5 require a great deal of strength, methods that match the level of each patient and incremental increases in strength would therefore be necessary (Gamelin et al. 2009).

References

Barker PJ, Guggenheimer KT, Grkovic I, Briggs CA, Jones DC, Thomas CD, Hodges PW (2006) Effects of tensioning the lumbar fasciae on segmental stiffness during flexion and extension. Spine (Phila Pa 1976) 31:397–405

Gamelin FX, Baquet G, Berthoin S, Thevenet D, Nourry C, Nottin S, Bosquet L (2009) Effect of high intensity intermittent training on heart rate variability in prepubescent children. Eur J Appl Physiol 105:731–738. doi:10.1007/s00421-008-0955-8

Hangai M, Kaneoka K, Kuno S, Hinotsu S, Sakane M, Mamizuka N, Sakai S, Ochiai N (2008) Factors associated with lumbar intervertebral disc degeneration in the elderly. Spine J 8(5):732–740

Hangai M, Kaneoka K, Hinotsu S, Shimizu K, Okubo Y, Miyakawa S, Mukai M, Sakane M, Ochiai N (2009) Lumbar intervertebral disc degeneration in athletes. Am J Sports Med 37:149–155

Hangai M, Kaneoka K, Okubo Y, Miyakawa S, Hinotsu S, Mukai N, Sakane M, Ochiai N (2010) Relationship between low back pain and competitive sports activities during youth. Am J Sports Med 38:791–796

Murakami E (2007) Low back pain derived from sacroiliac joint (Japanese). J Jpn Soc Lumbar Spine Disord 13:40–47

Okubo Y, Kaneoka K, Imai A, Shiina I, Tatsumura M, Izumi S, Miyakawa S (2010) Electromyographic analysis of transversus abdominis and lumbar multifidus using wire electrodes during lumbar stabilization exercises. J Orthop Sports Phys Ther 40:743–750

Takaki S, Kaneoka K, Okubo Y, Otsuka S, Miyamoto K, Tatsumura M, Shiina I, Miyakawa S (2010) Analysis of the muscle activity during active pelvic tilting motion. Abstract of 37th International society for the study of the lumbar spine

Terai T, Sairyo K, Goel VK, Ebraheim N, Biyani A, Faizan A, Sakai T, Yasui N (2010) Spondylolysis originates in the ventral aspect of the pars interarticularis: a clinical and biomechanical study. J Bone Joint Surg Br 92(8):1123–1127

Chapter 32
The Epidemiology of Low Back Disorders in Athletes

Mika Hangai

Abstract Low back disorders may have a serious effect on an athlete's performance. In previous reports, low back pain has been particularly associated with sports activities during youth. In addition, low back disorders such as lumbar disk degeneration have been associated with competitive sports activities. Furthermore, we reported that the incidence of lumbar disk degeneration among baseball players and swimmers was significantly higher than that of non-athletes. The relationships between positions in sports and low back disorders have also been also examined.

In this chapter we review these relationships, and also examine low back disorders of Japanese top athletes.

Keywords Lumbar disk degeneration • Lumbar disk herniation • Spondylolysis • Competitive sports

32.1 Introduction

The lifetime prevalence of low back pain (LBP) in the general adult population is estimated to be 60–90 % (Hangai et al. 2008; Long et al. 1996; Trainor and Wiesel 2002). Accordingly, it is difficult to demonstrate that LBP is the result of sports injuries. However, LBP strongly affects an athlete's performance, and proper recovery allows the athletes to return to their preinjury level of competition. There is no doubt that LBP is a major problem for athletes. Lumbar disk herniation, disk degeneration, and spondylolysis are the most common low back disorders associated with LBP in athletes. On the other hand, despite athletes typically being highly motivated to return to sports activity, a specific pain generator is not always found. Recently, such disorders were categorized as nonspecific LBP. It is reported that more than 80 % of low back disorders are included in nonspecific LBP.

In this chapter, we review the literature concerning LBP and low back disorders in athletes.

M. Hangai (✉)
Department of Orthopaedic Surgery, Medical Center, Japan Institute of Sports Sciences, Tokyo, Japan

K. Kanosue et al. (eds.), *Sports Injuries and Prevention*,
DOI 10.1007/978-4-431-55318-2_32

32.2 The Relationships Between Sports Activities and Low Back Disorders

The relationship between sports activities and LBP has largely focused on the effect of sports during youth (Harreby et al. 1999; Kovacs et al. 2003; Kujala et al. 1996; Mattila et al. 2008; Skoffer and Foldspang 2008). Sato et al. reported that in a total of 43,630 Japanese pupils, the no-sports group reported a 21.3 % incidence of LBP, while, in the sports group, 34.9 % experienced LBP (Sato et al. 2010). Sward et al. reported that the incidence of LBP in elite gymnasts (79 %) was significantly higher than that of the control group (38 %) (Sward et al. 1990). We also documented a relationship between LBP and competitive sports activities during youth (Hangai et al. 2010a). The participants in our study were 4667 new university students who, from 2004 to 2006, answered a questionnaire concerning LBP and their participation in competitive sports. The participants were divided into a "no" group (NO), a middle group (MID), and a high group (HI) based on the duration of their participation in competitive sports. There were statistically significant linear associations in the NO, MID and HI groups with 50.0, 61.8, and 71.7 % of the students experiencing LBP, respectively (Fig. 32.1).

It has been reported that lumbar spondylolysis is a fatigue fracture that causes pars interarticularis, and occurs during youth (Sairyo et al. 2003; Sakai et al. 2010; Wiltse 1962). Imamoto et al. reported that an abnormality such as spondylolysis is the most significant risk factor for LBP in high school and college football players, and that disk space narrowing and spinal instability are significant risk factors for

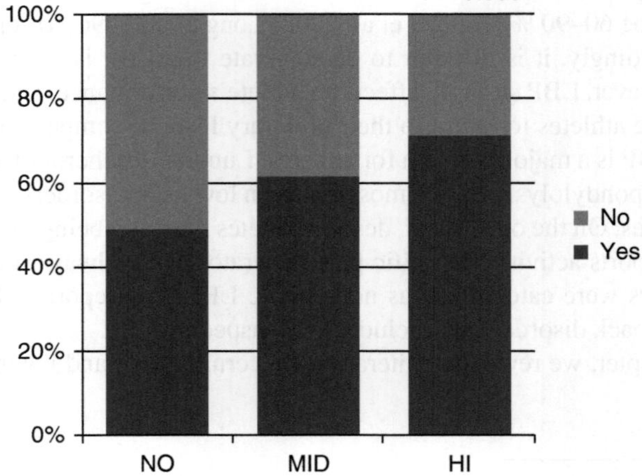

Fig. 32.1 Linear associations between low back pain experienced during the students' lifetime and 3 groups based on the duration of sports experience (Hangai et al. 2010a). *No* no sports, *MID* middle sports experience, *HI* high sports experience

LBP in athletes with high levels of athletic activity, such as a college football players (Iwamoto et al. 2004).

Therefore, we suggest that excessive exposure to competitive sports activities is associated with LBP and low back disorders. This influence appears to be particularly strong during youth. On the other hand, some studies have found that competitive sports activities are not associated with LBP (Balague et al. 1993; Feldman et al. 2001; Mogensen et al. 2007).

32.3 Low Back Disorders of Top Athletes

We investigated the disorders of top athletes who visited the orthopedics department of the Japan Institute of Sports Sciences (JISS) from 2008 to 2012 (Hangai et al. 2013). In Fig. 32.2, the incidence by anatomical region is shown. Of 7,873 disorders reported, disorders of the knee joint were the most frequent (22.6 %), followed by low back disorders (10.1 %). Of the low back disorders, nonspecific LBP was reported most frequently at 33.6 %, followed by lumbar disc herniation at 18.9 %

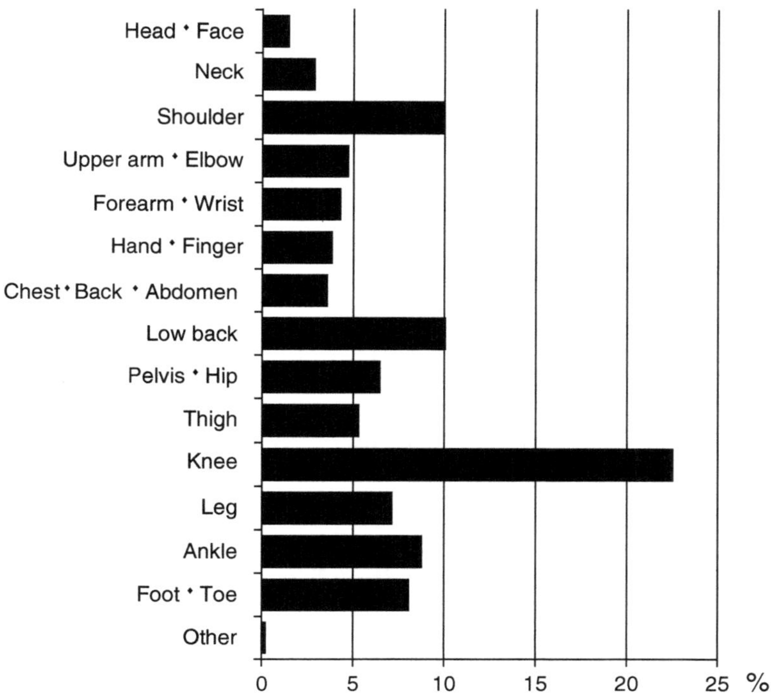

Fig. 32.2 Incidence by anatomical region of disorders in athletes who visited the Japan Institute of Sports Sciences (Hangai et al. 2013)

and lumbar disc degeneration at 17.0 %. Thus, lumbar disk disorders are strongly associated with LBP in athletes.

Other studies report that the incidence of spondylolysis for Japanese professional soccer and baseball players with or without pain is much higher (30 %), at >5 times the incidence that occurs in the general Japanese population (Matsumoto 2006; Sakai et al. 2009, 2010). However, in our study of JISS, spondylolysis was reported by 12.7 % of participants with low back disorders. The mean age of our participants (22.9 years) was older than the age considered susceptible to the development of spondylolysis. When spondylolysis is complete, the pain often disappears, which may have influenced our findings.

32.4 The Relationships Between Type of Sport, Position, and Low Back Disorders

In our study of new university students mentioned above, we selected students who had participated in the same sport for 5 or more years, and categorized the students according to the type of sport. Differences in LBP among the groups were analyzed using logistic regression with the NO as the reference group. All 8 sports groups that we analyzed experienced LBP at a significantly higher rate than the NO group, with the highest odds ratios in the volleyball group (3.8) (Fig. 32.3).

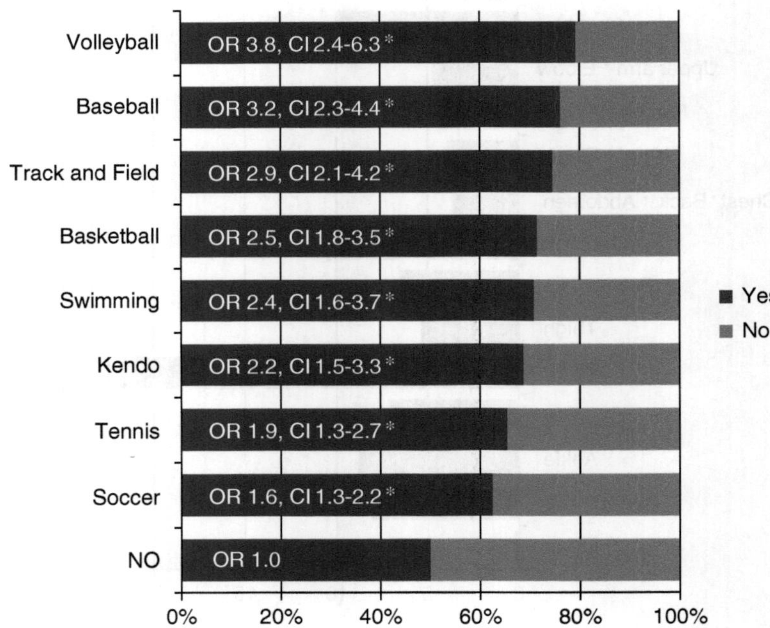

Fig. 32.3 Logistic regression analysis of low back pain during the students' lifetime by sport group (Hangai et al. 2010a). *OR* odds ratio, *CI* 95 % confidence interval. * P<.05

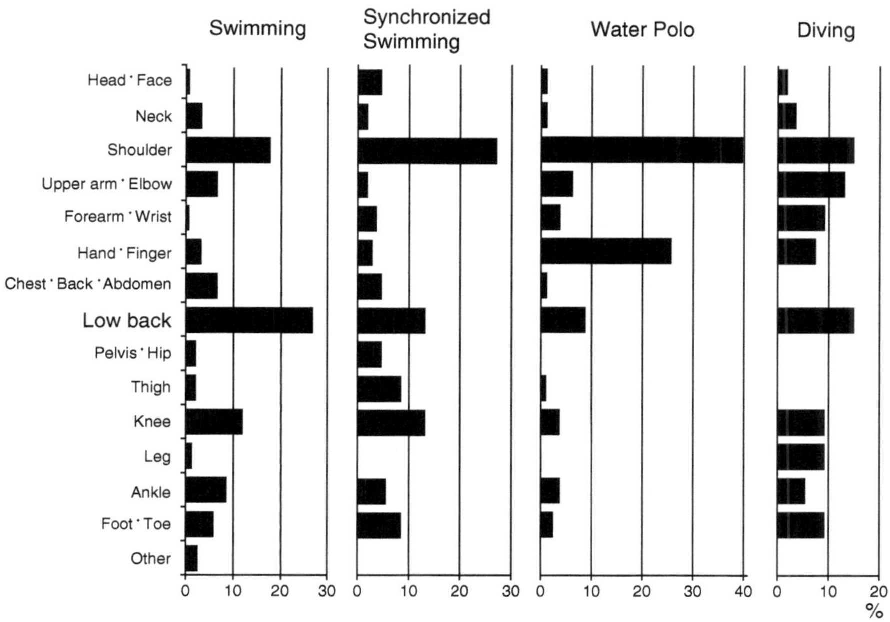

Fig. 32.4 Incidence by anatomical region of orthopedic disorders and type of sports in elite aquatic athletes (Hangai et al. 2010b)

Several other studies have also reported an association between the type of competitive sport activity and low back disorders. In the investigation of athletes at the JISS, swimmers reported a high incidence of low back disorders (22.8 %). On the other hand, few soccer athletes (2.8 %) and track and field athletes (5.9 %) reported low back disorders. We also compared the difference in the anatomical region of the disorders by type of sport in aquatic athletes who visited the JISS from 2001 to 2007 (Hangai et al. 2010b). Swimmers showed the highest incidence of low back disorders (Fig. 32.4). Morikita reported that the proportion of low back disorders in young volleyball athletes was 71.4 %, while the ankle disorders were even more frequent at 77.6 % (Morikita 2002). In badminton, low back disorders (19 %) and shoulder disorders (18 %) were reported as the two major disorders (Uchiyama 2007).

We examined the relationship between lumbar disk degeneration, which is a cause of LBP, and the type of sport. Our subjects were 308 well-trained university athletes that included baseball players, basketball players, kendo competitors, runners, soccer players, swimmers and 71 non-athletic university students (reference group). The proportions of lumbar disk degeneration among baseball players (odds ratio, 3.23) and the swimmers (odds ratio, 2.95) were significantly higher than those of non-athletes (Hangai et al. 2009). The results did not change by logistic regression adjusted for gender and BMI (Fig. 32.5). In Spanish elite athletes, the percentages

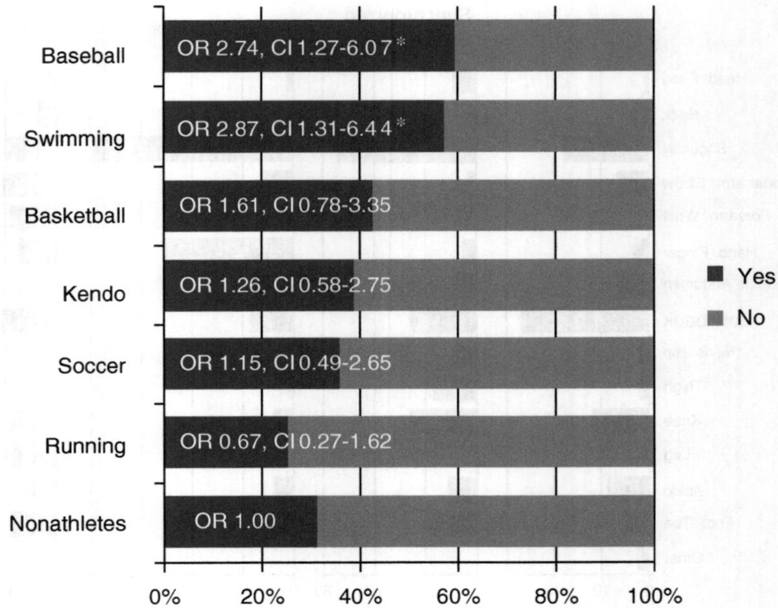

Fig. 32.5 Logistic regression analysis of lumbar disk degeneration by sports groups (Hangai et al. 2009). *OR* odds ratio, *CI* 95 % confidence interval. * P<.05. OR and CI were adjusted by gender and BMI

of spondylolysis were 26.67 % in throwing sports, 16.96 % in artistic gymnastics, and 16.88 % in rowing (Soler and Calderon 2000). The incidence of lumbar spondylolysis in Japanese amateur athletes participating in various sports has been reported (Sakai et al. 2010).

It is difficult to make precise comparisons of the relationships between sports activities and low back disorders in the preceding studies because the studies were designed differently. However, we are able to conclude that the frequency of low back disorders varies according to the type of sport.

We also investigated whether the proportion of low back disorders was different among the specialties of swimmers who visited the JISS. Figure 32.6 shows that the incidence was different among their specialties (Hangai et al. 2010b). Tajima et al. showed that outfielders frequently reported low back disorders, while pitchers often reported elbow injuries (Tajima and Kuwahara 2000). It was also reported that elite cross-country skiers who used the “classical method” experienced LBP more than the skiers who used the “skating method” only (Eriksson et al. 1996). These reports make it clear that, within a particular sport, position or style of performance also affect the frequency of low back disorders.

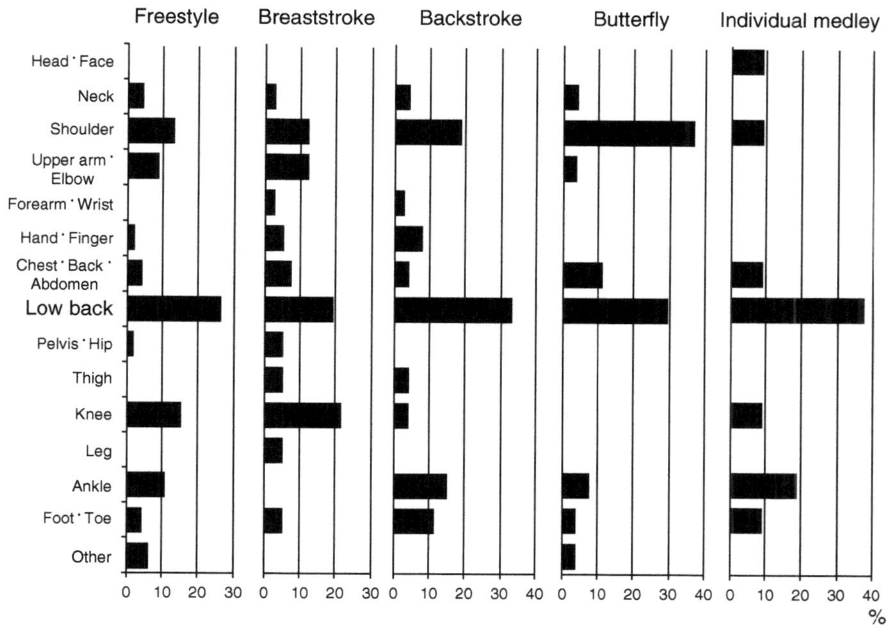

Fig. 32.6 Incidence by anatomical region of orthopedic disorders and specialty in elite swimmers (Hangai et al. 2010b)

32.5 Conclusions

In this review, we confirmed the association between excessive exposure to competitive sports and both LBP and low back disorders. The incidence of LBP and low back disorders is also affected by type of sport and position or style of performance. Future studies are needed to further examine the relationships between specific postures and actions and low back disorders.

References

Balague F, Damidot P, Nordin M et al (1993) Cross-sectional study of the isokinetic muscle trunk strength among school children. Spine 18:1199–1205

Eriksson K, Németh G, Eriksson E (1996) Low back pain in elite cross-country skiers. A retrospective epidemiological study.Scand. J Med Sci Sports 6(1):31–35

Feldman DE, Shrier I, Rossignol M et al (2001) Risk factors for the development of low back pain in adolescence. Am J Epidemiol 154:30–36

Hangai M, Kaneoka K, Kuno S et al (2008) Factors associated with lumbar intervertebral disc degeneration in the elderly. Spine J 8:732–740

Hangai M, Kaneoka K, Hinotsu S et al (2009) Lumbar intervertebral disk degeneration in athletes. Am J Sports Med 37:149–155
Hangai M, Kaneoka K, Okubo Y et al (2010a) Relationship between low back pain and competitive sports activities during youth. Am J Sports Med 38:791–796
Hangai M, Kaneoka K, Okuwaki T (2010b) Sports injuries in elite swimmers. Jpn J Orthop Sports Med 30:161–166
Hangai M, Nakajima K, Nakamura K et al (2013) Orthopedics medical treatment in Japan Institute of Sports Sciences (JISS) – second report. Jpn J Clin Sports Med 21:S227
Harreby M, Nygaard B, Jessen T et al (1999) Risk factors for low back pain in a cohort of 1389 Danish school children: an epidemiologic study. Eur Spine J 8:444–450
Iwamoto J, Abe H, Tsukimura Y et al (2004) Relationship between radiographic abnormalities of lumbar spine and incidence of low back pain in high school and college football players a prospective study. Am J Sports Med 32:781–786
Kovacs FM, Gestoso M, Gil del Real MT et al (2003) Risk factors for non-specific low back pain in schoolchildren and their parents: a population based study. Pain 103:259–268
Kujala UM, Taimela S, Erkintalo M et al (1996) Low-back pain in adolescent athletes. Med Sci Sports Exerc 28:165–170
Long DM, BenDebba M, Torgerson WS et al (1996) Persistent back pain and sciatica in the United States: patient characteristics. J Spinal Disord 9:40–58
Matsumoto M (2006) Conservative therapy of low back disorder. J Clin Sports Med 11:1301–1309
Mattila VM, Saarni L, Parkkari J et al (2008) Predictors of low back pain hospitalization -a prospective follow-up of 57,408 adolescents. Pain 139:209–217
Mogensen AM, Gausel AM, Wedderkopp N et al (2007) Is active participation in specific sport activities linked with back pain? Scand J Med Sci Sports 17:680–686
Morikita I (2002) Sports disorders in volleyball. Cent Jpn J Orthop Traumat 45:1021–1022
Sairyo K, Katoh S, Sakamaki T et al (2003) Three successive stress fractures at the same level in an adolescent baseball player. Am J Sports Med 31:606–610
Sakai T, Sairyo K, Takao S et al (2009) Incidence of lumbar spondylolysis in the general population in Japan based on multidetector computed tomography scans from two thousand subjects. Spine 34:2346–2350
Sakai T, Sairyo K, Takao S et al (2010) Incidence and etiology of lumbar spondylolysis; review of the literature. J Orthop Sci 15:281–288
Sato T, Ito T, Hirano T et al (2010) Low back pain in childhood and adolescence: assessment of sports activities. Eur Spine J 20:94–99
Skoffer B, Foldspang A (2008) Physical activity and low-back pain in schoolchildren. Eur Spine J 17:373–379
Soler T, Calderon C (2000) The prevalence of spondylolysis in the Spanish elite athlete. Am J Sports Med 28:57–62
Sward L, Hellstrom M, Jacobsson B et al (1990) Back pain and radiologic changes in the thoraco-lumbar spine of athletes. Spine 5:124–129
Tajima N, Kuwahara S (2000) Low back pain related with sports activities. J Jpn Orthop Assoc 74:545–552
Trainor TJ, Wiesel SW (2002) Epidemiology of back pain in the athlete. Clin Sports Med 21:93–103
Uchiyama E (2007) Badminton. Sports injuries in different athletic events, Authored by National Team Physicians. Nankodo Co. Ltd, Tokyo
Wiltse LL (1962) The etiology of spondylolosthesis. J Bone Joint Surg Am 44:539–560

Chapter 33
The Prevention of Low Back Disorders in Divers

Takaya Narita and Koji Kaneoka

Abstract During competitive diving events, divers jump up from 1 to 3 m springboards or 5–10 m platforms and dive into the water. The impact forces are very large in the water entry phase, and therefore, the divers experience a great deal of physical stress. This chapter is comprised of four sections on the prevention of low back pain(LBP) in divers. The first section is an introduction. The second section describes the very high incidence rate of LBP in Japanese elite junior divers. The third section details the importance of shoulder flexibility and age as critical factors for the prevention of LBP in elite junior divers. The fourth section describes investigations on the influence of the shoulder angle on entry alignment and compares the differences in entry phase alignment between the divers in the LBP diver group and non-LBP diver group. The results showed a negative correlation ($r=-0.623$) between the shoulder flexion angle and trunk extension angle for the entry phase. Four of 13 male divers experienced LBP. The trunk angle showed a significant difference between the LBP group ($210.4\pm5.7°$) and no-LBP group ($199.9\pm9.0°$). Our results suggested that shoulder flexibility is an important factor in the prevention of LBP in elite male junior divers.

Keywords Low back pain • Diving • Rate of incidence • Risk factors

33.1 Introduction

Competitive diving involves jumping off a 10 m-high platform or a 1 or 3 m-high springboard and doing somersaults and twists in the air before entering the water. Divers jumping from a 10 m platform reach velocities of 51 km/h before entering the water and quickly decelerate to 33 km/h upon impact with the water. This amounts to a force of about 400 kg N (Le Viet et al. 1993). Microtraumatic injuries

T. Narita (✉)
Department of Physical Therapy, Health Science University, Yamanashi, Japan
e-mail: t-narita@kenkoudai.ac.jp

K. Kaneoka
Faculty of Sport Sciences, Waseda University, Saitama, Japan

K. Kanosue et al. (eds.), *Sports Injuries and Prevention*,
DOI 10.1007/978-4-431-55318-2_33

are common (Narita et al. 2011; Badman and Rechtine 2004) because of this tremendous physical stress that is placed on a diver. Injury prevention, then, becomes very important. Based on scientific evidence, van Mechelen et al. (1992) formulated a four-step practical model as an effective injury prevention plan. First, investigations into diving injuries should be implemented to understand the incidence rate and severity of trauma. Second, the cause, mechanism and risk factors for the injuries should be determined. Third, a prevention plan should be introduced to counteract the risk factors. Finally, a post-implementation investigation should be conducted to assess the effectiveness of the prevention plan. In this chapter we describe implementation of the above steps in the development of a program to prevent low back pain in divers.

33.2 Second Section: The LBP Incidence Rate in Japanese Elite Junior Divers (First Step)

There are few investigations about injury occurrence in divers, especially for junior divers. In this section we describe our investigations in this area. We sought to describe the prevalence of injuries among elite junior divers over a 7-year period. One hundred and sixteen (60 male, 56 female) Japanese elite junior divers were included in this study (Table 33.1). Details of the pain and injury incidence that were experienced by the divers were assessed by the administration of questionnaires given during the national training camps. Low back pain(LBP)was also assessed by interviews and physical examinations.

33.2.1 Site of Pain in Training Camp

A total of 112 reports of pain were recorded. Among all types of injuries, the highest incidence was for LBP (38.4 %) in elite junior divers in Japan (Fig. 33.1).

33.2.2 Dimension of LBP

LBP in the divers became more prevalent as the divers' age increased (Fig. 33.2).

Table 33.1 Characteristics of the participants

	Male (n=60)	Female (n=56)
Age (years)	14.8±1.5	14.7±1.8
Height (cm)	162.0±9.3	154.6±5.9
Weight (kg)	54.3±10.0	47.0±7.1
Athletic career (years)	6.4±1.8	6.3±2.0

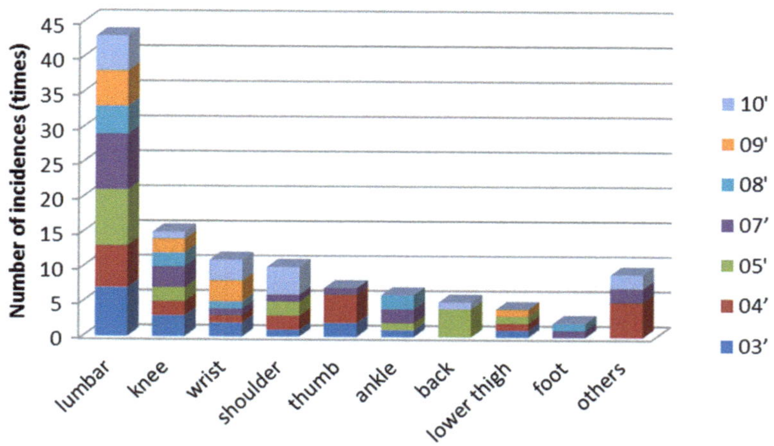

Fig. 33.1 Body parts associated with pain during training camp

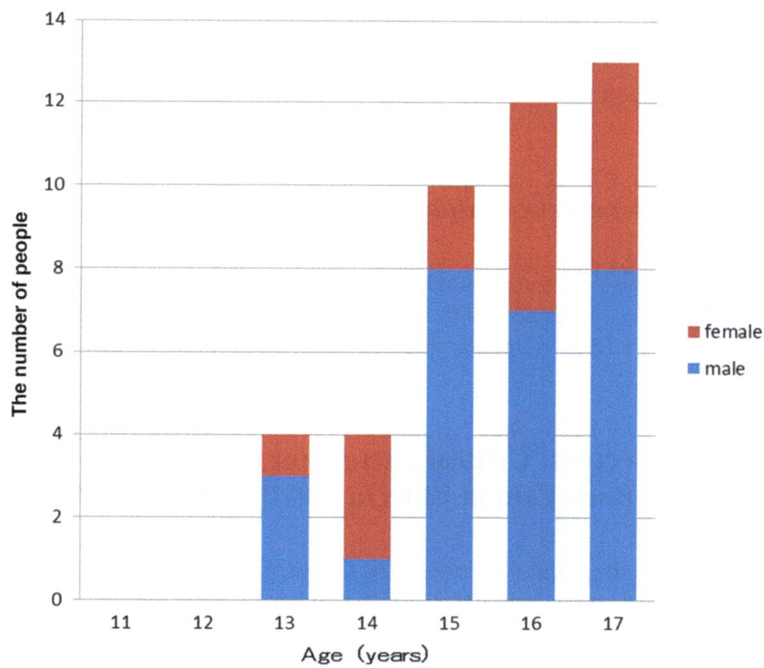

Fig. 33.2 The number of low back pain divers of each age

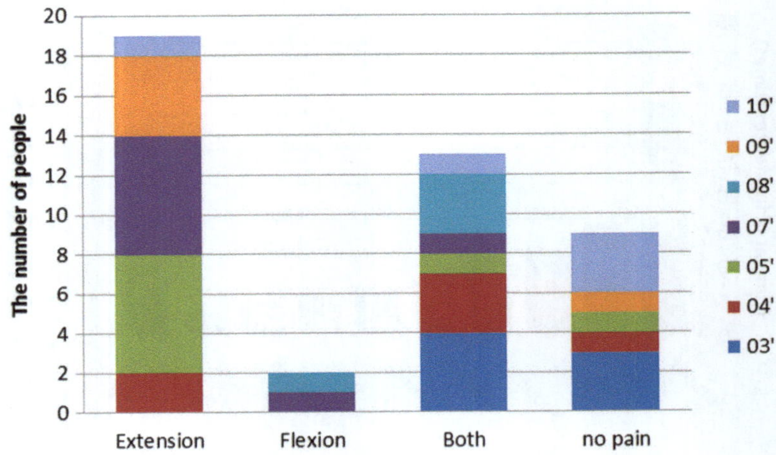

Fig. 33.3 Type of motion associated with low back pain

33.2.3 Direction of Motion with LBP

Thirty four of the 43 divers with LBP (79.1 %) experienced the pain during movement; of the 34, 19 (55.9 %) experienced pain only during extension, 13 during both extension and flexion (38.2 %), and 2 only during flexion (5.9 %) (Fig. 33.3).

33.2.4 The Occurrence Situation of LBP

24LBP reported for pain questionnaire before for 1 year training camp. The nine (38.0 %)LBP have generated with hyperextension at the entry phase. The 11 (46.0 %) cause of LBP was unclear and that occurred gradually (Fig. 33.4).

33.3 Third Section: Critical Factors for the Prevention of Low Back Pain in Elite Junior Divers (Second Step)

Our description in the Second section noted that the incidence of LBP in junior divers aged 13–17 was quite high (38.4 %). A number of other researchers have also found that LBP is the most frequently reported symptom in divers (Badman and Rechtine 2004; Narita et al. 2011; Ito 2007; Rubin 1999; Keene 1985; Carter 1986). However, these divers were still able to continue to practice with LBP (Badman and Rechtine 2004). Meeuwisse suggested that there are potential intrinsic and extrinsic risk factors for sports-related injuries; age, sex, physical characteristics, aptitude, psychological characteristics, health status, and history of injury are all considered intrinsic factors (Meeuwisse 1994). However, Steffen and colleagues (2010)

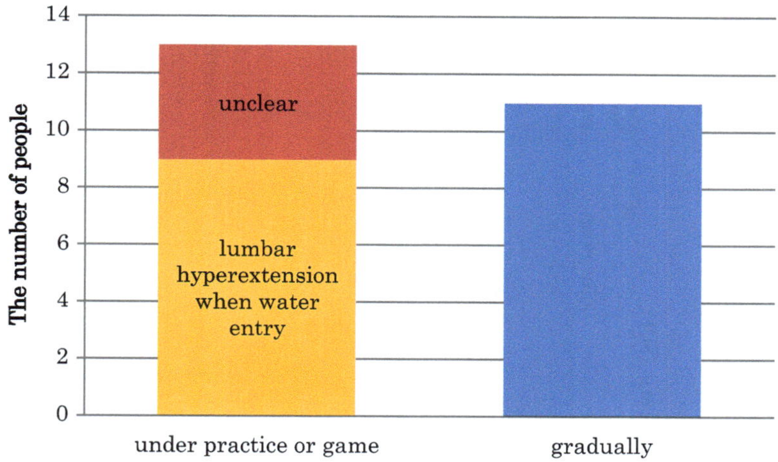

Fig. 33.4 The occurrence situation of LBP

Table 33.2 Characteristics of the participants

	Male, mean (SD)	Female, mean (SD)
Age, year	14.5 (1.6)	14.3 (1.8)
Height, cm	161.8 (9.1)	154.8 (6.2)
Weight, kg	54.1 (9.8)	46.5 (7.2)
Body mass index, kg/m^2	20.7 (2.1)	19.4 (2.1)
Diving experience, year	6.1 (1.8)	6.0 (2.2)

acknowledged that few studies have been conducted regarding the risks for sports-related injuries, and that there is a need to elucidate the risk factors, mechanisms, and prevention methods for high-level athletes. This lack is also true for competitive diving, since relatively few reports exist regarding the risks for injury. Understanding the athlete's intrinsic risk factors for injury incidence is necessary to establish a viable prevention plan. Therefore, in this section we investigate the intrinsic factors that correlate with the risk factors for LBP in competitive divers.

Eighty-three elite junior divers (42 males, 41 females) in Japan were included in this study (Table 33.2). LBP was assessed by questionnaire, interview, and physical examination during a national training camp. Morphologic data, physical fitness, and diving skills were also evaluated. The factors related to LBP were extracted by using logistic-regression analysis and the forward selection method (likelihood ratio).

33.3.1 Comparison of the LBP and No-LBP Groups

Table 33.3 shows the physical and fitness characteristics of both groups. Eighteen of 42 male divers reported LBP. The following characteristics showed a significant difference between the LBP and no-LBP groups: age, weight, body mass index,

Table 33.3 Comparison of LBP and no-LBP groups

	Male, mean (SD)			Female, mean (SD)		
	Overall	LBP group	No-LBP group	Overall	LBP group	No-LBP group
Age, year	14.5 (1.6)	15.6 (1.4)**	13.8 (1.3)**	14.3 (1.8)	15.5 (1.6)**	13.8 (1.7)**
Height, cm	161.8 (9.1)	164.9 (7.1)	159.4 (9.8)	154.8 (6.2)	157.5 (4.6)	153.5 (6.5)
Weight, kg	54.1 (9.8)	57.8 (8.8)*	51.2 (9.7)*	46.5 (7.2)	50.1 (5.5)*	44.8 (7.4)*
Body mass index, kg/m^2	20.7 (2.1)	21.3 (1.9)*	20.1 (2.2)*	19.4 (2.1)	20.2 (1.6)	19.0 (2.2)
Diving experience, year	6.1 (1.8)	6.5 (1.7)	5.8 (1.8)	6.0 (2.2)	6.8 (2.3)	5.5 (2.3)
Back muscle strength, kg	113.2 (35.2)	125.7 (32.9)*	103.8 (34.5)*	81.0 (16.9)	89.0 (13.3)*	77.3 (17.3)*
30-s sit-ups, repetitions	41.1 (3.6)	42.2 (3.2)	40.3 (3.8)	39.6 (4.4)	40.4 (4.9)	39.3 (4.2)
Body anteflexion, cm	55.3 (7.7)	59.1 (5.8)*	52.4 (7.8)*	57.4 (6.6)	60.7 (5.2)*	55.8 (6.6)*
Body retroflexion, cm	53.8 (6.4)	55.8 (6.6)*	52.3 (5.8)*	55.6 (6.6)	56.7 (6.4)	55.0 (6.7)
Vertical jump, cm	62.0 (7.4)	65.9 (6.2)**	59.1 (6.9)**	50.8 (6.0)	53.5 (5.1)	49.5 (6.0)
Standing long jump, cm	235.1 (20.3)	245.8 (18.4)**	227.0 (18.0)**	205.1 (14.3)	208.6 (10.6)	203.5 (15.7)
Shoulder rotation width, cm	52.2 (14.2)	57.2 (11.9)**	48.5 (14.9)**	29.3 (16.8)	29.5 (13.3)	29.3 (18.5)
Hand stand posture, points	6.2 (2.4)	7.1 (1.6)**	5.1 (2.3)**	5.6 (2.3)	5.3 (2.5)	5.7 (2.3)
Hand stand duration, seconds	25.0 (17.3)	31.6 (14.5)	19.8 (17.9)	9.7 (13.6)	7.2 (11.8)	10.8 (14.4)

*p<0.05
**p<0.01

back muscle strength, body anteflexion, body retroflexion, vertical jump, standing long jump, shoulder rotation width, and hand stand posture.

Thirteen of 41 female divers reported LBP. The following characteristics showed a significant difference between the LBP and no-LBP groups: age, weight, back muscle strength, and body anteflexion.

33.3.2 Factors Associated with LBP

According to the logistic-regression analysis, LBP correlated significantly with age ($P=0.009$; OR, 0.441; 95 % confidence interval [CI], 0.239–0.814) and shoulder rotation width ($P=0.031$; OR, 0.919; 95 % CI, 0.851–0.992) in male divers (Table 33.4). The following regression equation estimating the rate of incidence was derived from the factors above:

$$Rate\ of\ Incidence\ P1 = \frac{1}{1 = \exp\left[-16.608 + (-0.818) \times age + (-0.085) \times shoulder\ rotation\ width\right]}$$

The regression model was a good fit at $P=0.961$, as assessed by the Hosmer-Lemeshow test, with a discriminant accuracy of 81.1 %.

For female divers, the only characteristic that showed a significant correlation with LBP was age ($P=0.009$; OR, 0.536; 95 % CI, 0.335–0.856) (Table 33.5). The regression equation estimating the rate of incidence was:

Table 33.4 Results of logistic-regression analysis for male divers

		P		95 % CI	
	B (coefficient)	value	OR	Lower limit	Upper limit
Age	–0.818	.009	0.441	0.239	0.814
Shoulder rotation width	–0.085	.031	0.919	0.851	0.992

Discriminant accuracy, 81.1 %
Abbreviations: *CI* confidence interval, *OR* odds ratio

Table 33.5 Results of logistic-regression analysis for female divers

				95 % CI	
	B (coefficient)	*P* value	OR	Lower limit	Upper limit
Age	–0.624	.009	0.536	0.335	0.856

Discriminant accuracy, 75.0 %
Abbreviations: *CI* confidence interval, *OR* odds ratio

$$Rate\,of\,Incidence\,P2 = \frac{1}{1+\exp\left[-9.790+(-0.624)\times age\right]}$$

33.3.3 Summary

We investigated the risk factors associated with LBP in elite junior competitive divers (42 male, 41 female) during a national training camp. Risk factors in male divers included age and shoulder rotation width, whereas the only significant risk factor in female divers was age.

33.4 Fourth Section: Entry Phase Alignment in Men's Backward Dive Pike (201B) from 3 m Springboard Dives (Second Step)

In the previous section we demonstrated that there was a high prevalence of LBP in elite junior divers (Narita et al. 2011) and that less shoulder flexibility in male junior divers was a risk factor for LBP (Narita et al. 2014). This factor needs to be analysed in terms of motion during performance. In this section, we investigate the influence of the shoulder angle on entry alignment and compare the differences in entry phase alignment between divers with LBP divers and those with no-LBP.

Thirteen elite male divers in Japan were included in this study (Table 33.6). LBP was assessed by questionnaire, interview, and physical examination during a national training camp. The backward pike dive (201B)entry phase was filmed laterally using a high-speed camera (shutter speed: 1/200 s) (Fig. 33.5). The shoulder, trunk and hip angles for the entry phase (5 frames) were derived from the digitized film (Fig. 33.6). The relevance of the shoulder, trunk, and hip angles was examined using the Pearson product–moment correlation coefficient. We divided the divers into two groups on the basis of LBP status. An intergroup comparison was carried out using the Mann–Whitney's *U* test on the basis of entry phase alignment.

We measured the average angle of the diver in the 5th frame (0.025 s), before his palm touched the water's surface.

Table 33.6 Characteristics of the participants

	Male, mean (SD)
Age, year	16.8 (2.9)
Height, cm	165.4 (6.6)
Weight, kg	56.7 (7.7)
Diving experience, year	9.5 (2.5)

Fig. 33.5 Backward dive pike (201B). ① Take off, ② Pike position: maximum hip flexion, ③ Come-out (hip extension), ④ line-up (maximum shoulder flexion), ⑤ entry

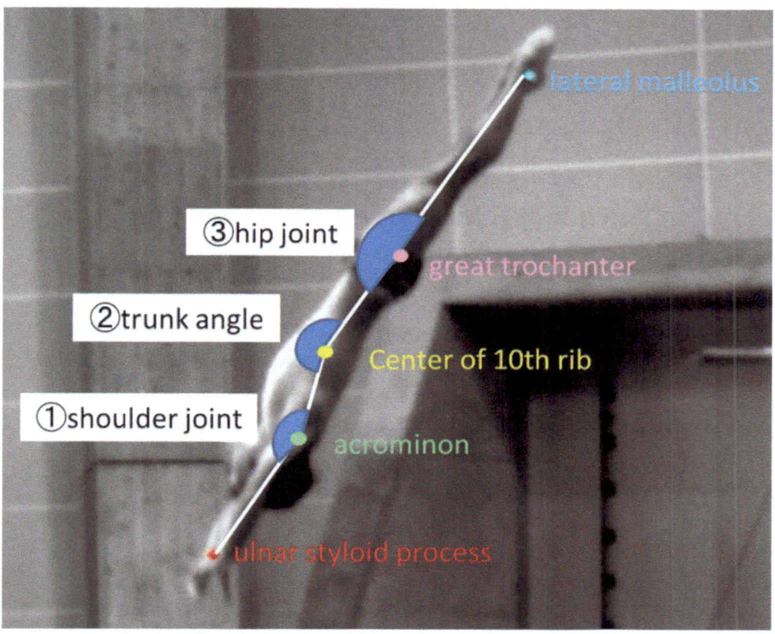

Fig. 33.6 The measurement angle

33.4.1 Shoulder Joint Angle and Trunk Angle at the Entry Phase

A negative correlation (r = −0.623) was observed in the shoulder flexion angle and trunk extension angle for the entry phase (Fig. 33.7).

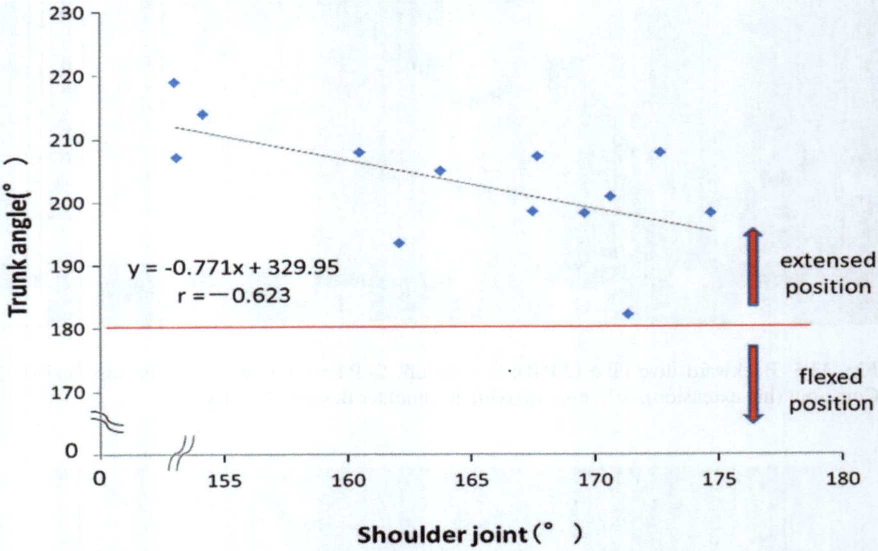

Fig. 33.7 Shoulder joint angle and trunk angle at the entry phase

33.4.2 Comparison of Each Angle at the Entry Phase in the LBP Group and No-LBP Group

Four of 13 male divers experienced LBP. The trunk angle showed a significant difference between the LBP group (210.4±5.7°) and no-LBP groups (199.9±9.0°) (Fig. 33.8).

33.4.3 Summary

The shoulder flexion angle was related to the trunk extension angle for the entry phase. The LBP group had a larger trunk extension angle than did the no-LBP group. Therefore retaining shoulder flexibility should help to prevent LBP in divers. This finding is likely relevant not only to diving, but also across throwing sports, swimming sports(i.e., the streamline position) and gymnastics.

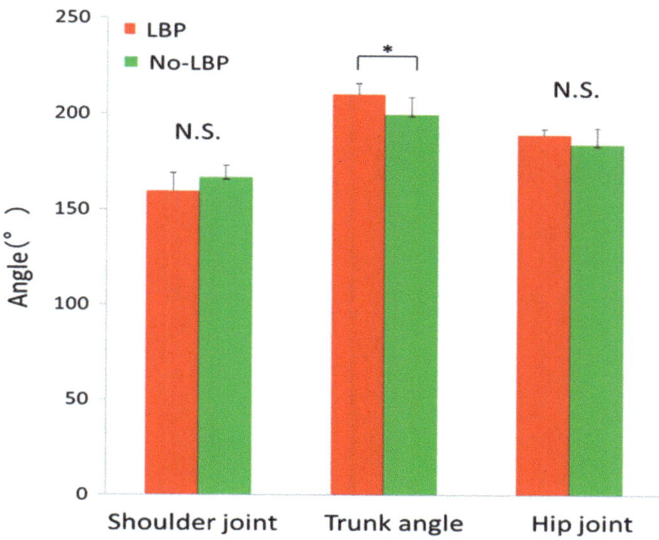

Fig. 33.8 Comparison of each angle at the entry phase in the LBP group and no-LBP group

References

Badman BL, Rechtine GR (2004) Spinal injury considerations in the competitive diver: a case report and review of the literature. Spine J 4:584–590

Carter RL (1986) Prevention of springboard and platform diving injuries. Clin Sports Med 5:185–194

Ito Y (2007) A result of investigation about injuries and disorders of divers and old players. Jpn J Sci Swim Water Exerc 10:11–18

Keene JS (1985) Mechanical back pain in the athlete. Compr Ther 11:7–14

Le Viet DT, Lantieri LA, Loy SM (1993) Wrist and hand injuries in platform diving. J Hand Surg 18:876–880

Meeuwisse WH (1994) Assessing causation in sport injury: a multifactorial model. Clin J Sport Med 4:166–170

Narita T, Kaneoka K, Takemura M, Nomura T, Sakata K, Miyakawa S (2011) Injury incidence in Japanese elite junior divers. Jpn J Sci Swim Water Exerc 14:1–6

Narita T, Kaneoka K, Takemura M, Nomura T, Sakata K, Miyakawa S (2014) Critical factors for the prevention of low back pain in Elite Junior Divers. Br J Sports Med 48:919–923

Rubin BD (1999) The basics of competitive diving and injuries. Clin Sports Med 18:293–303

Steffen K, Engebretsen L (2010) More data needed on injury risk among young elite athletes. Br J Sports Med 44:485–489

van Mechelen W, Hlobil H, Kemper HC (1992) Incidence, severity, aetiology and prevention of sports injuries. A review of concepts. Sports Med 14:82–99

Chapter 34
Electromyographic Analysis of Deep Trunk Muscles During Sports Activities

Yu Okubo

Abstract An understanding of the function of trunk muscles is important for the treatment of low back pain as well as for the improvement of athletic performance. In particular, the importance of deep trunk muscles (local muscles) has been documented in the recent literature. In this chapter we report on investigation about activity patterns of deep trunk muscles during lumbar stabilization exercises and sports activities. We recorded muscle activity using intramuscular electrodes. The results provide information that allows us to propose effective exercises for enhancing deep trunk muscle activity and improve the complex coordination of trunk muscle activity that is important during sports.

Keywords Lumbar spinal stability • Trunk muscle exercise • Coordination

34.1 Introduction

An understanding of the function of trunk muscles is important for the treatment of low back pain as well as for the improvement of athletic performance. In this regard, trunk muscle exercises are commonly performed and taught both in the clinic and on the athletic field. Conventional trunk exercises comprised of repeated trunk flexion and extension, such as sit-ups or back extension, have been widely performed for improving trunk strength. Recently, lumber stabilization exercises, which keep the lumbar spine in a neutral position with minimal trunk movements, such as the prone bridge or side bridge are frequently performed. The aim of lumbar stabilization exercises is to enhance trunk stability by co-activation of superficial and deep trunk muscles in the neutral zone.

Muscle function in the neutral zone is important for stabilization of the lumbar spine (Panjabi 1992a, b). Bergmark (1989) classified the trunk muscles into a local and a global system in order to better understand how the muscular system acts

Y. Okubo (✉)
School of Physical Therapy, Saitama Medical University, Saitama, Japan
e-mail: yokubo@saitama-med.ac.jp

K. Kanosue et al. (eds.), *Sports Injuries and Prevention*,
DOI 10.1007/978-4-431-55318-2_34

on the stability of the lumbar spine. The local muscle system includes the deeper muscles which have their origin or insertion points either directly or indirectly on the lumbar vertebrae; this system directly controls lumbar segmental stabilization. The transversus abdominis and lumbar multifidus are examples of local muscles. The global muscle system includes those muscles that do not directly attached to the lumbar vertebrae, such as the rectus abdominis and external oblique muscles. The global muscles generate a large torque across multiple segments and control trunk movement. It has been reported that increased vertebral stiffness by co-activation of the local and global muscles is important in improving the stability of the lumbar spine (Stanton and Kawchuk 2008). Although conventional trunk exercises have focused mainly on the global muscles, lumbar stabilization exercises are intended to train motor control of both the local and global muscles.

Important functions of the local muscles include their roles in spinal stability and postural responses. Hodges and Richardson report that the onset of the transversus abdominis (TrA) is earlier than that of agonist muscles (feedforward contraction) during upper limb movement (Hodges and Richardson 1997). It has also been documented that feedfrorward contraction of the TrA is significantly delayed in patients with low back pain (Hodges and Richardson 1996). Moreover, the local muscles are reported to contribute to lumbar spinal stability during walking and running (Saunders et al. 2004, 2005). It is also documented that spinal stability is important not only for disorders of the lumbar region but also for the prevention of lower extremity injuries (Leetun et al. 2004). Zazulak et al.(2007) report that deficient trunk core stability is a risk factor for anterior cruciate ligament injury in female athletes.

Therefore, enhancing motor coordination of local and global muscles is important for athletic performance as well as the prevention of sports injury. In the next portion of this chapter, I will introduce electromyographic (EMG) data of trunk muscles during various lumbar stabilization exercises and sports activities.

34.2 Intramuscular EMG Setup

It is difficult to measure local muscle activity because such muscles are located in a deep layer. To record EMG data from these muscles, we used fine-wire electrodes. The wire was threaded into hypodermic needles (23 gauge × 60 mm) with 2 mm of urethane removed and tips bent back to form 1- and 2-mm hooks. The wire electrodes were sterilized via autoclave. We inserted the electrodes into the TrA and multifidus (MF, L5 level) under the guidance of ultrasound imaging (Fig. 34.1). Once the electrodes reached the targeted muscle, placement was confirmed by applying electrical stimulation and using ultrasound imaging to confirm contraction of the proper muscle.

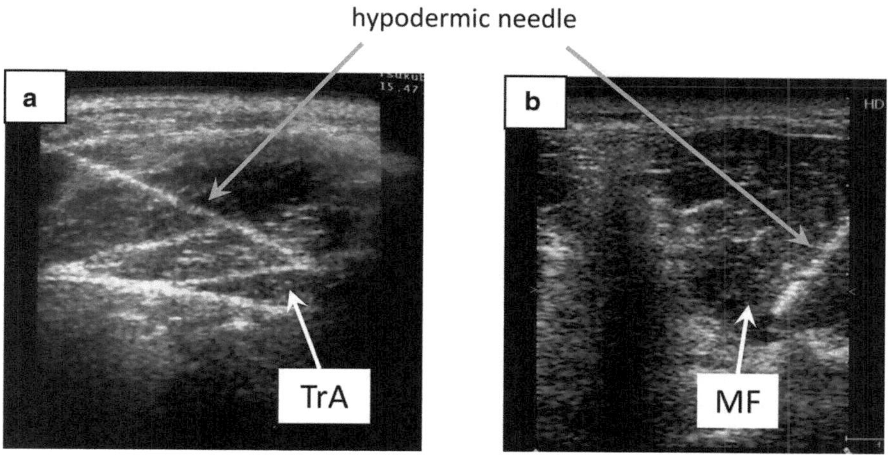

Fig. 34.1 Ultrasound imaging of the transversus abdominis (*TrA*, **a**) and multifidus (*MF*, **b**) with insertion of hypodermic needles

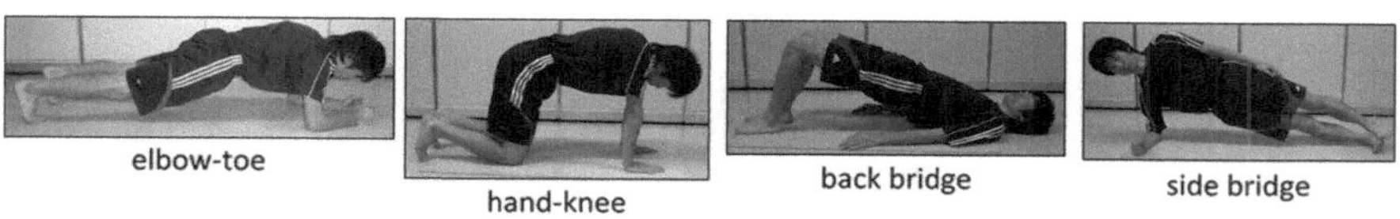

Fig. 34.2 Basic posture for each lumbar stabilization exercise

34.3 EMG Analysis of Trunk Mucles During Lumbar Stabilization Exercises

34.3.1 The Activity of the Trunk Muscles During Various Lumbar Stabilization Exercises

Bridge exercises, such as the elbow-toe, hand-knee, back bridge and side bridge exercise, are performed as basic lumbar stabilization exercises in clinics and are illustrated in Fig. 34.2. We measured EMG amplitudes during the ten lumbar stabilization exercises to determine if there were exercises that were more effective for the activation of local muscles. The elbow-toe with contralateral arm and left leg lift exercise significantly produced higher activity level for the TrA than did the other exercises (Fig. 34.3) (Okubo et al. 2010). The activity levels of the right TrA showed 41.8 ± 20.2 % of the maximum voluntary contraction (MVC) (mean ± SD) during the elbow-toe with right arm and left leg lift exercise, and activity levels of the left TrA showed 50.6 ± 28.4 % of the MVC during the elbow-toe with left arm and right leg lift exercise. For the MF, back bridge exercises and the hand-knee with contralateral arm and leg lift exercise produced significantly higher activity levels than did

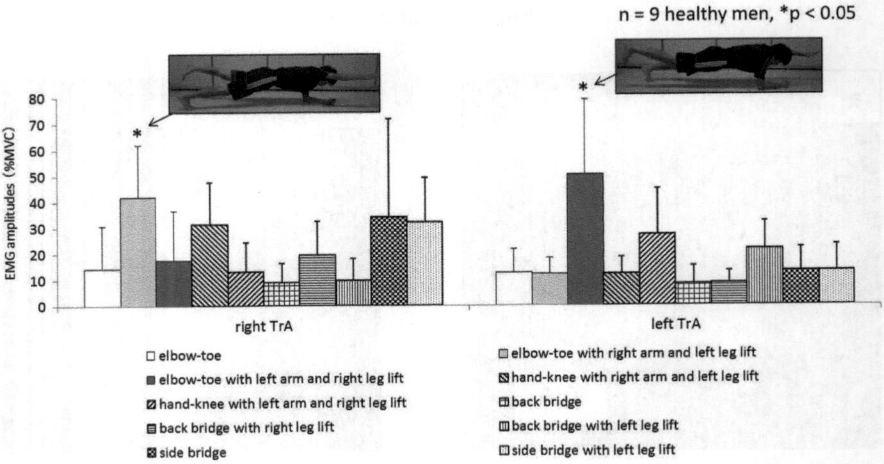

Fig. 34.3 EMG amplitudes of the tranversus abdominis (*TrA*) during each lumbar stabilization exercise

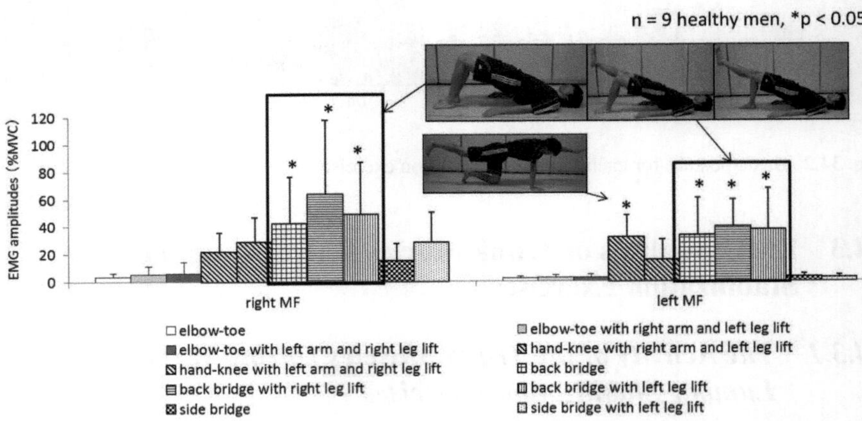

Fig. 34.4 EMG amplitudes of the multifidus (*MF*) during each lumbar stabilization exercise

the other exercises (Fig. 34.4) (Okubo et al. 2010). In general, the results showed a tendency for the antigravity muscles facing the floor, such as the abdominal muscles, to have a greater activation during bridge exercises and during elbow-toe exercises. The back muscles were more active during back bridge exercises and the external oblique (EO) during side bridge exercises (Fig. 34.5).

Additionally, we compared trunk muscle activity during three different prone bridge exercises, the elbow-toe, elbow-knee, and hand-knee exercises. These exercises

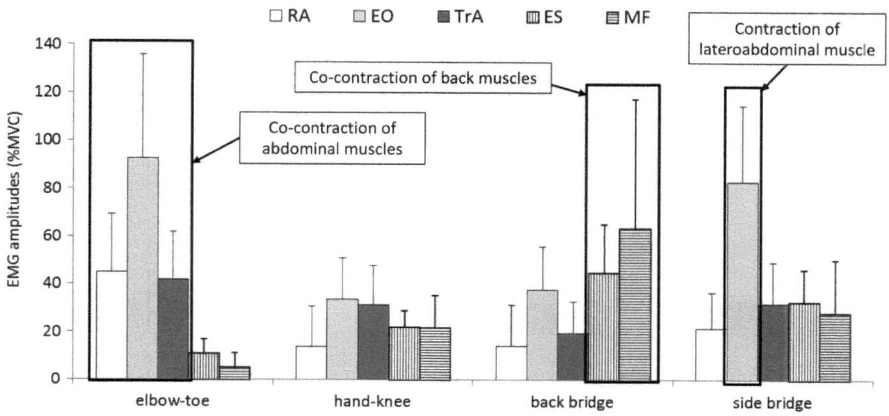

Fig. 34.5 Activity pattern of trunk muscles for each lumbar stabilization exercise. *RA* rectus abdominis, *EO* external oblique, *TrA* transversus abdominis, *ES* erector spinae, *MF* multifidus

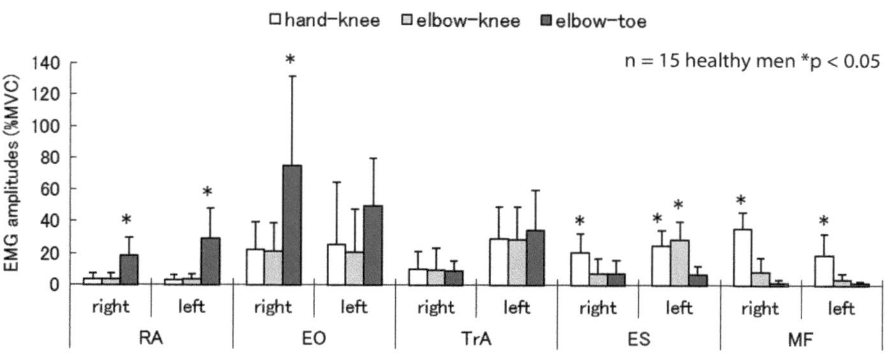

Fig. 34.6 EMG amplitudes during 3 prone bridge with left arm and right leg lift. *RA* rectus abdominis, *EO* external oblique, *TrA* transversus abdominis, *ES* erector spinae, *MF* multifidus

all involve lifting of the contralateral arm and leg. The activity of the rectus abdominis (RA) and EO was significantly greater during the elbow-toe exercise than during the elbow-knee or hand-knee exercises. The activity of the erector spinae (ES) and MF was significantly greater during the hand-knee exercise than during the elbow-toe or elbow-knee exercises. It was remarkable that there was no significant difference in the activity of the TrA during the three exercises (Fig. 34.6) (Okubo et al. 2012). These results indicate that the elbow-toe exercise activated the superficial abdominal muscles, and the hand-knee exercises activated the back muscles. On the other hand, the TrA activity level remained similar during the different prone bridge exercises.

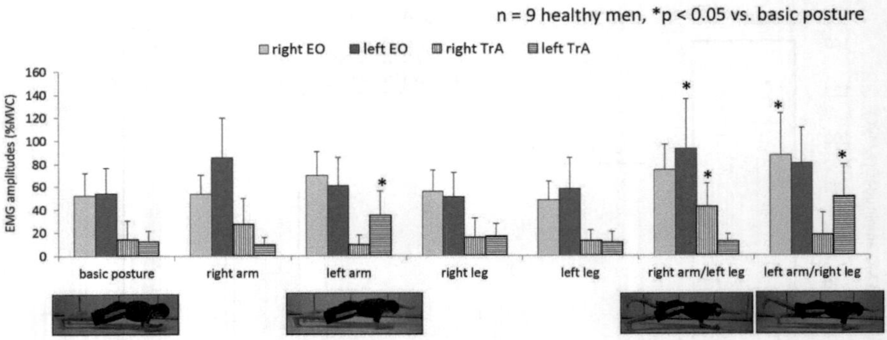

Fig. 34.7 Change in muscle activity by lifting up the extremities during elbow-toe exercise. *EO* external oblique, *TrA* transversus abdominis

34.3.2 Changes in Trunk Muscle Activity by Lifting of the Extremities During Lumbar Stabilization Exercise

The lifting of the extremities during lumbar stabilization exercises is thought to enhance the deep trunk muscle contraction. We examined changes in trunk muscle activity elicited by lifting of the extremities during elbow-toe and hand-knee exercises. Activity in the left TrA was significantly greater during the elbow-toe exercise with the left arm lift. Moreover, in the lifted arm, activity of the TrA and contralateral EO increased significantly during the elbow-toe with contralateral arm and leg lift (Fig. 34.7) (Okubo et al. 2011). These results suggest that activity of the ipsilateral TrA and contralateral EO in the lifted arm increase during the elbow-toe with arm lift. Therefore, there is thought to be disfunction of the right TrA or the left EO if an athlete has difficulty in lifting the right arm during the elbow-toe exercise. During the hand-knee exercise, the left MF activity significantly increased with the left leg lift, while the activity of the MF in the lifted leg and contralateral ES was significantly greater with the contralateral arm and leg lift. These results suggest that trunk stability is enhanced by co-contraction of deep and superficial trunk muscles during elbow-toe and hand-knee exercises involving the contralateral arm and leg lift.

34.3.3 Trunk Muscle Activity During Lumbar Stabilization Exercises on Stable and Unstable Surfaces

Lumbar stabilization exercises on an unstable surface are performed widely. However, there is little evidence in the literature as to whether performing lumbar stabilization exercises on an unstable surface actually increases muscular demand.

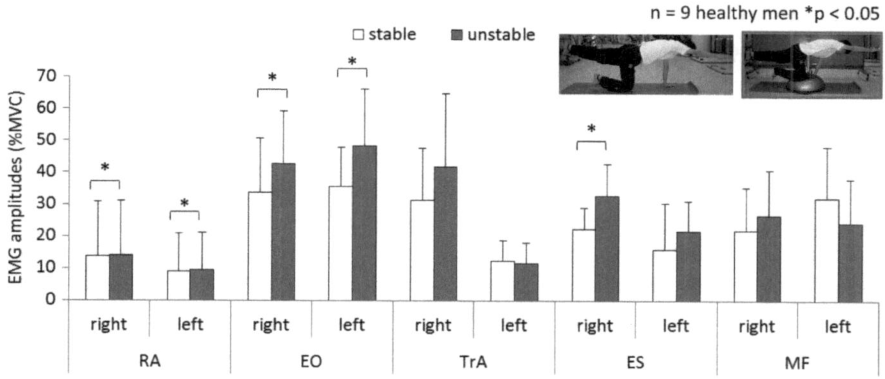

Fig. 34.8 Comparison of muscle activity for hand-knee exercise between stable and unstable surface. *RA* rectus abdominis, *EO* external oblique, *TrA* transversus abdominis, *ES* erector spinae, *MF* multifidus

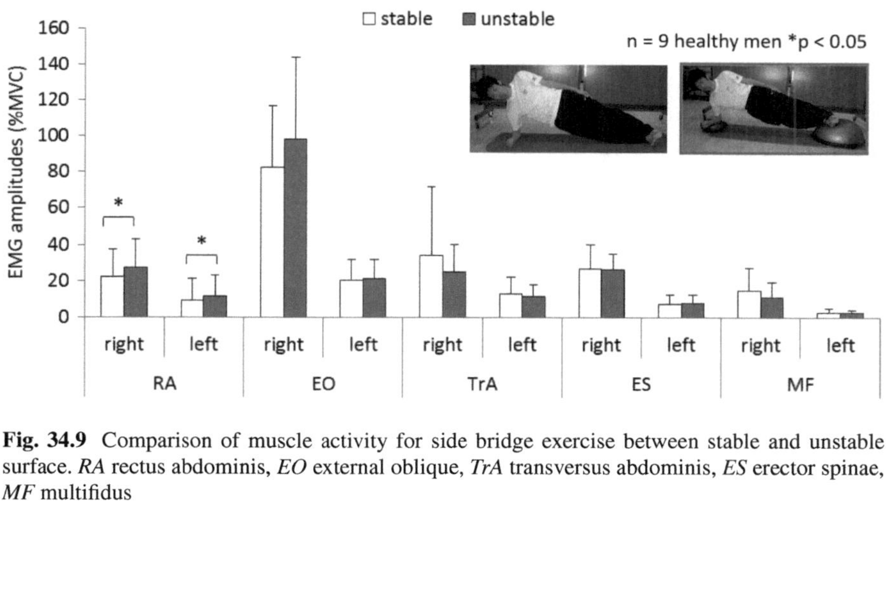

Fig. 34.9 Comparison of muscle activity for side bridge exercise between stable and unstable surface. *RA* rectus abdominis, *EO* external oblique, *TrA* transversus abdominis, *ES* erector spinae, *MF* multifidus

To evaluate the value of this exercise, we compared muscle activity for hand-knee and back bridge exercises between stable and unstable conditions. The activity of the RA, EO, and ES during hand-knee exercise was significantly greater when performed on an unstable surface as compared to a stable surface (Fig. 34.8) (Imai et al. 2010). With the side bridge exercise, only activity of the RA was significantly greater in the unstable condition (Fig. 34.9) (Imai et al. 2010). These results indicate that activity of the global muscles, such as that in the RA and EO, is greater when lumbar stabilization exercises are performed on an unstable surface.

34.4 EMG Analysis of Trunk Muscles During Sports Activities

34.4.1 Abdominal Muscle Activity During Jumping

Jumping is a fundamental movement of sports that requires complex motor coordination, but few studies have examined the neuromuscular control of the trunk during this task. Therefore, we quantified the activation patterns (onset and magnitude) of the abdominal muscles during a standing long jump using wire and surface EMG recordings. The onset of the TrA, EO, and RA relative to the toe-off were -0.35 ± 0.06 s, -0.28 ± 0.06 s, and -0.21 ± 0.07 s respectively. The onset of the TrA activation was earlier than that of the RA and EO (Fig. 34.10). The activation levels of the abdominal muscles were significantly greater during the push-off phase (from heel-off to toe-off) than during the preparation (from 200 ms prior to heel-off to heel-off) and float phases (from toe-off to 200 ms after toe-off) (Fig. 34.11) (Okubo et al. 2013b). These results imply that the abdominal muscles involved in the standing long jump are sequentially activated from deep muscle to superficial muscle, and that the abdominal muscles' activity levels increase more during the push-off phase than during the preparation and float phases.

34.4.2 Abdominal Muscle Activity During Volleyball Block Jump

Volleyball players are instructed to contract their abdominal muscles during the block jump. We measured activity in the abdominal muscles (RA, EO, and IO). We compared those results with results obtained during a vertical jump using

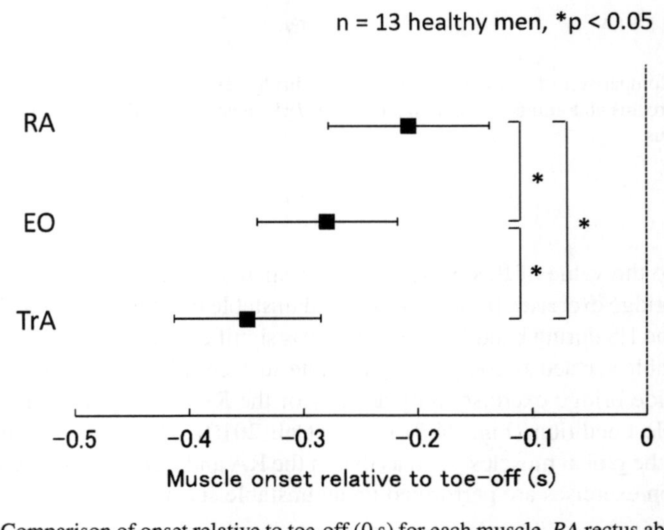

Fig. 34.10 Comparison of onset relative to toe-off (0 s) for each muscle. *RA* rectus abdominis, *EO* external oblique, *TrA* transversus abdominis

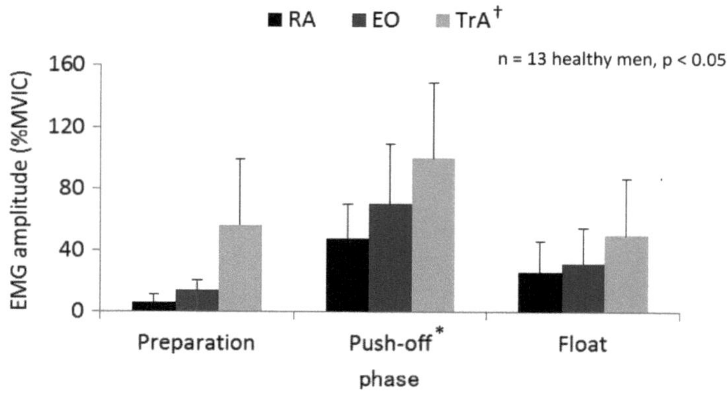

Fig. 34.11 EMG amplitudes for each phase during standing long jump. *When averaged across all muscles, EMG amplitude was significantly greater during the push-off phase compared to the preparation and float phases. †Activity of the TrA was significantly greater than that of the RA or the EO when averaged across all phases. *RA* rectus abdominis, *EO* external oblique, *TrA* transversus abdominis

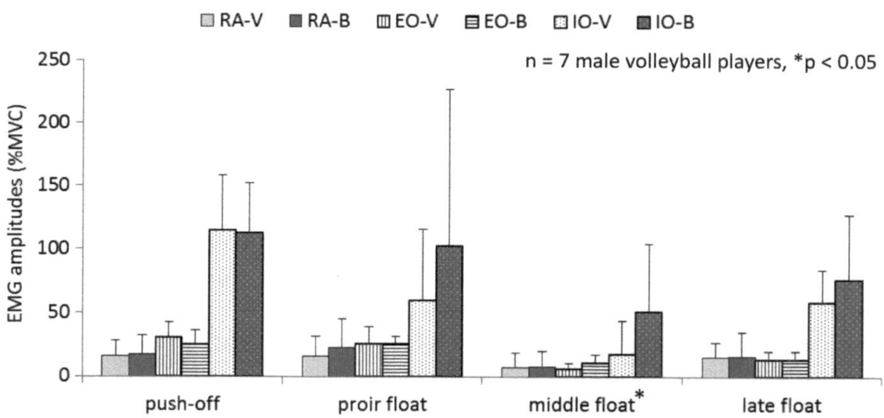

Fig. 34.12 EMG amplitudes of abdominal muscles for each phase during vertical jump and block jump. *EMG amplitude was significantly lower during middle float phase compared to the other phases. *RA* rectus abdominis, *EO* external oblique, *IO* internal oblique, *V* vertical jump, *B* block jump

surface EMG recordings in volleyball players. There was no significant interaction between the activities of the various abdominal muscles activity for the block jump and the vertical jump. Abdominal muscle activities were significantly lower during the middle float phase, when the jump reached its peak (Fig. 34.12). These results suggest that the activity pattern of the abdominal muscles during the volleyball block jump is the same as that during a normal vertical jump, and activity levels in the abdominal muscles are low when a player blocks a spike.

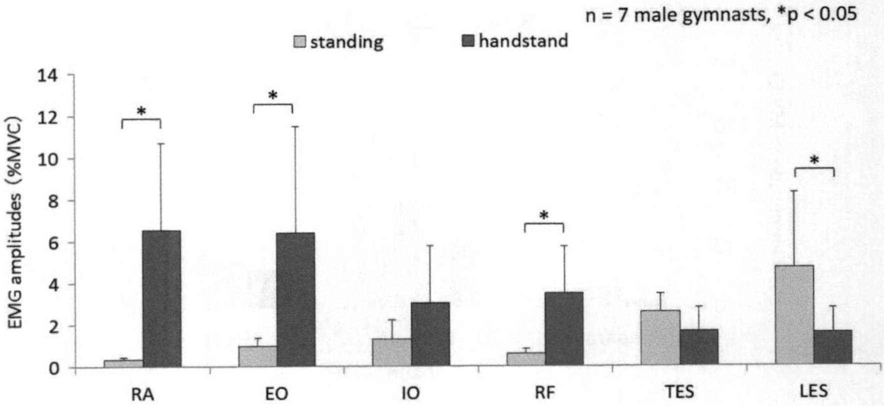

Fig. 34.13 Comparison of muscle activity between standing and handstand. *RA* rectus abdominis, *EO* external oblique, *IO* internal oblique, *RF* rectus femoris, *TES* thoracic erector spinae, *LES* lumbar erector spinae

34.4.3 Trunk Muscle Activity During a Handstand Posture

Handstands are frequently performed by competitive gymnasts and divers. Muscle activity during a handstand posture differs from that which occurs during normal standing posture because an individual's center of mass is controlled by activity in their upper extremities. We compared trunk muscle activity between handstand and normal standing posture in gymnasts. The RA and EO activity levels during the handstand were 6.5 ± 4.2 % of MVC and 6.4 ± 5.1 % of MVC, respectively, and were significantly higher than the values recorded while standing. On the other hand, the LES activity during the handstand (1.6 ± 1.2 % MVC) was significantly lower than that during standing (4.7 ± 3.6 %MVC) (Fig. 34.13) (Okubo et al. 2013a). These results indicate that the anterior surface muscles (RA and EO) are activated as antigravity muscles to maintain the handstand posture. In contrast, the LES muscles are activated as antigravity muscles while standing. Antigravity muscles are thus counterchanged during handstands as compared with their activation during normal standing.

References

Bergmark A (1989) Stability of the lumbar spine: a study in mechanical engineering. Acta Orthop Scand 230(Suppl):20–24

Hodges PW, Richardson CA (1996) Inefficient muscular stabilization of the lumbar spine associated with low back pain. A motor control evaluation of transversus abdominis. Spine 21:2640–2650

Hodges PW, Richardson CA (1997) Feedforward contraction of transversus abdominis is not influenced by the direction of arm movement. Exp Brain Res 114:362–370

Imai A, Kaneoka K, Okubo Y, Shiina I, Tatsumura M, Izumi S, Shiraki H (2010) Trunk muscle activity during lumbar stabilization exercises on both a stable and unstable surface. J Orthop Sports Phys Ther 40:369–375

Leetun DT, Ireland ML, Willson JD, Ballantyne BT, Davis IM (2004) Core stability measures as risk factors for lower extremity injury in athletes. Med Sci Sports Exerc 36:926–934

Okubo Y, Kaneoka K, Imai A, Shiina I, Tatsumura M, Izumi S, Miyakawa S (2010) Electromyographic analysis of transversus abdominis and lumbar multifidus using wire electrodes during lumbar stabilization exercises. J Orthop Sports Phys Ther 40:743–750

Okubo Y, Kaneoka K, Imai A, Shiina I, Tatsumura M, Izumi S, Miyakawa S (2011) Changes in trunk muscle activity by lifting of the extremities during lumbar stabilization exercises. Jpn J Clin Sports Med 19:94–101

Okubo Y, Kaneoka K, Shiina I, Tatsumura M, Miyakawa S (2012) Comparison of trunk muscle activity during three different prone bridge exercises. Spineweek 2012

Okubo Y, Kaneoka K, Matsunaga N, Hasebe K, Tanaka Y (2013) Trunk muscle activity and body alignment during a handstand posture. 18th annual congress of the European College of Sport Science. Barcelona

Okubo Y, Kaneoka K, Shiina I, Tatsumura M, Miyakawa S (2013b) Abdominal muscle activity during a standing long jump. J Orthop Sports Phys Ther 43:577–582

Panjabi MM (1992a) The stabilizing system of the spine. Part I: function, dysfunction, adaption, and enhancement. J Spinal Disord 5:383–389

Panjabi MM (1992b) The stabilizing system of the spine. Part II: neutral zone and stability hypothesis. J Spinal Disord 5:390–397

Saunders SW, Rath D, Hodges PW (2004) Postural and respiratory activation of the trunk muscles changes with mode and speed of locomotion. Gait Posture 20:280–290

Saunders SW, Schache A, Rath D, Hodges PW (2005) Changes in three dimensional lumbo-pelvic kinematics and trunk muscle activity with speed and mode of locomotion. Clin Biomech 20:784–793

Stanton T, Kawchuk G (2008) The effect of abdominal stabilization contractions on posteroanterior spinal stiffness. Spine 33:694–701

Zazulak BT, Hewett TE, Reeves NP, Goldberg B, Cholewicki J (2007) Deficits in neuromuscular control of the trunk predict knee injury risk: a prospective biomechanical-epidemiologic study. Am J Sports Med 35:1123–1130

Index

K. Kanosue et al. (eds.), *Sports Injuries and Prevention*,
DOI 10.1007/978-4-431-55318-2

MIX
Papier aus verantwortungsvollen Quellen
Paper from responsible sources
FSC® C105338

If you have any concerns about our products,
you can contact us on
ProductSafety@springernature.com

In case Publisher is established outside the EU,
the EU authorized representative is:
Springer Nature Customer Service Center GmbH
Europaplatz 3, 69115 Heidelberg, Germany

Printed by Libri Plureos GmbH
in Hamburg, Germany